T0318644

Theory and Modeling of Dispersed Multiphase Turbulent Reacting Flows

Theory and Modeling of Dispersed Multiphase Turbulent Reacting Flows

Lixing Zhou
Tsinghua University, Beijing, China

Butterworth-Heinemann
An imprint of Elsevier

British Library Cataloguing-in-Publication Data
A catalogue record for this book is available from the British Library

Library of Congress Cataloging-in-Publication Data
A catalog record for this book is available from the Library of Congress

ISBN: 978-0-12-813465-8

For Information on all Butterworth-Heinemann publications
visit our website at https://www.elsevier.com/books-and-journals

Working together
to grow libraries in
developing countries

www.elsevier.com • www.bookaid.org

Publisher: Matthew Deans
Acquisition Editor: Glyn Jones
Editorial Project Manager: Naomi Robertson
Production Project Manager: Vijayaraj Purushothaman
Cover Designer: Miles Hitchen

Typeset by MPS Limited, Chennai, India

清華大學出版社
TSINGHUA UNIVERSITY PRESS

Contents

Preface

Multiphase, turbulent, and reacting flows are widely encountered in engineering and the natural environment. The basic theory, phenomena, mathematical models, numerical simulations, and applications of multiphase (gas or liquid flows with particles/droplets or bubbles), turbulent reacting flows are presented in this book. The special feature of this book is in combining the multiphase fluid dynamics with the turbulence modeling theory and reacting fluid dynamics (combustion theory). There are nine chapters in this book, namely: "Fundamentals of Dispersed Multiphase Flows"; "Basic Concepts and Description of Turbulence"; "Fundamentals of Combustion Theory"; "Basic Equations of Multiphase Turbulent Reacting Flows"; "Modeling of Single-Phase Turbulent Flows"; "Modeling of Dispersed Multiphase Turbulent Flows"; "Modeling of Turbulent Combustion"; "The Solution Procedure for Modeling Multiphase Turbulent Reacting Flows"; and "Simulation of Flows and Combustion in Practical Fluid Machines, Combustors and Furnaces." The main difference between this book and previous books written by the author is that more much better descriptions of basic equations and closure models of multiphase turbulent reacting flows are introduced, and recent advances made by the author and other investigators between 1994 and 2016 are included.

This book serves as a reference book for teaching, research, and engineering design for faculty members, students, and research engineers in the fields of fluid dynamics, thermal science and engineering, aeronautical, astronautical, chemical, metallurgical, petroleum, nuclear, and hydraulic engineering.

The author wishes to thank Prof. F.G. Zhuang, H.X. Zhang, and C.K. Wu for their valuable comments and suggestions. Thanks also go to colleagues and former students: Prof. W.Y. Lin, R.X. Li, X.L. Wang, J. Zhang, B. Zhou, Y.C. Guo, H.Q. Zhang, L.Y. Hu, Y. Yu, F. Wang, Z.X. Zeng, K. Li, Y. Zhang; Drs. Gene X.Q. Huang, T. Hong, C.M. Liao, W.W. Luo, K.M. Sun, Y. Li, T. Chen, Y. Xu, G. Luo, M. Yang, L. Li, H.X. Gu, X.L. Chen, X. Zhang, and Y. Liu. Their research results under the direction and cooperation of the author contributed to the context of this book.

Finally, the author's gratitude is given to the editors from Elsevier and the Executive Editor, Dr. Qiang Li from the Tsinghua University Press for their hard work in the final editing and publishing of this book.

Any comments and suggestions from the experts and readers would be highly appreciated.

Lixing Zhou
Tsinghua University, Beijing, China
February, 2017

Nomenclature

A	area
B	preexponential factor
c	empirial constants, specific heat
c_d	drag coefficient
d	diameter
D	diffusivity
E	activation energy
e	internal energy
F	force
f	mixture fraction
G	production term
g	gravitational acceleration; mean squire value of concentration fluctuation
H	stagnation enthalpy
h	enthalpy
J	diffusion fux
k	turbulent kinetic energy; reaction-rate coefficient
l	turbulent scale; length
M	molecular weight
m	mass
N	total particle number flux; particle number density
n	fluctuation of particle number density; exponent in particle-size distribution function; reaction order; mole number density
Nu	Nusselt number
p	pressure; probability density distribution function
Pr	Prandtl number
Q	heat; heating effect
q	heat flux
R	universal gas constant; weight fraction in particle-size distribution
r	radius; radial coordinate
Re	Reynolds number
R_f	flux Richardson number
S	source term
Sc	Schmidt number
Sh	Sherwood number
T	temperature
t	time
u,v,w	velocity components
V	volume; drift velocity

w	reaction rate
x,y,z	coordinates
X	combined mass fraction; mole fraction
Y	mass fraction

GREEK ALPHABETS

α	volume fraction
μ	dynamic viscosity
ν	kinematic viscosity
λ	heat conductivity
ε	dissipation rate of turbulent kinetic energy; emissivity
ϕ	generalized dependent variable
θ	dimensionless temperature
τ	shear stress
σ	Stefan-Boltzmann constant; generalized Prandtl number

SUBSCRIPTS

A,a	air
c	raw coal, reaction
ch	reaction; char
d	diffusion
e	effective; exit
F,fu	fuel
f	flame; fluid
g	gas
h	char; heterogeneous
hr	heterogeneous
i, in	initial; inlet
i,j,k	coordinate directions
k	*k*-th particle group
l	liquid
m	mixture

Introduction

Dispersed multiphase turbulent reacting flows are widely encountered in thermal, aeronautical, astronautical, nuclear, chemical, metallurgical, petroleum, and hydraulic engineering, and in water and atmosphere environments. As early as the 1950s, Von Karman and H.S. Tsien suggested using continuum mechanics to study laminar gas reacting flows and combustion, called "aerothermochemistry" or "dynamics of chemically reacting fluids." Multiphase fluid dynamics was first proposed by S.L. Soo in the 1960s for studying nonreacting multiphase flows. The classical reacting fluid dynamics and multiphase fluid dynamics do not include the theory of turbulence modeling. On the other hand, the theory of turbulence modeling was first proposed by P.Y. Chou in the 1950s, and was fully realized by B.E. Launder and D.B. Spalding in the 1970s. Within the last 40 years, through worldwide study and application, it has become the only reasonable and economical method to solve complex turbulent flows in engineering problems. However, up until the 1980s, the theory of turbulence modeling was limited to only single-phase fluid flows themselves, and did not concern the dispersed phase, i.e., particles/droplets/bubbles in multiphase flows.

Since the 1980s, the author has combined multiphase fluid dynamics with the theory of turbulence modeling, and proposed the concept of multiphase (two-phase) turbulence models, in particular the turbulence models of the dispersed phase, i.e., particles/droplets/bubbles. Furthermore, we developed the turbulence-chemistry models for single-phase and two-phase combustion using a method similar to turbulence modeling. Hence, the dynamics of multiphase turbulent reacting flows was developed, where the modeling theory, numerical simulation, measurements, and their application in combustion systems were systematically studied. The comprehensive models, basic conservation equations, the relationships between slip and diffusion, the energy distribution between the continuum and dispersed phases, the fluid-particle/droplet/bubble turbulence interactions, the interactions between particle turbulence and particle reaction, the gas-phase turbulence-chemistry interaction, and the particle—wall interaction were thoroughly studied. A series of new closure models were proposed, many 2-D and 3-D computer codes were developed based on the proposed models and some of the simulation results were validated using the laser Doppler velocimeter (LDV), phase Doppler particle anemometer (PDPA), and particle imaging velocimeter (PIV) measurements and direct numerical simulation (DNS). The research results were applied to develop innovative swirl combustors, cement kilns, oil—water hydrocyclones, gas—solid cyclone separators, and innovative cyclone coal combustors. This book is written based on the research results of the author, as well as those obtained by other investigators in recent years. In the following sections some basic definitions and descriptions are discussed.

TURBULENT DISPERSED MULTIPHASE FLOWS

Gas/liquid flows containing a vast amount of particles/droplets/bubbles are called dispersed multiphase flows. This terminology is widely accepted by the academic and engineering communities in the fields of fluid dynamics, thermal science and engineering, aeronautical, astronautical, metallurgical, chemical, petroleum, nuclear, and hydraulic engineering. Frequently, the concept of "phase" is considered as a thermodynamic state, so multiphase flows are divided into gas—solid (gas—particle), liquid—solid (liquid—particle), gas—liquid (gas—spray or bubble—liquid), liquid—liquid (oil—water) two-phase flows and gas—solid—liquid, oil—water—gas three-phase flows. Also, sometimes the terminologies "suspension flows" and "dispersed flows" are adopted. Besides, there are nondispersed two-phase flows, such as stratified and annular gas—liquid flows. However, from the multiphase fluid dynamic point of view, in particular in multifluid models, particles/droplets/bubbles with different sizes, velocities, and temperatures may constitute different phases. This is the reason why the terminology "multiphase fluid dynamics" was first proposed by S.L. Soo in the 1960s. In short, although different academic and engineering communities have different understanding of the above-listed terminologies, nowadays "multiphase flow" as a general concept of a branch of science and technology is widely accepted without disagreement.

Most practical fluid flows, maybe more than 99% of flows in the natural environment and engineering, are laden with particles, droplets, or gas bubbles. Pure single-phase flows exist only in a few cases such as flows in artificial ultraclean environment. There are a variety of multiphase flows, such as cosmic dust in cosmic space, cloud and fog (rain droplets), dusty-air flow, sandy rivers, blood flows in biological bodies, pneumatic/hydraulic conveying, dust separation and collection, spray coating, drying and cooling, spray/pulverized-coal combustion, plasma chemistry, fluidized bed, flows in gun barrels, solid-rocket exhaust, steam-droplet flows in turbines and gas-fiber flows, steam-water flows in boilers and nuclear reactors, oil—water and gas—oil—water flows in petroleum pipes, and gas—liquid—solid flows in steel making furnaces.

Most fluid flows in engineering facilities, such as flows in hydraulic channels, gas pipes, heat exchangers, fluid machines, chemical reactors, combustors, and furnaces, are turbulent flows due to the size of the geometric system, the velocity range and the presence of various barriers or expansions leading to flow separation, and frequently they are complex turbulent flows, such as recirculating flows, swirling flows, and buoyant flows.

MULTIPHASE TURBULENT REACTING FLOWS AND COMBUSTION

In many cases we are dealing with nonisothermal multiphase turbulent flows with heat and mass transfer and chemical reactions (exothermic or endothermic), and even electrostatic effects (for gas flows laden with fine particles in electrostatic dust separators, or metal or plastic pipes) or magnetic effects (for gas flows in plasma torch and magneto-hydrodynamic (MHD) generators). Spray or pulverized-coal combustion is a typical case of multiphase turbulent reacting flows.

The word "combustion" denotes a class of chemical reactions with high heat release and light radiation. These reactions in the first place are oxidation of solid fuel (nonmetals or metals), liquid fuel or gaseous fuel, but chlorination, fluoridation, nitridation, dissociation, and substitution reactions (e.g., sodium−water reaction) and self-propagating reactions in the synthesis of solid materials can also be considered as combustion, if they have a high heating effect. As a matter of fact, combustion is not simply reactions themselves, but a complex process of gas flows or gas−particle flows with heat and mass transfer and chemical reactions. Just the interactions between heat and mass transfer and chemical reactions control the processes of ignition, extinction, flame propagation, and combustion rate.

Frequently, "flame" is considered as the high-temperature combustion products. The word "flame" in its scientific sense can be defined as a zone with a sharp change of temperature and concentration. An important flame property is that the flame zone can propagate automatically. The flame propagation velocity is the flame velocity relative to that of the cold fresh combustible mixture, which is equal to the difference between the flame displacement velocity and the mixture flow velocity, such as that observed in flame propagation in a long tube with one closed end. For stationary flames such as those in Bunsen burners or flat burners, the displacement velocity is zero and the propagation velocity should be equal to the flow velocity. Two possible regimes of flame propagation were observed in experiments: deflagration with flame velocity of 0.2−1 m/s and detonation with flame velocity near 3000 m/s. Other flame properties are carbon formation, flame radiation (due to carbon or soot particles), ionization with concentration up to 10^{12} ions/cm^3 in laminar premixed flames and noise in turbulent flames.

DIFFERENT FLOW REGIMES OF DISPERSED MULTIPHASE TURBULENT REACTING FLOWS

To understand physically the general features of dispersed multiphase turbulent reacting flows, it is proper to judge their flow regimes. At first, the following characteristic times and nondimensional numbers are defined as:

Flow time (residence time) $\qquad \tau_f = L/v$

Particle relaxation time $\qquad \tau_r = d_p^2 \rho_{pm}/(18\mu)$

Particle relaxation time for mean motion $\qquad \tau_{r1} = \tau_r(1 + \mathrm{Re}_p^{2/3}/6)^{-1}$

Fluid fluctuation time (diffusion time) $\qquad \tau_T = l/u' = k/\varepsilon$

Particle collision time $\qquad \tau_p = l_p/u_p' = (c\pi n_p r_p^2)^{-1}(u_p')^{-1}$

Fluid reaction time $\qquad \tau_c = w_{s\infty}Q_s/(\rho_\infty c_p T_\infty)$

Stokes number $\qquad St = \tau_{r1}/\tau_f$

Hinze−Tchen number $\qquad Ht = \tau_r/\tau_T$

Soo number $\qquad Sl = \tau_{r1}/\tau_p$

First Damköhler number $\qquad D_1 = \tau_f/\tau_c$

Second Damköhler number $\qquad D_2 = \tau_T/\tau_c$

According to the magnitude of nondimensional numbers, the following regimes for limiting cases of dispersed multiphase turbulent reacting flows can be identified as:

$St \ll 1$ No-slip or dynamic equilibrium flows
$St \gg 1$ Strong-slip or dynamic frozen flows
$Ht \ll 1$ Diffusion-equilibrium flows
$Ht \gg 1$ Diffusion-frozen flows
$Sl \ll 1$ Dilute suspension flows
$Sl \gg 1$ Dense suspension flows
$D_1 \ll 1$ Reacting-frozen flows
$D_1 \gg 1$ Reacting-equilibrium flows
$D_2 \ll 1$ Kinetics-controlled combustion
$D_2 \gg 1$ Diffusion-controlled combustion

DEVELOPMENT OF VARIOUS DISPERSED-PHASE MODELS

There are two basic approaches to studying dispersed multiphase turbulent flows. One is the Eulerian−Eulerian or multifluid (two-fluid) approach, in which the fluid phase (liquid/gas) is treated as a continuum and the dispersed phase (particles/droplets/bubbles) is treated as a pseudo-fluid (PF) or pseudo-continuum. Different phases occupy the same space, interpenetrate into each other, and all are described in the Eulerian coordinate system. The other is the Eulerian−Lagrangian approach, in which only the fluid phase is treated as a continuum in the Eulerian coordinate system and the particles/droplets/bubbles are treated as a dispersed system in the Lagrangian coordinate system. This approach for the dispersed phase is also called the trajectory approach. Early studies are limited to particle/droplet/bubble motion in a known flow field, called one-way coupling, neglecting the effect of particles/droplets/bubbles on the fluid flows. One of the main features of modern multiphase fluid dynamics is to fully account for the mass, momentum, and energy interactions between the continuous phase and the dispersed phase, called two-way coupling. For dense dispersed multiphase flows or the flow region where the particle/droplet/bubble concentration is sufficiently large, it is necessary to account for the particle—particle interaction, called four-way coupling. In the last case, both two-phase turbulence and interparticle collision will dominate the coupling effects.

Different multiphase-flow models are identified by how to treat the dispersed-phase modeling. Table 1 gives the development of typical dispersed-phase models from the 1960s to now. The earliest model is the single-particle-dynamics (SPD) model, developed in the 1950s to 1960s, in which the single-particle trajectory (PT) of mean motion or convection in a known flow field (frequently uniform velocity and temperature field) and the particle velocity and temperature change along the trajectory are considered, neglecting the effect of particles on the flow field. This is an oversimplified model for actual complex multiphase flows and is no longer used. However, it is useful for understanding the basic features of particle motion, such as

TABLE 1 Development of Dispersed-Phase Models

Items Models	Approach	Coupling	Slip	Coordinate	Particle Diffusion
Single-particle-dynamics model	Dispersed system	One-way	Yes	Lagrangian	No
Small-slip model	Continuum	One-way	Slip = diffusion	Eulerian	Diffusion = slip
No-slip model	Continuum	Two-way	No	Eulerian	Yes
Particle-trajectory model	Dispersed system	Two-way	Yes	Lagrangian	No (deterministic) Yes (stochastic)
Pseudo-fluid model (two-fluid model)	Continuum	Two-way	Yes	Eulerian	Yes

the particle relaxation time, particle terminal velocity, etc. In the meantime, Tchen, Hinze, and others studied the particle behavior in a fluctuating flow field, i.e., turbulent diffusion just based on the SPD approach.

In the late 1960s, Soo first proposed a PF model of particle phase based on a continuum concept, and the slip between fluid and particles was taken into account. However, the slip was considered to be a result of particle diffusion, so this model can be called a "small-slip" model. The particles are described in the Eulerian coordinates with one-way coupling by using analytical solutions.

In the early 1970s, due to the preliminary use of computational fluid dynamics in multiphase flows, the single-fluid model or no-slip model was proposed. This model is a dynamic and thermal equilibrium model, in which the particle velocity and temperature are assumed to be equal to those of the gas phase everywhere, and the particle diffusion is also assumed to be equal to the gas species diffusion. This is another kind of oversimplified model, which cannot describe practical complex flow well and is now seldom used to solve practical problems.

From 1980s till now, more advanced models have been developed. These models account for the velocity slip, mass, momentum, and energy interaction between fluid and dispersed phases, in particular, the effect of dispersed phase on fluid flows. These models are the Lagrangian PT model (including deterministic trajectory model and stochastic trajectory model) and the PF model of dispersed phase with simultaneous phase slip and particle diffusion. Actually, all of the different models are an approximation of the real process from different points of view. The feasibility of these models depends on the particle volume fraction, relaxation time, flow time, and particle–particle impaction time. Unlike the SPD model, the modern PT model is a two-way coupling model, accounting for the effect of particles on the fluid mass, momentum, and energy. The stochastic trajectory model accounts for particle diffusion caused by particle fluctuation, while the deterministic trajectory model does not account for this effect. The two-fluid model is used to treat the dispersed phase as a PF in the Eulerian coordinate system. Unlike the single-fluid model and the small-slip model, the two-fluid model slip is independent of particle diffusion, and the particle diffusion is caused by its turbulent fluctuation, hence the two-fluid model is a full PF model, and a two-way coupling model. The trajectory model is widely used in engineering applications, and nowadays the two-fluid models find wider and wider application. It is also a two-way coupling model.

The above-discussed models are based on the Reynolds averaged N-S equations and were recently called Reynolds Averaged N-S equations (RANS) modeling. Alternatively, more elaborate models, such as the direct numerical simulation (DNS, including the so-called "point-particle DNS, PDNS" and "fully resolved DNS, FDNS," which accounts for the finite volume of particles/droplets/bubbles) and large-eddy simulation (LES), have been developed. However, due to the large computation time, DNS is mainly used for understanding the turbulence and flame structure and validating the LES and RANS modeling. It still cannot be directly used to solve engineering problems.

Chapter 1

Some Fundamentals of Dispersed Multiphase Flows

1.1 PARTICLE/SPRAY BASIC PROPERTIES

To characterize gas-particle or gas-spray flows, it is necessary first to describe the particle/spray basic properties [1−4] as follows.

1.1.1 Particle/Droplet Size and Its Distribution

The particle/droplet size distribution is frequently expressed by the semiempirical Rosin−Rammler formula as:

$$R(d_k) = \exp[-(d_k/\overline{d})^n] \tag{1.1}$$

where $R(d_k)$ is the weight fraction of particles with sizes larger than d_k, n is the index of nonuniformness, and $\overline{d}$ is a characteristic size. Both n and $\overline{d}$ are determined by experiments. The derivative of $R(d_k)$ is

$$\frac{dR}{d(d_k)} = n(d_k)^{n-1}(\overline{d})^{-n}\exp\left[-(d_k/\overline{d})^n\right] \tag{1.2}$$

which expresses the differential particle size distribution, and $R(d_k)$ is the integral size distribution. The mean particle sizes can be defined as:

$$
\begin{aligned}
d_{10} &= \sum n_k d_k / \sum n_k \\
d_{20} &= \left(\sum n_k d_k^2 / \sum n_k\right)^{1/2} \\
d_{30} &= \left(\sum n_k d_k^3 / \sum n_k\right)^{1/3} \\
d_{32} &= \sum n_k d_k^3 / \left(\sum n_k d_k^2\right)
\end{aligned}
\tag{1.3}
$$

where d_{10}, d_{20}, d_{30}, and d_{32} are diameter-averaged, surface-averaged, volume-averaged, and Sauter mean sizes, respectively. The Sauter diameter is most widely used in engineering. The typical particle sizes are:

Coal particles in fluidized beds	1−10 mm
Liquid spray	10−200 μm
Pulverized coal	1−100 μm
Soot particles	1−5 μm

Theory and Modeling of Dispersed Multiphase Turbulent Reacting Flows.
DOI: https://doi.org/10.1016/B978-0-12-813465-8.00001-6

1.1.2 Apparent Density and Volume Fraction

For gas-particle/droplet flows there are differently defined densities. The relationships among them are:

$$\rho_m = \rho + \rho_p = \rho + \sum \rho_k = \rho + \left(\sum n_k \pi d_k^3 / 6\right) \bar{\rho}_p \qquad (1.4)$$

where $\rho_m, \rho, \rho_p, \rho_k,$ and $\bar{\rho}_p$ are mixture density, fluid apparent density, particle total apparent density, k-th size particle apparent density and particle material density, respectively. The particle volume fraction and fluid volume fraction are defined as:

$$\alpha_p = \rho_p / \bar{\rho}_p; \quad \alpha_f = 1 - \alpha_p = 1 - \rho_p / \bar{\rho}_p \qquad (1.5)$$

For dilute gas-particle flows we have:

$$\rho = \bar{\rho}(1 - \rho_p / \bar{\rho}_p) \approx \bar{\rho}$$

where $\bar{\rho}$ is the fluid material density. Obviously, the fluid apparent density in dilute gas-particle flows is almost equal to the fluid material density. The so-called mass loading, which is the ratio of particle mass flux to fluid mass flux, is defined as $\rho_{p0} u_{p0} / (\rho_0 u_0)$. When the fluid initial velocity is equal to the particle initial velocity, the mass loading is equal to the ratio of apparent densities. For example, in spray or pulverized-coal flames the typical value of the mass loading is:

$$\rho_p / \rho = 1/15 = \frac{\bar{\rho}_p}{\rho} \frac{\alpha_p}{1 - \alpha_p} \approx 1000 \frac{\alpha_p}{1 - \alpha_p}$$

namely, $\alpha_p < 0.01\%$, hence the spray flame and pulverized-coal flame are dilute gas-particle flows. Other examples are: pneumatic transport $\alpha_p \approx 0.1\%$ (mass loading ≈ 1), fluidized beds and flows in gun barrels $\alpha_p \approx 0.8-1$. It can be seen that when $\alpha_p = 0.1\%$, due to $1 = 1000 n \pi d^3 / 6$, the average inter-particle size will be:

$$\Delta \approx n^{-1/3} = (1000\pi/6)^{1/3} d_p = 8.1 d_p. \quad \Delta > 20 d_p.$$

1.2 PARTICLE DRAG, HEAT, AND MASS TRANSFER

For different ranges of particle Reynolds number the particle drag is given as:
Newton drag formula: $c_d = 0.44 \quad (\text{Re}_p > 1000)$
Wallis–Kliachko drag formula: $c_d = (1 + \text{Re}_p^{2/3}/6)24/\text{Re}_p (1 < \text{Re}_p < 1000)$

$$\text{Stokes drag formula: } c_d = 24/\text{Re}_p (\text{Re}_p < 1) \qquad (1.6)$$

where Re_p is the particle Reynolds number of particle motion relative to fluid. When the particle temperature is higher than the gas temperature, the particle drag will increase according to the so-called 1/3 law. The gas viscosity in the particle Reynolds number will be:

$$\nu = \nu_p/3 + 2\nu_g/3 \tag{1.7}$$

where the subscripts p and g denote the gas viscosity under the particle temperature and gas temperature, respectively. The particle mass loss due to evaporation, devolatilization, or heterogeneous combustion will reduce the particle drag to:

$$c_d = c_{d0}\ln(1 + B)/B \tag{1.8}$$

where B is a dimensionless parameter given by

$$\ln(1 + B) = \dot{m}/(\pi d_p Nu D \rho) \tag{1.9}$$

The particle heat and mass transfer are given by the Ranz−Marshell formula:

$$\begin{aligned} Nu &= 2 + 0.6Re_p^{0.5} \; Pr^{0.33} \\ Sh &= 2 + 0.6Re_p^{0.5} \; Sc^{0.33} \end{aligned} \tag{1.10}$$

where Nu, Sh, Re, Pr, and Sc are the Nusselt number, Shewood number, Reynolds number, Prandtl number, and Schmidt number, respectively. The droplet mass, diameter, and temperature change during evaporation and solid-fuel particle mass and temperature change during moisture evaporation, devolatilization, and char combustion are given in the combustion theory, see Chapter 3, Fundamentals of Combustion.

1.3   SINGLE-PARTICLE DYNAMICS

Consider the single-particle motion in a known simple flow field and neglect the effect of particles on the fluid flow; this is single-particle dynamics [6]. For turbulent gas-particle flows single-particle dynamics is a basic phenomenon observed in practical cases.

1.3.1   Single-Particle Motion Equation

Taking into consideration only the drag and gravitational forces, the simplest single-particle motion equation can be given as:

$$\frac{dv_{pi}}{dt_p} = (v_i - v_{pi})/\tau_r + g_i \tag{1.11}$$

where τ_r is the particle relaxation time, expressing the ratio of particle inertia to particle drag, determined by the drag law.

1.3.2   Motion of a Single Particle in a Uniform Flow Field

Assuming a particle with initial velocity v_{P0} and Stokes' drag law, moving in a uniform flow field (Fig. 1.1), when neglecting the gravitational force, the particle momentum equation in the x direction is

$$\frac{du_p}{dt} = (u_\infty - u_p)/\tau_r \tag{1.12}$$

where $\tau_r = d_p^2 \bar{\rho}_p/(18\mu)$. Integration of Eq. (1.12) with an initial condition of $u_p = u_{p0}$ at $t = 0$ gives the particle longitudinal velocity

$$u_p = u_\infty - (u_\infty - u_{p0})\exp(-t/\tau_r) \tag{1.13}$$

The particle lateral velocity can be obtained in a similar way as

$$v_p = v_{p0}\exp(-t/\tau_r) \tag{1.14}$$

Integration of Eqs. (1.13) and (1.14) with respect to t gives the particle trajectory equations as

$$x_p = u_\infty t - (u_\infty - u_{p0})\tau_r(1 - e^{-t/\tau_r})$$
$$y_p = v_{p0}\tau_r(1 - e^{-t/\tau_r}) \tag{1.15}$$

Similar equations can also be derived for non-Stokes' particle drag. Eqs. (1.13, 1.14, 1.15) point out that as the time approaches ∞, the particle longitudinal velocity approaches the fluid velocity, the particle lateral velocity approaches zero and the particle lateral displacement approaches $y = v_{p0}\tau_r$. When $t = \tau_r$, we have $v_p = v_{p0}/\tau_r$. Hence the physical meaning of the particle relaxation time is the time needed for the fluid-particle velocity slip to decrease to $1/e$ of its initial value. It expresses the easiness with which particles follow the fluid.

1.3.3   Particle Gravitational Deposition

For an initially stagnant particle acting only by Stokes' drag and gravity, the motion equation is:

$$\frac{dv_p}{dt} + \frac{v_p}{\tau_r} - g = 0 \tag{1.16}$$

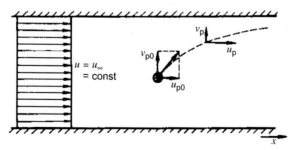

FIGURE 1.1   Motion of a single particle in uniform flow field.

For the initial condition of $v_{p0} = 0$ at $t = 0$, its solution is:

$$v_p = \tau_r g(1 - e^{-t/\tau_r}) \qquad (1.17)$$

As the time approaches infinity, v_p approaches $\tau_r g = v_{pr}$, the particle acceleration becomes zero and the gravity and drag force will be in equilibrium. In this case the particle velocity is called the terminal velocity.

1.3.4   Forces Acting on Particles in Nonuniform Flow Field

1.3.4.1   Magnus Force

As a nonspherical particle moves in the flow field with velocity gradient, in particular after its impact on the wall, it may rotate, causing a lifting force perpendicular to the direction of relative velocity, called the Magnus force. Its magnitude is:

$$F_M = \pi d_p^3 \bar{\rho} |v - v_p| |\omega_p - \Omega| \qquad (1.18)$$

where ω_p is the angular velocity of particle rotation, and Ω is the half of fluid vorticity. It has been estimated that the ratio of Magnus force to the drag force is 0.04 for a 1-μm particle and 3 for a 10-μm particle. However, experimental studies have shown that in most regions of the flow field, particles do not rotate due to fluid viscosity. Therefore, except in the region adjacent to the wall, the Magnus force is not important.

1.3.4.2   Saffman Force

If the particle is sufficiently large and there is a large velocity gradient in the flow field (for example, near the wall), there will be a particle-lifting force called the Saffman force. Its magnitude is

$$F_s = 1.6(\mu\bar{\rho})^{1/2} d_p^2 \left| v - v_p \right\| \left\| \frac{\partial v}{\partial y} \right\|^{1/2} \qquad (1.19)$$

The ratio of the Saffman force to the Magnus force is much greater than unity; hence the Saffman force may play an important role, in particular in the region of a large velocity gradient, such as in the recirculation region and the near-wall region.

1.3.4.3   Particle Thermophoresis, Electrophoresis, and Photophoresis

Tiny particles smaller than 1 μm may move under the effects of so-called "thermophoresis," "electrophoresis," and "photophoresis," caused by a large temperature gradient, electric field gradient, and nonuniform light radiation, respectively. The forces of thermophoresis and electrophoresis can be estimated by

$$F_{Tj} = -4.5\nu^2(\rho/T)d_p[\lambda(2\lambda + \lambda_p)]\frac{\partial T}{\partial x_j}$$

$$F_E = (\pi/6)\overline{\rho}_p d_p^3 qE \tag{1.20}$$

where λ and λ_p are the gas and particle thermoconductivities, respectively, and E and q are electric field strength and particle electric charge, respectively. All of these forces are significant merely for submicron or ultrafine particles.

1.3.5 Generalized Particle Motion Equation

Eq. (1.11) is a very simple particle motion equation. C.M. Tchen [7], using a method of intuitive superposition of various possible forces, proposed a generalized particle motion equation, with Stoke drag and accounting for the Magnus force, Saffman force, thermophoresis, and electrophoresis forces, as

$$m_p\frac{dv_{pi}}{dt_p} = F_{di} + F_{vmi} + F_{pi} + F_{Bi} + F_{Mi} + F_{si} + F_{Ti} + F_{Ei} +$$

$$\dots = 3\pi d_p\mu(v_i - v_{pi}) + 0.5(\pi d_p^3/6)\rho\frac{d}{dt_p}(v_i - v_{pi}) +$$

$$(\pi d_p^3/6)\rho\frac{dv_i}{dt} + 1.5(\pi\rho\mu)^{1/2}d_p^2\int_{-\infty}^{t}\frac{d}{d\tau}(v_i - v_{pi})(\tau - t)d\tau + \tag{1.21}$$

$$F_{Mi} + F_{si} + F_{Ti} + F_{Ei} + \dots$$

where the first, second, third, and fourth terms on the right-hand side of Eq. (1.21) denote the drag force, virtual-mass force, pressure-gradient force, and Basset force (due to unsteady flow), respectively. It should be noted that in most cases the forces other than the drag force are of minor importance, so the approximation made in Eq. (1.11) is still valid.

1.3.6 Recent Studies on Particle Dynamics

Sommerfeld and Kussin [8] studied the forces acting on particles of irregular shapes. Zhang and Lin [9] studied the motion, its orientation, and forces acting on elliptical particles. Bagchi and Balachandar [10] give the detailed flow field around a single particle using direct numerical simulation (DNS). Sundaresan and Cate [11] show the detailed flow field around several particles using a Lattice−Boltzmann simulation. From these simulation results the exact forces acting on the particles can be obtained. For example, it is found that the virtual mass force can be neglected, if the ratio of the fluid material density to the particle material density is small. The effect of small-scale turbulence on the forces acting on particles is also studied. Michaelides [12,13] systematically summarized the research results of forces and heat and mass transfer acting on particles and proposed a more comprehensive particle motion equation. A comparison was made between the classical analytic solutions and the recent results of DNS and Lattice−Boltzmann simulation results. For example, the effect of particle concentration on particle drag force was discussed.

Alternatively, in some cases the electric forces and van der Waals forces are also considered when particles are located in the electric field and are very near to each other. The contact force and collision forces between particles should be considered for dense gas-particle flows. For further details the reader should refer to Refs [14,15].

REFERENCES

[1] L.X. Zhou, Theory and Numerical Modeling of Turbulent Gas-Particle Flows and Combustion., Science Press, and Florida, CRC Press, Beijing, 1993.

[2] L.X. Zhou, Numerical Simulation of Turbulent Two-Phase Flows and Combustion., Tsinghua University Press, Beijing, 1991.

[3] S.L. Soo, Fluid Dynamics of Multiphase Systems. Blaisdell (Ginn), New York, 1967.

[4] S.L. Soo, Multiphase Fluid Dynamics, Science Press, and Hong Kong, Gower Technical, Beijing, 1990.

[6] N.A. Fuchs, Mechanics of Aerosols., MacMillan, New York, 1964.

[7] C.M. Tchen, Mean value and correlation problems connected with the motion of small particles in a turbulent field. Ph.D. Dissertation, Delft University, Hague, Martinus Nijhoff, 1947.

[8] M. Sommerfeld, J. Kussin, On the behavior of non-spherical particles in pneumatic conveying. In: Guo Liejin Ed., Proceedings of the 5th International Symposium on Multiphase Flow, Heat Transfer and Energy Conversion, Xi'an, China, 2005-07-03-06, Abstract p1, 2005.

[9] W.F. Zhang, J.Z. Lin, Forces on cylindrical particles in suspension flows. In: Zhou Lixing Ed., Proceedings of the 5th Chinese Symposium on Multiphase Fluid, Non-Newtonian Fluid and Physical-Chemical Fluid Flows, 2000-10-31-11-01 Wuhan, China, pp. 8−13, 2000.

[10] P. Bagchi, S. Balachandar, Unsteady motion and forces on a spherical particle in non-uniform flows. In: Stock D.E., Ed., Proceedings of 2000 ASME Fluids Engineering Summer Conference, Boston, 2000-06-11-06-15, Paper FEDSM2000-11128, 2000.

[11] S. Sundaresan, A.T. Cate, Analysis of unsteady forces in ordered arrays of mono-disperse spheres., J. Fluid Mech. 552 (2006) 257−287.

[12] E.E. Michaelides, Hydrodynamic forces and heat/mass transfer from particles, bubbles and drops—The Freeman Scholar Lecture, Transactions of the ASME., J. Fluids Eng. 125 (2003) 209−238.

[13] E.E. Michaelides, Transient equations for the motion, heat and mass transfer of bubbles, drops and particles. In: Matsumoto Y Ed., Proceedings of 5th International Conference on Multiphase Flow, Yokohama, Japan, 2004-05-30-06-04, 2004, Abstracts, Paper No. K11.

[14] L.S. Fan, C. Zhu, Principles of Gas-Solid Flows., Cambridge University Press, Cambridge, 1998.

[15] C.T. Crowe, M. Sommerfeld, Y. Tsuji, Multiphase Flows with Droplets and Particles., CRC Press, Florida, 1998.

FURTHER READING

L.X. Zhou, Combustion Theory and Reacting Fluid Dynamics., Science Press, Beijing, 1986.

Chapter 2

Basic Concepts and Description of Turbulence

2.1 INTRODUCTION

As this book is related to multiphase turbulent reacting flows, some knowledge of the fundamental concepts and description of turbulence are introduced. More than 100 years ago, Osborn Reynolds (1883−94) first indicated that flows can be either laminar or turbulent by observing injected dyed water flow in a tube. A parameter called the Reynolds number $Re = vd/\nu$ (where v is the velocity, d is the size, and ν is the kinematic viscosity) is used to identify these two flow regimes. He first suggested the decomposition of the flow variables into time-averaged and fluctuation quantities for mathematically analyzing turbulent flows.

Turbulent flows widely occur in nature and engineering, particular in astronomy and natural water bodies. In engineering facilities, turbulent flows are encountered in fluid machines, heat exchangers, and combustors, because frequently in these facilities the fluid velocity is higher, and the geometrical sizes are large, in other words, the Reynolds number is large. However, in some cases, even if the Re number is not large, but there are rough walls and flow separation by obstacles, turbulence may also be produced.

Examples of turbulent flows are the discharge of smoke from a stack, as shown in Fig. 2.1, and a turbulent gas jet flame, as shown in Fig. 2.2. The flows of small soot particles and combustion products show on one hand the irregular behavior of instantaneous gas flows, but on the other hand, some organized structures—the so-called "coherent structures." Therefore, turbulent flows have both random and organized structures. Fig. 2.3 shows the vorticity map of a sudden-expansion flow by large-eddy simulation (LES). It gives the detailed turbulence structures. It can be seen that there are different length scales of eddies in turbulent flows.

2.2 TIME AVERAGING

Let us consider a variable $\tilde{a}(x;t)$ changing with time at a given spatial location of the turbulent flow field. O. Reynolds first introduced the concept of time averaging of a variable ϕ [1] as:

$$\overline{\phi}(x,y,z) = \lim_{\tau \to \infty} \frac{1}{\tau} \int_0^\tau \phi(x,y,z;t)\mathrm{d}t \qquad (2.1)$$

Theory and Modeling of Dispersed Multiphase Turbulent Reacting Flows.
DOI: https://doi.org/10.1016/B978-0-12-813465-8.00002-8

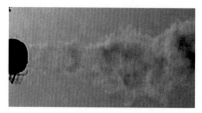

FIGURE 2.1   Stack smoke.

FIGURE 2.2   Turbulent CH_4-air flame.

FIGURE 2.3   Vorticity map of a sudden-expansion flow.

Here ϕ may be any variable in the turbulent flow field, such as the velocity component v_i, or temperature T (enthalpy h), or species concentration Y_s. The time-averaging period τ must be much larger in comparison with the integral time scale of turbulent fluctuation, but should be smaller than the macroscopic time period of unsteady flows, such as the wavy flows. The so-called Reynolds' expansion is defined as:

$$\phi = \overline{\phi} + \phi'; \quad \phi = v_i, \quad T(h), \quad Y_s$$
$$\overline{\overline{\phi}} = \overline{\phi}; \qquad \overline{\phi'} = 0; \quad \overline{\phi'\psi'} \neq 0 \tag{2.2}$$

For compressible flows, the so-called Favre averaging or density-weighed averaging usually is used, and is defined as:

$$\phi = \tilde{\phi} + \phi''; \quad \tilde{\phi} = \overline{\rho\phi}/\overline{\rho} \quad \overline{\phi''} \neq 0 \quad \overline{\rho\phi''} = 0 \tag{2.3}$$

2.3   PROBABILITY DENSITY FUNCTION

A different description of the fluctuation of a variable is the so-called probability density function (PDF). The PDF-$p(f)$ is defined as follows: the

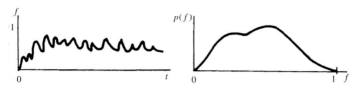

FIGURE 2.4   The probability density function.

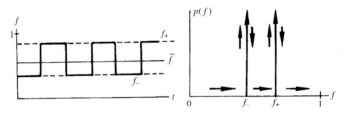

FIGURE 2.5   The top-hat PDF.

probability of a stochastic function f (changing from 0 to 1) in the range of f to $f + df$ is $p(f)df$ (Fig. 2.4).

Alternatively, it can be defined as:

$$p(\tilde{a}, x) = \lim_{\Delta \to 0} \frac{1}{\Delta} \left[\lim_{T \to \infty} \left(\frac{T_a}{T} \right) \right] \tag{2.4}$$

Then, for a variable f we have:

$$1 = \int_0^1 p(f)df \quad \bar{f} = \int_0^1 fp(f)df$$

$$\overline{f'^2} = \overline{f^2} - (\bar{f})^2 = \int_0^1 f^2 p(f)df - \left(\int_0^1 fp(f)df \right)^2 \tag{2.5}$$

For any other variable $\phi(f)$, it is statistically averaged and the fluctuation values are:

$$\overline{\phi(f)} = \int_0^1 \phi(f)p(f)df$$

$$\overline{\phi'^2} = \int_0^1 \phi^2(f)p(f)df - \left(\int_0^1 \phi(f)p(f)df \right)^2 \tag{2.6}$$

One of the typical forms of PDF is the top-hat PDF (Fig. 2.5):

$$p(f) = \alpha \delta(f_-) + (1 - \alpha)\delta(f_+) \tag{2.7}$$

When $\alpha = 0.5$, we have

$$\bar{f} = (f_- + f_+)/2g = \overline{f'^2} = (\bar{f} - f_-)^2 = (f_+ - \bar{f})^2 \tag{2.8}$$

$$f_- = \bar{f} - g^{1/2}, \quad f_+ = \bar{f} + g^{1/2} \tag{2.9}$$

Other useful PDF forms are clipped Gaussian PDF:

$$p(f) = \alpha\delta(f_-) + (1 - \alpha)\delta(f_+) + \int_{f_-}^{f_+} \frac{1}{(2\pi\sigma^2)^{1/2}} \exp\left[\frac{-(f - \bar{f})^2}{2\sigma^2}\right] \tag{2.10}$$

β-function PDF:

$$p(f) = \frac{f^{a-1}(1-f)^{b-1}}{\int_0^1 f^{a-1}(1-f)^{b-1}df} \tag{2.11}$$

and the joint PDF for species concentrations and temperature is:

$$P_c(Y_f, Y_{\alpha x}, T) = \frac{1}{(2\pi)^{3/2}|B_m|^{1/2}} \exp\left[-\frac{1}{2}(X - \bar{X})'B_m^{-1}(X - \bar{X})\right] \tag{2.12}$$

2.4 CORRELATIONS, LENGTH, AND TIME SCALES

The above-stated description is only about a variable at a fixed location, there is no information on the length or time scales. The time autocorrelation coefficient of a variable at the same location but at different time instants is defined as [2]:

$$R_\phi(\tau, x) = \frac{\overline{\phi(t, x)\phi(t + \tau; x)}}{\overline{\phi^2(x)}} \tag{2.13}$$

The integral time scale is

$$T_\phi(x) = \int_0^\infty R_\phi(\tau, x)d\tau \tag{2.14}$$

The correlation of two variables at the same spatial location and same time instant is

$$R_{ab}(t; x) = \frac{\overline{a(t; x)b(t, x)}}{\left[\overline{a^2(x)}\overline{b^2(x)}\right]} \tag{2.15}$$

The correlation of a variable at the same time but at different spatial locations separated by h is

$$R_\phi(h, x) = \frac{\overline{\phi(t, x)\phi(t; x + h)}}{\overline{\phi^2(x)}} \tag{2.16}$$

The integral length scale is

$$L_\phi(x) = \int_{h_{1m}}^{\infty} R_\phi(h_1; x) \mathrm{d}h_1 \tag{2.17}$$

where h_{1m} is the separation length, where the correlation is a maximum. For more knowledge on the fundamentals of turbulence, readers are referred to Refs. [1,2].

REFERENCES

[1]   P.A. Libby, Introduction to Turbulence, Taylor and Francis, Oxford, UK, 1996.
[2]   S.B. Pope, Turbulent Flows., Cambridge University Press, Cambridge, UK, 2000.

Chapter 3

Fundamentals of Combustion Theory

3.1 COMBUSTION AND FLAME

Combustion, in the first place, is chemical reactions with high heat release and light emission. These reactions may be oxidation of solid fuels (coal, carbon, silica, boron and metals, solid propellants), liquid fuels (petroleum products and liquid hydrocarbons), or gaseous fuels (natural gas and gaseous hydrocarbons), oxidation reactions of nitrogen, chlorine, and fluorine, dissociation of N_2H_4 into nitrogen and hydrogen, and substitution reactions (e.g., $Na + H_2O$). However, generally speaking, combustion is not simply chemical reactions, but the interaction of fluid flow, heat, and mass transfer with chemical reactions. Flame is the external appearance of combustion. According to the flow regimes, flames may be laminar or turbulent. According to the mixing state of fuel with oxidizer before combustion, flames may be premixed or nonpremixed (diffusion flame). In chemically reacting flows there are three typical characteristic times—flow time (residence time), reaction time, and diffusion time:

$$\tau_f = L/v$$
$$\tau_c = \rho c_p (T_f - T_\infty)/(w_f Q_f)$$
$$\tau_d = L^2/D_T = \tau_T = l/u' = k/\varepsilon$$

The dimensionless parameters, called Damköhler numbers D_I and D_{II} are the ratios of these characteristic times:

$$D_I = \tau_f/\tau_c \quad D_{II} = \tau_d/\tau_c$$

There are two limiting cases of reacting flows:

$$D_I \ll 1 \quad \text{Reaction Frozen Flows}$$
$$D_I \gg 1 \quad \text{Reaction Equilibrium Flows}$$

Combustion is the case between these two limiting cases. There are two limiting cases of combustion:

$$D_{II} \gg 1 \quad \text{Diffusion-controlled combustion}$$
$$D_{II} \ll 1 \quad \text{Kinetics-controlled combustion}$$

Theory and Modeling of Dispersed Multiphase Turbulent Reacting Flows.
DOI: https://doi.org/10.1016/B978-0-12-813465-8.00003-X

15

The premixed flame includes the cases between these two limiting cases—diffusion-kinetic combustion, $D_{II} \approx 1$.

The adiabatic flame temperature is determined by:

$$T_m = T_\infty + Q_f / [(1 + \alpha L_0)c_p] \quad (\alpha > 1)$$
$$T_m = T_\infty + 0.23\alpha L_0 Q_{ox} / [(1 + \alpha L_0)c_p] \quad (\alpha < 1)$$

where α is the coefficient of excess air.

3.2   BASIC EQUATIONS OF LAMINAR MULTICOMPONENT REACTING FLOWS AND COMBUSTION

Since combustion is the phenomenon of interactions among fluid flow, heat and mass transfer, and chemical reactions, for theoretical analysis, we must start from the basic conservation equations.

3.2.1   Thermodynamic Relationships of Multicomponent Gases

For multicomponent gases, the total mass density should be the summation of species mass concentration, and the total pressure should be the summation of species partial pressure:

$$\rho = \sum_s \rho_s \tag{3.1}$$

$$p = \sum_s p_s \tag{3.2}$$

The mass fraction (relative concentration) of species s and mole fraction are:

$$Y_s = \rho_s / \rho \tag{3.3a}$$

$$X_s = p_s / p \tag{3.3b}$$

The relationships among the mixture density, s species concentration, and mole number density are:

$$\rho = nM \tag{3.4}$$

$$\rho_s = n_s M_s \tag{3.5}$$

Except a few cases of very high temperature and high pressure, it is recognized that in most reacting flows the gas mixture and its species obey the state of perfect gases. Hence we have

$$p_s = \rho_s RT / M_s = n_s RT \tag{3.6a}$$

$$p = \rho RT / M = nRT \tag{3.6b}$$

and

$$X_s = p_s / p = n_s / n$$

Due to

$$n = \sum n_s \quad \rho = \sum \rho_s = \sum n_s M_s = nM$$

The relationship between the mixture molecular weight and the species molecular weight, and that between the mass fraction and mole fraction can be obtained as

$$M = \sum X_s M_s \qquad X_s = Y_s M / M_s$$
$$\sum (Y_s M / M_s) = 1 \quad M = \left(\sum Y_s / M_s\right)^{-1} \qquad (3.7)$$
$$M / M_s = \left(M_s \sum Y_s / M_s\right)^{-1}$$

Let us consider the motion and diffusion of multicomponent reacting gases. There are three kinds of macroscopic velocities:

v-mixture velocity with reference to the laboratory coordinate;
v_s-s species velocity with reference to the laboratory coordinate;
V_s-s species velocity with reference to the mixture velocity, i.e., the diffusion drift velocity.

Obviously there should be

$$V_s = v_s - v \qquad (3.8)$$

Correspondingly, there are three kinds of mass fluxes:

$\rho v = g$—mixture mass flux;
$\rho_s v_s = g_s$—s species mass flux;
$\rho_s V_s = J_s$—s species diffusion flux.

The s species mass flux with reference to the laboratory coordinate should be equal to the summation of its diffusion flux and its mass flux carried by the gas mixture, namely

$$g_{sj} = \rho_s v_{sj} = J_{sj} + Y_s \rho v_j = \rho_s V_{sj} + \rho_s v_j \qquad (3.9)$$

The mixture mass flux should be the summation of species mass fluxes:

$$g_j = \rho v_j = \sum g_{sj} = \sum \rho_s v_{sj} = \sum \rho_s V_{sj} + \rho v_j$$

Hence we have

$$\sum \rho_s V_{sj} = \sum J_{sj} = 0 \qquad (3.10)$$

Eqs. (3.9) and (3.10) point out that species mass fluxes do not equal their diffusion fluxes. The summation of diffusion fluxes is zero. The summation of diffusion fluxes does not affect the mixture flows. However, the summation of diffusion velocities is not zero:

$$\sum V_{sj} \neq 0$$

3.2.2 Molecular Transport Laws of Multicomponent Reacting Gases

For two-component gas mixtures the molecular diffusion due to concentration gradient obeys the Fick law:

$$J_{sj} = -\rho D_{12}\frac{\partial Y_s}{\partial x_j} \tag{3.11}$$

For multicomponent gas mixtures the Fick law still keeps a similar form as

$$J_{sj} = -\rho D_s\frac{\partial Y_s}{\partial x_j} \tag{3.12}$$

Here the molecular diffusivity D_s is a function of species concentrations. In some cases further assumptions are made. For example, the assumption of equality of species diffusivity as

$$D_1 = D_2 = \cdots = D_s = D$$

For molecular heat conduction, the total heat flux is determined by the generalized Fourier law, accounting for the net enthalpy carried by the species diffusion fluxes entering and leaving the elementary volume of the gas mixture:

$$q_j = -\lambda\frac{\partial T}{\partial x_j} + \sum \rho_s V_{sj} h_s \tag{3.13}$$

where λ is the molecular thermoconductivity, and h_s is the s species enthalpy, including the chemical enthalpy (enthalpy of formation).

$$h_s = h_{0s} + \int_{T_0}^{T} c_{ps}\mathrm{d}T \tag{3.14}$$

The mixture enthalpy is

$$h = \sum Y_s h_s = \sum Y_s h_{0s} + \int_{T_0}^{T}\sum Y_s c_{ps}\mathrm{d}T = h_0 + \int_{T_0}^{T} c_p \mathrm{d}T \tag{3.15}$$

where h_{0s} is the enthalpy of formation of s species, c_{ps} is the specific heat of s species, which is a function of temperature T and pressure p. Attention should be paid to the fact that both species enthalpy and mixture enthalpy include thermal enthalpy and chemical enthalpy. The species formation enthalpy is constant, but the mixture formation enthalpy is a function of species concentration. The species specific heat is independent of species concentration, but the mixture specific heat depends on species concentration. There are three kinds of mixture enthalpies: thermal enthalpy; thermal enthalpy + chemical enthalpy; stagnant enthalpy = thermal enthalpy + chemical enthalpy + kinetic energy. For momentum transfer in multicomponent gas mixtures, the relationship between the stress and the deformation can still be described by the generalized Newton–Stokes law for single-component nonreacting flows:

$$p_{ij} = -p\delta_{ij} + \mu\left(\frac{\partial v_i}{\partial x_j} + \frac{\partial v_j}{\partial x_i}\right) - \frac{2}{3}\mu\left(\frac{\partial v_j}{\partial x_j}\right)\delta_{ij} \tag{3.16}$$

Here the molecular viscosity μ is a function of not only temperature and pressure, but also of species concentration.

3.2.3   Basic Relationships of Chemical Kinetics

The stoichiometric relationships of chemical reactions can be expressed by

$$\sum \nu_s A_s \rightarrow \sum \nu'_s A'_s$$

where A_s, A'_s are symbols denoting the reactants and products respectively, and ν_s, ν'_s are corresponding stoicheometric coefficients. The reaction rate of s species w_s is the consumed or produced mass by reactions per unit time per unit volume, namely

$$w_s = -(d\rho_s/dt)_{chem}$$

Obviously, the stoicheometric relationships give

$$w_1/\nu_1 = w_2/\nu_2 = \cdots = -w'_1/\nu'_1 = -w'_2/\nu'_2 = \cdots$$

The law of mass action is

$$w_s = k_s \prod_{s=1}^{z} C_s^{\nu_s}$$

where C_s may be the mass concentration ρ_s, or the mole concentration $X_s M_s$, k_s is the reaction rate coefficient, z is the total number of species, and $\nu = \nu_1 + \nu_2 + \cdots = \sum \nu_s$ is the apparent reaction order. In most cases the actual reaction order is not the apparent order, i.e.,

$$w_s = k_s \prod_{s=1}^{z} C_s^{m_s} \tag{3.17}$$

where $m_s \neq \nu_s$, $m = \sum m_s \neq \nu = \sum \nu_s$, and m is the actual reaction order. For simultaneous elementary reactions, the r-th elementary reaction rate is

$$\sum_{s=1}^{z} \nu_{sr} A_s \rightarrow \sum_{s=1}^{z} \nu'_{sr} A'_s$$

So, the total reaction rate is

$$w_s = \sum_r w_{sr} = \sum_r k_{sr} \prod_{s=1}^{z} C_s^{m_{sr}} \tag{3.18}$$

For reversible reactions

$$\sum \nu_s A_s \Leftrightarrow \sum \nu'_s A'_s$$

There are forward reaction $w_{s+} = k_s \prod_s C_s^{m_s}$ and backward reaction $w_{s-} = k'_s \prod_s C'_s m'_s$ so, the net reaction rate is

$$w_s = w_{s+} - w_{s-} = k_s \prod_s C_s^{m_s} - k'_s \prod_s C_s^{'m'_s} \tag{3.19}$$

In the case that there are several reversible elementary reactions, the total reaction rate is

$$w_s = \sum_r w_{sr} = \sum_r (k_{sr} \prod_s C_s^{m_{sr}} - k'_{sr} \prod_s C_s^{'m'_{sr}} \tag{3.20}$$

The reaction rate coefficient k_s is a strong nonlinear function of temperature, and can be expressed by the Arrhenius law

$$k = B \exp(-E/RT) \tag{3.21}$$

where E is called the activation energy, expressing the energy barrier of reaction; R is the universal gas constant; and B is called the pre-exponential factor. In theoretical analysis or numerical simulation, frequently the reactions are considered as a one-step global reaction and its reaction rate is

$$w_s = B \rho^m \exp(-E/RT) \prod_s Y_s^{m_s} \tag{3.22}$$

where the pre-exponential factor B, activation energy E, and reaction order m are all empirical kinetic constants, based on experimental results.

3.2.4 The Reynolds Transport Theorem

The relationship between the change in Lagrangian coordinate and Eulerian coordinate for any variables is

$$\frac{d\phi}{dt} = \frac{\partial \phi}{\partial t} + v_j \frac{\partial \phi}{\partial x_j}$$

From here we obtain the Reynolds transport theorem

$$\frac{d\Phi}{dt} = \frac{\partial \Phi}{\partial t} + \int_S \phi v_n dS = \frac{\partial \Phi}{\partial t} + \int_V \frac{\partial}{\partial x_j}(\phi v_j)\delta V \tag{3.23}$$

where

$$\Phi = \int_V \phi \delta V$$

is the integral of the generalized variable φ in the control volume. The physical meaning of Eq. (3.23) is that the change of the integral of any variable φ in the control volume Φ in Lagrangian coordinate is equal to the static changing rate in time plus the changing rate due to convection in the Eulerian coordinate.

The generalized conservation law is

$$\frac{d\Phi}{dt} + \int_V S_\phi \delta V = 0 \tag{3.24}$$

where S_φ is the source term, i.e., the production rate or destruction rate of φ per unit volume.

3.2.5 Continuity and Diffusion Equations

For the conservation or continuity of the gas mixture mass, taking

$$\phi = \rho, \quad S_\varphi = 0, \quad \frac{d\Phi}{dt} = 0$$

we have the continuity equation

$$\frac{\partial \rho}{\partial t} + \frac{\partial}{\partial x_j}(\rho v_j) = 0 \tag{3.25}$$

or

$$\frac{d\rho}{dt} + \rho \frac{\partial v_j}{\partial x_j} = 0 \tag{3.26}$$

It can be seen that there is no difference between the form of continuity equation of multicomponent reacting gases and that of single-component nonreacting gases.

For species mass conservation or diffusion equation, taking

$$\phi = \rho_s = Y_s \rho$$

and accounting for the species mass change in the moving control volume due to reactions and the net diffusion fluxes through the surrounding surface of the control volume,

$$\int_V S_\phi \delta V = \int_S j_s dS + \int_V w_s \delta V$$

after using the Reynolds transport theorem and Fick law, the mass conservation equation or diffusion equation of species s can be obtained as

$$\frac{\partial}{\partial t}(\rho Y_s) + \frac{\partial}{\partial x_j}(\rho Y_s v_j) = \frac{\partial}{\partial x_j}\left(D\rho\frac{\partial Y_s}{\partial x_j}\right) - w_s \qquad (3.27)$$

or by combining with the continuity equation, we have

$$\rho\frac{dY_s}{dt} = \frac{\partial}{\partial x_j}\left(D\rho\frac{\partial Y_s}{\partial x_j}\right) - w_s \qquad (3.28)$$

3.2.6 Momentum Equation

For the momentum conservation equation, taking $\phi = \rho v_i$, and using Newton's second law, that is, the momentum change in the moving control volume should be the summation of all external forces, including surface forces and body forces, acting on the control volume, we have

$$\frac{d}{dt}\int_V \rho v_i \delta V = -\int_V S_v \delta V = \int_V \left(\frac{\partial p_{ij}}{\partial x_j} + \sum \rho_s F_{si}\right)\delta V$$

By using the Reynolds transport theorem and the generalized Newton–Stokes law

$$\frac{d}{dt}\int_V \rho v_i \delta V = \int_V \left[\frac{\partial}{\partial t}(\rho v_i) + \frac{\partial}{\partial x_j}(\rho v_j v_i)\right]\delta V; \quad p_{ij} \approx \mu\left(\frac{\partial v_i}{\partial x_j} + \frac{\partial v_j}{\partial x_i}\right) - p\delta_{ij}$$

The momentum conservation equation can be obtained as

$$\frac{\partial}{\partial t}(\rho v_i) + \frac{\partial}{\partial x_j}(\rho v_j v_i) = -\frac{\partial p}{\partial x} + \frac{\partial}{\partial x_j}\left[\mu\left(\frac{\partial v_i}{\partial x_j} + \frac{\partial v_j}{\partial x_i}\right)\right] + \sum \rho_s F_{si} \qquad (3.29)$$

or

$$\rho\frac{dv_i}{dt} = -\frac{\partial p}{\partial x} + \frac{\partial}{\partial x_j}\left[\mu\left(\frac{\partial v_i}{\partial x_j} + \frac{\partial v_j}{\partial x_i}\right)\right] + \sum \rho_s F_{si} \qquad (3.30)$$

The body force in Eq. (3.29) or Eq. (3.30) may be the gravitational force ρg_i, electric force $\rho_e E_i$, or Lorenz magnetic force $(J \times B)_i$. It can be seen from these two equations that except for some differences in the body-force term and the effect of species concentration on the molecular viscosity, the form of the momentum equation for multicomponent reacting flows is the same as that for single-component nonreacting flows.

3.2.7 Energy Equation

The physical basis for the energy conservation equation is the first law of thermodynamics. The heat exchange from the outside to the control volume is equal to the summation of the energy change in this control volume and the mechanical work outside by various forces.

Using the Reynolds transport theorem and taking

$$\phi = \rho \left(e + \frac{v^2}{2} \right)$$

where e and $v^2/2$ denote the internal energy and the kinetic energy per unit mass, respectively, then the total energy change in the control volume is

$$\frac{d}{dt} \int_V \rho \left(e + \frac{v^2}{2} \right) \delta V = \int_V \left[\frac{\partial}{\partial t} \rho \left(e + \frac{v^2}{2} \right) + \frac{\partial}{\partial x_j} \rho v_j \left(e + \frac{v^2}{2} \right) \right] \delta V =$$

$$\int_V \rho \frac{d}{dt} \left(e + \frac{v^2}{2} \right) \delta V$$

The net heat exchange to the control volume by heat conduction, radiation, and diffusion of multicomponent gases is

$$\int_V \frac{\partial}{\partial x_j} \left(\lambda \frac{\partial T}{\partial x_j} + q_{rj} + \sum_s D\rho \frac{\partial Y_s}{\partial x_j} h_s \right) \delta V$$

The work includes the body-force work and surface-force work

$$\int_V \sum \rho_s F_{si} v_{si} \delta V = \int_V \left(v_i \sum \rho_s F_{si} + \sum \rho_s F_{si} V_{si} \right) \delta V; \quad \int_s \frac{\partial}{\partial x_j} (p_{ij} v_i) \delta V$$

Therefore, the total energy conservation equation is

$$\rho \frac{d}{dt} \left(e + \frac{v^2}{2} \right) = \frac{\partial}{\partial x_j} \left(\lambda \frac{\partial T}{\partial x_j} \right) + \frac{\partial q_{rj}}{\partial x_j} + \frac{\partial}{\partial x_j} \left(\sum_s D\rho \frac{\partial Y_s}{\partial x_j} h_s \right) + v_i \sum \rho_s F_{si}$$

$$+ \sum \rho_s F_{si} V_{si} + \frac{\partial}{\partial x_j} (p_{ij} v_i)$$

$$(3.31)$$

It can be seen that new terms for reacting flows appear: The net enthalpy carried by the species diffusion fluxes and the summation of work done by the species body forces with the diffusion drift velocities. Furthermore, the surface-force work includes the stress work and the deformation work

$$\frac{\partial}{\partial x_j} (p_{ij} v_i) = v_i \frac{\partial p_{ij}}{\partial x_j} + p_{ij} \frac{\partial v_i}{\partial x_j}$$

The first term on the right-hand side is the stress work, which is related to the kinetic energy and part of the body-force work via momentum equation. By multiplying v_i with each term of the momentum equation, the kinetic energy equation can be obtained as:

$$v_i \rho \frac{dv_i}{dt} = \rho \frac{d}{dt}\left(\frac{v^2}{2}\right) = v_i \frac{\partial p_{ij}}{\partial x_j} + v_i \sum \rho_s F_{si} \tag{3.32}$$

On the other hand, the deformation work $p_{ij}\dfrac{\partial v_i}{\partial x_j}$ can be divided into two parts:

$$p_{ij}\frac{\partial v_i}{\partial x_j} = (-p\delta_{ij} + \tau_{ij})\frac{\partial v_i}{\partial x_j} = -p\frac{\partial v_j}{\partial x_j} + \tau_{ij}\frac{\partial v_i}{\partial x_j} \tag{3.33}$$

where the first term on the right-hand side is the compression/expansion work and the second term is the shear deformation work or the viscous energy dissipation

$$\Phi \equiv \tau_{ij}\frac{\partial v_i}{\partial x_j} = \left[2\mu \dot{S}_{ij} - \left(\frac{2}{3}\mu \frac{\partial v_j}{\partial x_j}\right)\delta_{ij}\right]\frac{\partial v_i}{\partial x_j}$$

The tensor operation rule gives

$$\frac{\partial v_i}{\partial x_j} = \frac{1}{2}\left[\left(\frac{\partial v_i}{\partial x_j} + \frac{\partial v_j}{\partial x_i}\right) + \left(\frac{\partial v_i}{\partial x_j} - \frac{\partial v_j}{\partial x_i}\right)\right] = \dot{S}_{ij} + \xi_{ij}$$

$$\dot{S}_{ij}\cdot\xi_{ij} = 0, \quad \delta_{ij}\cdot\xi_{ij} = 0, \quad \tau_{ij}\cdot\xi_{ij} = 0, \quad \delta_{ij}\frac{\partial v_i}{\partial x_j} = \frac{\partial v_j}{\partial x_j}$$

Hence the final expression of the dissipation energy is

$$\Phi = 2\mu\dot{S}_{ij} - \frac{2}{3}\mu\left(\frac{\partial v_j}{\partial x_j}\right)^2 = 2\mu(\dot{S}_{11}^2 + \dot{S}_{22}^2 + \dot{S}_{33}^2 + \dot{S}_{12}^2 + \dot{S}_{23}^2 + \dot{S}_{31}^2) - \frac{2}{3}\mu(\dot{S}_{11} + \dot{S}_{22} + \dot{S}_{33})^2 \tag{3.34}$$

Putting the kinetic energy equation and the expression of the deformation work into the total energy equation, the energy equation in terms of the internal energy can be obtained as

$$\rho\frac{de}{dt} + p\frac{\partial v_j}{\partial x_j} = \frac{\partial}{\partial x_j}\left(\lambda\frac{\partial T}{\partial x_j}\right) + \left(\frac{\partial q_{rj}}{\partial x_j}\right) + \frac{\partial}{\partial x_j}\left(\sum D\rho\frac{\partial Y_s}{\partial x_j}h_s\right) + \Phi + \sum\rho_s F_{si}V_{si} \tag{3.35}$$

Attention should be paid to the fact that the total energy equation includes the change to internal and kinetic energies, heat transfer, and mechanical work. The conservation equation of the kinetic energy includes only the change to kinetic energy and mechanical work, but is not related to heat transfer. The conservation equation of internal energy is related to heat transfer and mechanical work, but is not related to kinetic energy. Frequently, in theoretical analysis and numerical simulation it is preferred to use the energy equation in terms of enthalpy. By using the thermodynamic relationships and continuity equation,

$$\frac{dh}{dt} = \frac{d}{dt}(e + p/\rho) = \frac{de}{dt} + \frac{1}{\rho}\frac{dp}{dt} - \frac{p}{\rho^2}\frac{d\rho}{dt} \quad p\frac{\partial v_j}{\partial x_j} = -\frac{p}{\rho}\frac{d\rho}{dt}$$

The energy equation in terms of enthalpy can be obtained as

$$\rho\frac{dh}{dt} - \frac{dp}{dt} = \frac{\partial}{\partial x_j}\left(\lambda\frac{\partial T}{\partial x_j}\right) - \frac{\partial q_{rj}}{\partial x_j} + \frac{\partial}{\partial x_j}\left(\sum D\rho\frac{\partial Y_s}{\partial x_j}h_s\right) + \Phi + \sum \rho_s F_{si} V_{si}$$

$$(3.36)$$

Using the definition of enthalpy

$$h = \sum Y_s h_s, \quad h_s = h_{0s} + \int c_{ps} dT, \quad \sum_s Y_s c_{ps} = c_p, \quad \frac{\partial h}{\partial x_j} = \sum_s h_s \frac{\partial Y_s}{\partial x_j} + c_p \frac{\partial T}{\partial x_j}$$

In the case of neglecting the body force and radiation, Eq. (3.36) can be rewritten as

$$\rho\frac{dh}{dt} - \frac{dp}{dt} = \frac{\partial}{\partial x_j}\left[\frac{\mu}{\Pr}\frac{\partial h}{\partial x_j} + \mu\left(\frac{1}{Sc} - \frac{1}{\Pr}\right)\sum_s h_s \frac{\partial Y_s}{\partial x_j}\right] \quad (3.37)$$

When $\Pr = Sc = Le = 1$, this equation can be reduced to

$$\frac{\partial}{\partial t}(\rho h) + \frac{\partial}{\partial x_j}(\rho v_j h) = \frac{dp}{dt} + \frac{\partial}{\partial x_j}(\mu\frac{\partial h}{\partial x_j}) + \Phi \quad (3.38)$$

In the classical combustion theory the energy equation in terms of temperature is usually used. The mixture enthalpy can be expanded into

$$h = \sum_s Y_s\left(h_{0s} + \int c_{ps} dT\right) = h_0 + \int c_p dT$$

Its derivative is

$$\rho\frac{dh}{dt} = \rho c_p\frac{dT}{dt} + \sum h_{0s}\rho\frac{dY_s}{dt} = \rho c_p\frac{dT}{dt} + \sum h_{0s}\left[\frac{\partial}{\partial x_j}(D\rho\frac{\partial Y_s}{\partial x_j}) - w_s\right]$$

$$= \rho c_p\frac{dT}{dt} + \frac{\partial}{\partial x_j}\left(\sum D\rho\frac{\partial Y_s}{\partial x_j}h_{0s}\right) - \sum w_s h_{0s}$$

Due to

$$h_s - h_{0s} = \int c_{ps} dT \quad \sum w_s h_{0s} = w_s Q_s$$

where $w_s Q_s$ is the heat release of reaction, the energy equation in terms of temperature can be obtained as

$$\rho c_p \frac{dT}{dt} - \frac{dp}{dt} = \frac{\partial}{\partial x_j}\left(\lambda \frac{\partial T}{\partial x_j}\right) + \frac{\partial q_{rj}}{\partial x_j} + \frac{\partial}{\partial x_j}\left(\sum D\rho \frac{\partial Y_s}{\partial x_j}\int c_{ps} dT\right) + w_s Q_s + \Phi$$
$$+ \sum \rho_s F_{si} V_{si}$$

(3.39)

If the Mach number is much smaller than unity, in comparison to the enthalpy change the pressure change and dissipation energy can be neglected. In case of neglecting the body force and radiative heat transfer, and assuming the equality of species specific heat,

$$c_{p1} = c_{p2} = c_{ps} = \cdots = c_p$$

then the energy equation can be simplified into the so-called Zeldovich−Schwab form

$$\rho c_p \frac{dT}{dt} - \frac{\partial p}{\partial t} = \frac{\partial}{\partial x_j}\left(\lambda \frac{\partial T}{\partial x_j}\right) + w_s Q_s$$

(3.40)

For the simplest laminar multicomponent reacting flows, assuming a global one-step second-order reaction, the basic equations can be summarized as

$$\frac{\partial \rho}{\partial t} + \frac{\partial}{\partial x_j}(\rho v_j) = 0$$

$$\frac{\partial}{\partial t}(\rho v_i) + \frac{\partial}{\partial x_j}(\rho v_j v_i) = -\frac{\partial p}{\partial x_i} + \frac{\partial}{\partial x_j}\left[\mu\left(\frac{\partial v_i}{\partial x_j} + \frac{\partial v_j}{\partial x_i}\right)\right] + \rho g_i$$

$$\frac{\partial}{\partial t}(\rho Y_s) + \frac{\partial}{\partial x_j}(\rho v_j Y_s) = \frac{\partial}{\partial x_j}\left(D\rho \frac{\partial Y_s}{\partial x_j}\right) - w_s$$

(3.41)

$$\frac{\partial}{\partial t}(\rho c_p T) + \frac{\partial}{\partial x_j}(\rho v_j c_p T) - \frac{\partial p}{\partial t} = \frac{\partial}{\partial x_j}\left(\lambda \frac{\partial T}{\partial x_j}\right) + w_s Q_s$$

$$w_s = B\rho^2 Y_1 Y_2 \exp(-E/RT)$$
$$p = \rho RT \sum Y_s/M_s$$

3.2.8   Boundary Conditions at the Interface and Stefan Flux

The gas−solid/liquid interface may be a channel wall, an ablating surface, a catalytic surface, or a solid/liquid-fuel surface. The physical/chemical changes may exist, such as evaporation/condensation, sublimation,

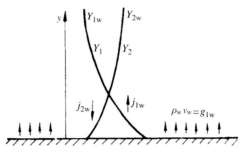

FIGURE 3.1 Water evaporation.

devolatilization, and gas–solid heterogeneous reactions at the interface. Alternatively, there may be no physical/chemical changes at the inert surfaces, such as at a ceramic wall. The physical/chemical processes at the interface determine the specific features of heat and mass transfer at the interface. First let us consider water evaporation, as shown in Fig. 3.1.

The mass fluxes at the interface can be written as

$$g_{1w} = J_{1w} + Y_{1w}\rho_w v_w = -D\rho \left(\frac{\partial Y_1}{\partial y}\right)_w + Y_{1w}\rho_w v_w$$

$$g_{2w} = J_{2w} + Y_{2w}\rho_w v_w = -D\rho \left(\frac{\partial Y_2}{\partial y}\right)_w + Y_{2w}\rho_w v_w = 0$$

(3.42)

$$g_{1w} = -D\rho \left(\frac{\partial Y_1}{\partial y}\right)_w + Y_{1w}\rho_w v_w = D\rho \left(\frac{\partial Y_2}{\partial y}\right)_w + Y_{1w}\rho_w v_w =$$

$$(Y_{2w} + Y_{1w})\rho_w v_w = \rho_w v_w - D\rho \left(\frac{\partial Y_1}{\partial y}\right)_w = (1 - Y_{1w})\rho_w v_w$$

(3.43)

It can be seen that the water evaporation rate is its total mass flux but not its diffusion flux. The water total mass flux is equal to its diffusion flux plus part of the mass flux carried by the mixture flows. The water evaporation rate is equal to the mixture mass flux, i.e., the Stefan flux [1,2]. The next case is carbon combustion in oxygen, as shown in Fig. 3.2.

The mass fluxes and diffusion fluxes at the carbon surface are

$$C + O_2 \rightarrow CO_2$$

$$\left(\frac{\partial Y_{ox}}{\partial y}\right)_w = -\left(\frac{\partial Y_{pr}}{\partial y}\right)_w, \quad J_{ox,w} = -J_{pr,w}$$

$$g_{ox,w} + g_{pr,w} = \rho_w v_w = -\frac{12}{32}g_{ox,w} = g_c \quad g_{ox,w} = -\frac{32}{44}g_{pr,w}$$

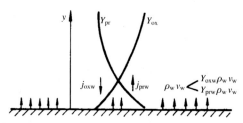

FIGURE 3.2   Carbon combustion in oxygen.

$$g_{\text{ox},w} = -D\rho\left(\frac{\partial Y_{\text{ox}}}{\partial y}\right)_w + Y_{\text{ox},w}\rho_w v_w \quad g_{\text{pr},w} = -D\rho\left(\frac{\partial Y_{\text{pr}}}{\partial y}\right)_w + Y_{\text{pr},w}\rho_w v_w$$

$$(3.44)$$

In this case, the oxygen and carbon-dioxide mass fluxes are not their diffusion fluxes, and none of them is equal to the total mass flux—the Stefan flux. From these two examples, it is concluded that the sufficient and necessary conditions for the occurrence of a Stefan flux are: there are physical/chemical process at the interface; there is multicomponent diffusion to or from the interface. Taking the Stefan flux into account, the boundary conditions at the liquid−gas interface can be described, as shown in Fig. 3.3.

The boundary conditions for the species mass fluxes and concentrations are

$$g_{sw} = -D\rho\left(\frac{\partial Y_s}{\partial y}\right)_w + Y_{sw}\rho_w v_w = \alpha\rho_w v_w \quad (s=F, \alpha=1; \ s=\text{ox},\text{pr},\text{iner},\ldots,\alpha=0)$$

$$\sum g_{sw} = \rho_w v_w = g_{Fw} \quad \sum Y_s = Y_F + Y_{\text{ox}} + Y_{\text{pr}} + Y_{\text{iner}} = 1$$

$$(3.45)$$

The boundary conditions for the heat flux is

$$\lambda\left(\frac{\partial T}{\partial y}\right)_w + \varepsilon\sigma(T_\infty^4 - T_w^4) = \rho_w v_w L + \lambda_l\left(\frac{\partial T}{\partial y}\right)_{lw} \quad Y_{Fw} = B_w \exp(-E_w/RT_w)$$

$$(3.46)$$

When neglecting the radiation and heat conduction inside the liquid, it becomes

$$\lambda\left(\frac{\partial T}{\partial y}\right)_w = \rho_w v_w L = \rho_w v_w q_e$$

$$(3.47)$$

The boundary conditions at the solid−gas interface are shown in Fig. 3.4.

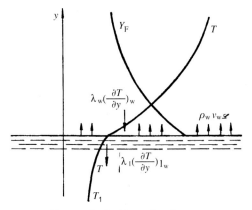

FIGURE 3.3 Boundary conditions at the liquid–gas interface.

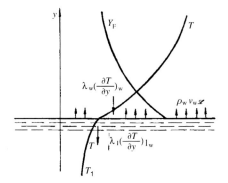

FIGURE 3.4 Boundary conditions at the solid–gas interface.

The boundary conditions for the species mass fluxes and concentrations are

$$g_{sw} = -D\rho\left(\frac{\partial Y_s}{\partial y}\right)_w + Y_{sw}\rho_w v_w = \sum_r w_{sr} \quad g_{iner} = -D\rho\left(\frac{\partial Y_{iner}}{\partial y}\right)_w + Y_{iner,w}\rho_w v_w = 0$$

$$\sum_s g_{sw} = \rho_w v_w, \quad \alpha g_{Fw} = \beta g_{ox,w} = \cdots$$

$$(3.48)$$

The boundary conditions for the heat flux are

$$\sum_r w_{sr}Q_{sr} = \lambda\left(\frac{\partial T}{\partial y}\right)_w + \lambda_1\left(\frac{\partial T}{\partial y}\right)_{1w} + \varepsilon\sigma(T_w^4 - T_\infty^4) \qquad (3.49)$$

3.3 IGNITION AND EXTINCTION

3.3.1 Basic Concept

Ignition is a critical phenomenon of transition from a slow reaction to a high-temperature fast reaction. Ignition is routinely classified into auto-ignition, forced ignition, thermal ignition, and chemical ignition. Actually, ignition is the interaction between heat release and heat loss, chemical kinetics, and heat and mass transfer. A false understanding is presented as ignition temperature being a constant property for gas/liquid/solid materials. In fact, the ignition temperature is a function of pressure, velocity, geometrical size, and mixing ratio $T_i = f(p, d, v, \alpha)$. Ignition condition is the initial or boundary condition when or where the reaction begins to automatically accelerate and rapidly reaches a high-temperature state. The combustion reactions have the following features: (1) the reactions have high heating effects: Carbon: $Q = 8800$ kcal/kg; Hydrocarbons: $Q = 10,000$ kcal/kg; Hydrogen: $Q = 33000$ kcal/kg; (2) the reaction rate rapidly increases with the increase of temperature. For example,

$w_s = B\rho^2 Y_1 Y_2 \exp(-E/RT)$ $E = 20,000$ kcal/kgmol $R = 2$ kcal/(kgmol $- k$)

$T_1 = 500$K, $\exp(-E/RT_1) \approx 2 \times 10^{-9}$

$T_2 = 1000$K, $\exp(-E/RT_2) \approx 4 \times 10^{-5}$

$T_2 = 2T_1$, $w_2 = 20,000\, w_1$

3.3.2 Dimensional Analysis

At first, ignition may be described by using the Frank-Kamenetsky's dimensional analysis [1,2]. The energy equation in Zeldovich−Schwab's form for a steady state is

$$\rho c_p \frac{dT}{dt} = \rho c_p v_j \frac{\partial T}{\partial x_j} = \frac{\partial}{\partial x_j}\left(\lambda \frac{\partial T}{\partial x_j}\right) + w_s Q_s \tag{3.50}$$

Dividing each term of Eq. (3.50) by $E/(RT_1^2)$, and setting

$$\tau_c = \frac{RT_1^2}{E}\frac{\rho c_p}{w_{s1} Q_s} \quad \tau_d = \rho c_p L^2/\lambda \quad D = \tau_d/\tau_c$$

$$\theta = (T - T_1)/(RT_1^2/E) \quad \bar{t} = t/\tau_d \quad \xi = x_j/L$$

the dimensionless energy equation is obtained as

$$\frac{d\theta}{d(t/\tau_c)} = \frac{1}{D}\frac{\partial}{\partial \xi}\left(\frac{\partial \theta}{\partial \xi}\right) - w_s/w_{s1} \tag{3.51}$$

The Frank-Kamenetsky's approximation gives

$$(T - T_1)/T_1 = \Delta T/T_1 \ll 1, \quad \exp(-E/RT) = \exp\left[-\frac{E}{RT_1(1 + \Delta T/T_1)}\right]$$

$$\approx \exp(-E/RT_1) \exp\left(\frac{E}{RT_1^2}\Delta T\right) = \exp(-E/RT_1)e^{\theta}$$

So, we have

$$w_s \approx w_{s1}e^{\theta}$$

$$\frac{d\theta}{d\bar{t}} = \frac{\partial}{\partial\xi_j}\left(\frac{\partial\theta}{\partial\xi_j}\right) + De \quad \theta = \theta(D, \xi_j, \bar{t}) \quad \text{Steady solution} \quad \theta = \theta(D, \xi_j)$$

Then, the ignition condition is

$$D = D_{cr} \quad\quad\quad\quad\quad (3.52)$$
$$\tau_d/\tau_c = D_{cr} = \text{const}$$

3.3.3 Ignition in an Enclosed Vessel—Simonov's Unsteady Model

Simonov [1,3] proposed an unsteady model of ignition in an enclosed vessel. Assume: (1) a premixed combustible mixture in a spherical vessel (Fig. 3.5); (2) uniformly distributed temperature increases with time; (3) no forced and natural convection in the vessel; (4) only heat conduction between the vessel and the surrounding environment; and (5) negligible concentration change before ignition. The simplified energy equation is:

$$\rho V c_v \frac{dT}{dt} = V w_s Q_s - 2\pi\lambda(T - T_\infty)$$

$$\rho c_v \frac{dT}{dt} = q_1 - q_2 = w_s Q_s - \frac{12}{d^2}(T - T_\infty) \quad\quad\quad (3.53)$$

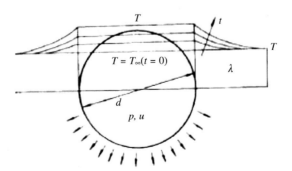

FIGURE 3.5 Ignition in an enclosed vessel.

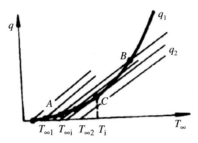

FIGURE 3.6 The ignition condition.

It can be seen from Fig. 3.6 that the equilibrium state may be a steady slow reaction or false equilibrium state. Increasing the initial temperature may lead to the heat-release curve being tangential to the heat-loss curve. The tangential point is the critical point. Increasing pressure, reducing heat conductivity, or increasing container size can also lead to ignition. Ignition temperature is the temperature satisfying the ignition condition, but not that at the tangent point. The final combustion temperature is much higher than the ignition temperature

Therefore, the ignition condition is

$$q_1 = q_2, \quad \frac{dT}{dt} = 0 \quad q_1 = q_2, \quad \frac{\partial q_1}{\partial T} = \frac{\partial q_2}{\partial T} \tag{3.54}$$

Then, the final expression for the ignition condition can be obtained from

$$q_1 = q_2 \rightarrow \quad w_s Q_s = 12\lambda(T_i - T_\infty)/d^2$$

$$\frac{\partial q_1}{\partial T} = \frac{\partial q_2}{\partial T} \rightarrow \quad E w_s Q_s/(RT_i^2) = 12\lambda/d^2$$

$$T_i - T_{\infty i} = RT_i^2/E \quad T_i \approx T_{\infty i} + RT_{\infty i}^2/E$$

Finally, we obtain the ignition condition in the enclosed vessal as

$$\frac{E}{RT_\infty^2} \, w_{s\infty} Q_s = 12\lambda/d^2$$

$$\tau_c = \rho c_v \, RT_\infty^2/(E w_{s\infty} Q_s) \quad \tau_d = \rho c_v d^2/\lambda$$

$$\tau_d/\tau_c = 12$$

From these results, it can be seen that ignition is the interaction of chemical kinetics with heat and mass transfer. Ignition condition is a functional relationship of temperature, pressure, mixing ratio, and container size. Ignition temperature decreases with an increase in pressure. Ignition temperature decreases with an increase in the container size. There are minimum ignition temperature and rich and lean concentration limits under fixed pressure and container size. The theoretical results are in agreement with the experimental results, as shown in Figs. 3.7–3.9.

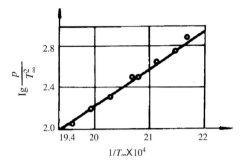

FIGURE 3.7 Ignition pressure versus temperature.

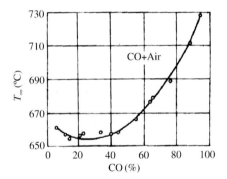

FIGURE 3.8 Ignition temperature versus CO concentration.

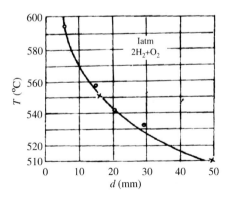

FIGURE 3.9 Ignition temperature versus vessel size.

3.3.4 Ignition Lag (Induction Period)

Ignition does not come immediately, even if the ignition condition is satisfied. Ignition lag is the time period from the instant of the initial state to the instant of a rapid rise in temperature. Ignition lag is a maximal finite value when the ignition condition is satisfied. Ignition lag is not zero even if the initial condition is superior to the ignition condition. Ignition lag and ignition condition are not the same thing, but they are closely related to each other. From the $T(t)$ curve, obtained by the energy equation, it can be seen that

$$T_\infty < T_{\infty i} \quad \frac{dT}{dt} \to 0 \quad \tau_i = \infty$$

$$T_\infty = T_{\infty i} \quad \frac{dT}{dt} > 0 \to \frac{dT}{dt} = 0 \quad t = \tau_i$$

$$T_\infty > T_{\infty i} \quad \frac{d^2 T}{dt^2} < 0 \to \frac{d^2 T}{dt^2} > 0 \quad \frac{d^2 T}{dt^2} = 0 \quad \frac{dT}{dt} \neq 0$$

Then, the ignition lag can be obtained from the following equation

$$c_v \frac{dT}{dt} = w_s Q_s - 12\lambda(T - T_\infty)/d^2 \tag{3.56}$$

Taking the following procedure, Eq. (3.56) is multiplied by $E/(RT_\infty^2)$ and divided by ρc_v

Taking $\dfrac{RT_\infty^2}{E}, \tau_c = \dfrac{RT_\infty^2}{E} \dfrac{\rho_\infty c_v}{w_{s\infty} Q_s}, \tau_d = \rho_\infty c_v d^2 / \lambda, \theta = E(T - T_\infty)/(RT_\infty^2), \bar{t} = t/\tau_c$

Then we obtain

$$\frac{d\theta}{d\bar{t}} = e^\theta - 12\theta/D \quad D = \tau_d/\tau_c \quad \text{Adiabatic Ignition Lag } q_2 = 0$$

$$\frac{d\theta}{dt} = e^\theta/\tau_c, \quad \int_0^\theta e^\theta d\theta = \int_0^t d\left(\frac{t}{\tau_c}\right) \quad t/\tau_c = 1 - e^\theta$$

$$t = \tau_i \quad \theta_i = 1 \quad \tau_i = \left(1 - \frac{1}{e}\right)\tau_c = \frac{RT_\infty^2}{E} \frac{\rho_\infty c_v}{w_{s\infty} Q_s}$$

The adiabatic ignition lag is

$$\tau_i = K p^{1-n} [\phi(\alpha)]^{-1} \exp(E/RT_\infty) \tag{3.57}$$

It can be seen that the ignition lag rapidly decreases with an increase in temperature; the ignition lag decreases with an increase in pressure; there are minimum ignition lag, higher and lower limits under fixed temperature and pressure when changing the mixing ratio. These theoretical results are validated by experiments, as shown in Figs. 3.10 and 3.11.

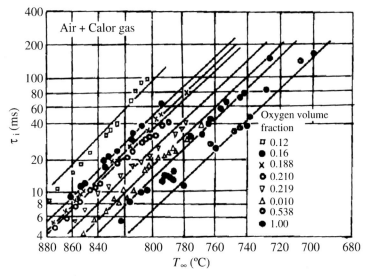

FIGURE 3.10   Ignition lag versus temperature.

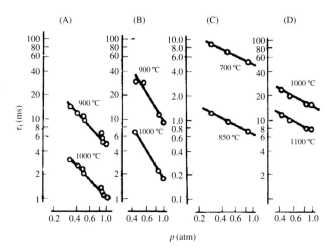

FIGURE 3.11   Ignition lag versus pressure.

3.3.5   Ignition by a Hot Plate—Khitrin–Goldenberg Model

Ignition may take place by a spark, a flamelet, or a hot surface. This is local ignition by a local high temperature. If the surface temperature is slightly higher than the gas temperature, the wall temperature gradient may become zero at a certain distance x, but there is no ignition (see Fig. 3.12). If the surface temperature is sufficiently high (e.g., 1273K), ignition may take place at a certain distance x, and after that there is a temperature peak. The location at which the wall temperature gradient is zero may be specified as the

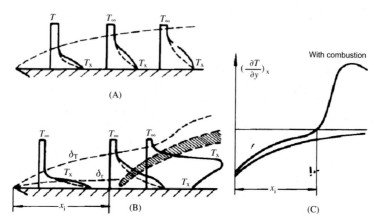

FIGURE 3.12 Ignition by a hot plate.

ignition distance, behind that the heat will be transferred from the gas to the wall. If the plate length L is smaller than x_i, ignition fails.

The energy equation for 2-D laminar boundary-layer reacting flows is

$$\rho u c_p \frac{\partial T}{\partial x} + \rho v c_p \frac{\partial T}{\partial y} = \frac{\partial}{\partial y}\left(\lambda \frac{\partial T}{\partial y}\right) + w_s Q_s \tag{3.58}$$

The simplified solution given by the Khitrin−Goldenberg model [1,4] divides the boundary layer into two zones:

Zone 1: Reaction zone, $\delta_r < y < \delta_T$, where the convection term can be neglected, and the energy equation is simplified to

$$\lambda_w \frac{\partial}{\partial y}\left(\frac{\partial T}{\partial y}\right) = -w_s Q_s \tag{3.59}$$

Zone 2: Preheating zone, $0 < y < \delta_r$, where the reaction term can be neglected, and the energy equation is simplified to

$$\rho u c_p \frac{\partial T}{\partial x} + \rho v c_p \frac{\partial T}{\partial y} = \frac{\partial}{\partial y}\left(\lambda \frac{\partial T}{\partial y}\right) \tag{3.60}$$

The solution to Eq. (3.50) gives

$$\frac{\partial}{\partial y}\left(\frac{\partial T}{\partial y}\right) = \frac{\partial T}{\partial y}\frac{\partial}{\partial T}\left(\frac{\partial T}{\partial y}\right) = \frac{1}{2}\frac{\partial}{\partial T}\left(\frac{\partial T}{\partial y}\right)^2$$

$$\left(\frac{\partial T}{\partial y}\right)_w^2 - \left(\frac{\partial T}{\partial y}\right)_1^2 = \frac{2Q_s}{\lambda_w}\int_{T_w}^{T_r} w_s dT \approx \frac{2Q_s}{\lambda_w}\int_{T_w}^{T_\infty} w_s dT$$

$$x = x_i, \quad \left(\frac{\partial T}{\partial y}\right)_w = 0 \quad \left(\frac{\partial T}{\partial y}\right)_1 = \sqrt{\frac{2Q_s}{\lambda_w} \int_{T_\infty}^{T_w} w_s dT}$$

The solution to Eq. (3.60) gives

$$\left(\frac{\partial T}{\partial y}\right)_2 \approx \frac{h^*(T_w - T_\infty)}{\lambda_w} = \frac{Nu_x^*}{x}(T_w - T_\infty)$$

The two-zone coupling gives

$$\left(\frac{\partial T}{\partial y}\right)_1 = \left(\frac{\partial T}{\partial y}\right)_2$$

So, the ignition condition can be obtained as

$$\left(\frac{Nu_x^*}{x}\right)^2 = \frac{1}{(T_w - T_\infty)^2} \frac{2Q_s}{\lambda_w} \int_{T_\infty}^{T_w} w_s dT \quad Nu_x^* = f''(0)Re_x^{0.5} = 0.332 Re_x^{0.5}$$

$$\tau_i = x_i/u_\infty = \rho c_p (T_w - T_\infty)^2 \Big/ \left(0.332 Q_s \int_{T_\infty}^{T_w} w_s dT\right)$$

$$\overline{\tau}_c = \rho c_p (T_w - T_\infty)^2 \Big/ \left(Q_s \int_{T_\infty}^{T_w} w_s dT\right) \quad \tau_i/\overline{\tau}_c = 0.332$$

$$(3.61)$$

These results indicate that the ignition distance is proportional to the oncoming velocity; it rapidly decreases with an increase in wall temperature, and there are a minimum ignition distance, a lean limit, and a rich limit when changing only the mixing ratio. These theoretical results are validated by experiments, as shown in Figs. 3.13−3.15.

3.3.6 Ignition and Extinction—Vulis Model

Although ignition and extinction are similar critical phenomena, they are however, not reversible; the ignition point and the extinction point do not coincide with each other. After Vulis [1,5], consider ignition and extinction in a simple opened vessel, as shown in Fig. 3.16. The assumptions are uniform T and Y in the vessel and also in the combustion products, inlet temperature, concentration, and mass flux are T_∞, Y_∞, and M, respectively; adiabatic wall and first-order reaction. It is seen from Fig. 3.17 that there are two intersection points (slow reaction and combustion) and two tangential points (ignition and extinction).

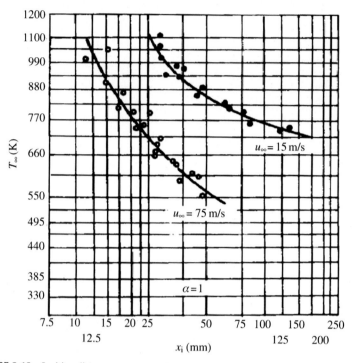

FIGURE 3.13 Ignition distance versus temperature.

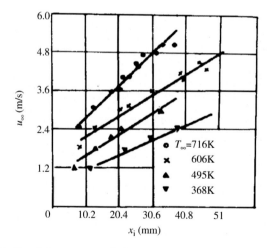

FIGURE 3.14 Ignition distance versus velocity.

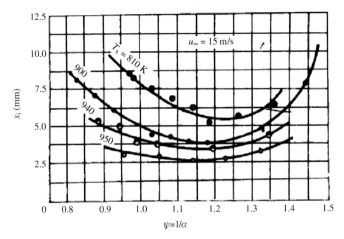

FIGURE 3.15 Ignition distance versus mixing ratio.

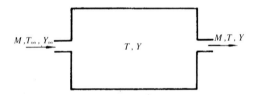

FIGURE 3.16 Ignition and extinction in an opened vessel.

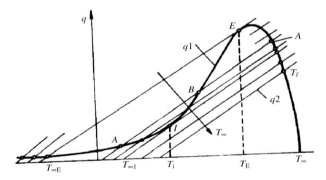

FIGURE 3.17 Two critical points.

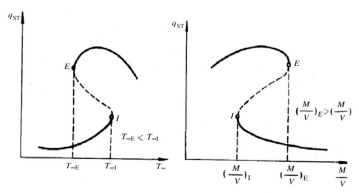

FIGURE 3.18   Ignition and extinction points.

In the present situation, the heat and mass balance between the heat release and heat loss, species mass consumption, and output are given by the following expressions:

$$Q_1 = VQ_sB\rho_\infty Y \exp(-E/RT) \quad Q_2 = Mc_p(T - T_\infty)$$

$$G_1 = VB\rho_\infty Y \exp(-E/RT) \quad G_2 = M(Y - Y_\infty)$$

$$Q_1 = Q_2 \quad G_1 = G_2$$

$$VQ_sB\rho_\infty Y \exp(-E/RT) = Mc_p(T - T_\infty) \quad VB\rho_\infty Y \exp(-E/RT) = M(Y - Y_\infty)$$

$$c_p(T - T_\infty) = Q_s(Y_\infty - Y) \quad c_p(T_m - T_\infty) = Q_sY_\infty$$

$$q_1 = Q_1/V = Q_sB\rho_\infty Y_\infty \left(\frac{T_m - T}{T_m - T_\infty}\right)\exp(-E/RT) \quad q_2 = Q_2/V = Mc_p(T - T_\infty)/V$$

It can be seen from Fig. 3.18 that changing T_∞ or M/V may lead to ignition or extinction. There are three regions: there is only combustion; there is only a slow reaction; there is no ignition but also no extinction if combustion already takes place. Therefore, the ignition and extinction conditions are expressed by:

Ignition-Extinction   Conditions

$$Q_1 = Q_2 \quad G_1 = G_2 \quad \frac{\partial Q_1}{\partial T} = \frac{\partial Q_2}{\partial T}$$

Ignition-Extinction   Expressions

$$1 + \frac{M}{\rho_\infty VB}\exp(E/RT_\infty) = \frac{Y_\infty QE}{c_pRT_\infty^2}$$

$$\frac{M}{\rho_\infty VB}\exp(E/RT_m)\left[1 + \frac{M}{\rho_\infty VB}\exp(E/RT_m)\right]^{-2} = \frac{Rc_pT_m^2}{EY_\infty Q}$$

(3.62)

From these expressions it can be concluded that both ignition and extinction are critical phenomena, similar to each other. Both depend on flow velocity, geometrical size, temperature, and initial concentration. However, the ignition condition does not equal the extinction condition. For ignition the temperature effect is larger and the concentration effect is smaller, and for extinction the concentration effect is larger and the temperature effect is smaller.

3.4 LAMINAR PREMIXED AND DIFFUSION COMBUSTION

3.4.1 Background

The aim of studying laminar premixed combustion is to understand the reaction kinetics. The laminar premixed flame is a thin layer with rapid temperature and concentration changes, where reaction, heat and mass transfer take place. At the hot boundary the fuel and oxygen concentration become zero and the temperature and combustion product concentration become maximal. The flame propagation is created by the continuous ignition of fresh mixture by heat conduction from the high-temperature combustion products. Reaction and heat conduction dominate flame propagation.

3.4.2 Basic Equations and Their Properties

The basic equations for one-dimensional laminar premixed reacting flows are:

$$\rho u = \rho_\infty u_\infty = \rho_\infty S_l = \text{const}$$

$$p \approx \text{const}$$

$$\rho u \frac{dY_s}{dx} = \frac{d}{dx}\left(D\rho \frac{dY_s}{dx}\right) - w_s$$

$$\rho u c_p \frac{dT}{dx} = \frac{d}{dx}\left(\lambda \frac{dT}{dx}\right) + w_s Q_s \qquad (3.63)$$

$$\rho_\infty S_l c_p \frac{dT}{dx} = \frac{d}{dx}\left(\lambda \frac{dT}{dx}\right) + w_s Q_s$$

For an adiabatic flame the boundary conditions are:

$$x \to -\infty \quad T = T_\infty, \quad \frac{dT}{dx} = 0$$

$$x \to +\infty \quad \frac{dT}{dx} = 0 \qquad (3.64)$$

There are three adiabatic boundary conditions for a second-order differential equation. The solution is possible only for a certain parameter S_l. Hence, this is an eigen-value problem. The energy equation can be transformed into

$$p = \frac{dT}{dx} \quad \frac{\lambda}{c_p \rho_\infty} p \frac{dp}{dT} - S_l p + \frac{w_s Q_s}{c_p \rho_\infty} = 0 \tag{3.65}$$

At the cold boundary, owing to a zero temperature gradient there will be no finite solution, i.e., an infinitive flame propagation velocity, if the reaction rate is not zero. The existence of a finite solution requires a zero reaction rate—no ignition at the cold boundary. The actual reaction rate at the cold boundary is not zero—this is called the "cold-boundary difficulty."

$$x \rightarrow -\infty \quad T = T_\infty \quad p = 0$$

Due to the low temperature at the cold boundary the reaction rate is close to zero, so, the mathematical formulation approximately reflects the physical process.

The similarity of temperature and concentration profiles and the law of conservation of enthalpy across the flame can be obtained from the basic equations without using their solution. Using the continuity equation, we have the energy and species equations in the following form:

$$\rho_\infty S_l c_p \frac{dT}{dx} = \frac{d}{dx} \left(\lambda \frac{dT}{dx} \right) + w_1 Q_1$$

$$\rho_\infty S_l \frac{dY_1}{dx} = \frac{d}{dx} \left(D\rho \frac{dY_1}{dx} \right) - w_1$$

Integration from the cold boundary to the hot boundary gives:

$$\rho_\infty S_l c_p (T_m - T_\infty) = \int_{-\infty}^{+\infty} w_1 Q_1 dx \quad \rho_\infty S_l Y_{1\infty} Q_1 = \int_{-\infty}^{+\infty} w_1 Q_1 dx \tag{3.66}$$
$$c_p (T_m - T_\infty) = Y_{1\infty} Q_1 \quad c_p T_m = Y_{1\infty} Q_1 + c_p T_\infty$$

It shows the conservation of enthalpy across the flame, that is, the cold-boundary enthalpy equals that of the hot boundary, whether the Le number is unity or not. The hot-boundary temperature is the adiabatic temperature. On the other hand, when $\mathrm{Le} = D\rho c_p / \lambda = 1$, we have

$$\theta = c_p (T_m - T_\infty)/(Y_1 Q_1) = (T_m - T)/(T_m - T_\infty)$$
$$F = Y_1/Y1_\infty$$

$$S_1 \frac{d\theta}{dx} = \frac{d}{dx} \left(D\rho \frac{d\theta}{dx} \right) - w_1/Y_{1\infty}$$

$$S_1 \frac{dF}{dx} = \frac{d}{dx} \left(D\rho \frac{dF}{dx} \right) - w_1/Y_{1\infty}$$

Hence, the result gives

$$\theta = F$$

$$Y_1/Y_{1\infty} = (T_m - T)/(T_m - T_\infty) \qquad (3.67)$$

$$c_p T + Y_1 Q_1 = c_p T_\infty + Y_{1\infty} Q_1 = c_p T_m$$

This result implies that when $Le = 1$, the temperature and concentration distributions across the flame are similar to each other, and when $Le = 1$, and only when $Le = 1$, the enthalpy conserves at each location across the flame.

3.4.3   Two-Zone Approximate Solution

Frank-Kamenetsky [1,2] proposed a two-zone approximate solution, according to which the flame can be divided into two zones: the preheating zone I and the reaction zone II, as shown in Fig. 3.19.

There is a negligible reaction in zone I, and negligible convection in zone II. For zone I, the integration from the cold boundary to its border with the reaction zone gives

$$\rho_\infty S_1 c_p \frac{dT}{dx} = \frac{d}{dx}\left(\lambda \frac{dT}{dx}\right) \qquad \rho_\infty S_1 c_p (T_1 - T_\infty) = \lambda \left(\frac{dT}{dx}\right)_1$$

For zone II, the integration from its border with the preheating zone to the hot boundary gives

$$\frac{d}{dx}\left(\lambda \frac{dT}{dx}\right) + w_1 Q_1 = 0 \qquad \left(\frac{dT}{dx}\right)_2 = \sqrt{\frac{2Q_1}{\lambda} \int_{T_2}^{T_m} w_1 \, dT}$$

$$T_1 = T_2 \qquad \left(\frac{dT}{dx}\right)_1 = \left(\frac{dT}{dx}\right)_2$$

$$S_l = \sqrt{\frac{2\lambda Q_1 \int_{T_1}^{T_m} w_1 \, dT}{\rho_\infty^2 c_p^2 (T_1 - T_\infty)^2}}$$

Further approximation was made as

$$T_1 - T_\infty \approx T_m - T_\infty \qquad \int_{T_1}^{T_m} w_1 \, dT \approx \int_{T_\infty}^{T_m} w_1 \, dT$$

Hence the final expression for the laminar flame propagation velocity is

$$S_1 = \sqrt{\frac{2\lambda Q_1 \int_{T_\infty}^{T_m} w_1 \, dT}{\rho_\infty^2 c_p^2 (T_m - T_\infty)^2}} \qquad (3.68)$$

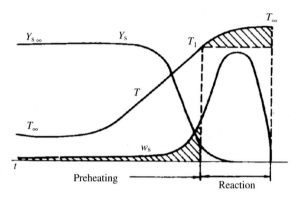

FIGURE 3.19   Two zones of the laminar premixed flame.

It can be seen that the laminar flame velocity is the interaction between the reaction heat release and the heat conduction. This depends on only reaction kinetics and thermophysical properties, but is independent of flow velocity and sizes. It slightly increases with the decrease of pressure. It has a peak value with the change of the mixing ratio. These relationships are:

$$S_l \sim \sqrt{\lambda} \quad S_l \sim \sqrt{w_1} \quad S_l \sim p^{n/2-1}; \quad S_l \sim \varphi(\alpha)$$

Tsukhanova [1] proposed a modified approximate solution for the laminar flame velocity, using the results of an exact solution of the energy equation, as shown in Fig. 3.20.

The flame can be divided into three zones: the preheating zone; and the front and rear reaction zones. The original two-zone model takes a coupling of zone 1 with zone 3. The modified model takes a coupling of zone 1 with zone 2. Therefore, we have

$$\text{Zone 1:} \quad w_1 Q_1 \approx 0 \quad S_l \frac{dT}{dx} = \frac{d}{dx}\left(\frac{\lambda}{c_p \rho_\infty} \frac{dT}{dx}\right)$$

$$\text{Zone 2:} \quad \frac{d^2 T}{dx^2} \approx 0 \quad S_l \frac{dT}{dx} = \frac{w_1 Q_1}{c_p \rho_\infty}$$

$$\text{Zone 3:} \quad S_l \frac{dT}{dx} \approx 0 \quad \frac{d}{dx}\left(\lambda \frac{dT}{dx}\right) = -w_1 Q_1$$

$$\text{Original} \quad \left(\frac{dT}{dx}\right)_{1,3} = \left(\frac{dT}{dx}\right)_{3,1}$$

$$\text{Modified} \quad \left(\frac{dT}{dx}\right)_{1,2} = \left(\frac{dT}{dx}\right)_{2,1}$$

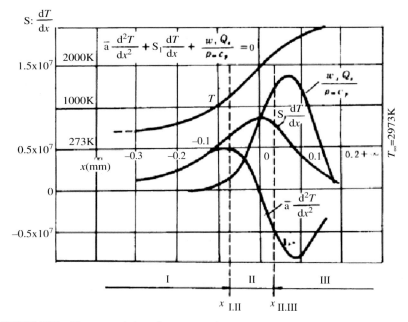

FIGURE 3.20 The exact solution of energy equation.

The flame velocity is obtained as

$$\left(\frac{dT}{dx}\right)_{1,2} = \rho_\infty c_p S_l (T_1 - T_\infty)/\lambda \qquad \left(\frac{dT}{dx}\right)_{2,1} = \frac{w_1 Q_1}{\rho_\infty c_p S_l}$$

$$S_l = \sqrt{\frac{\lambda_1 Q_1 w_{11}}{\rho_\infty^2 c_p^2 (T_1 - T_\infty)}}$$

To determine T_1, using the coupling condition, the result is

$$\frac{\partial}{\partial T}\left[\left(\frac{dT}{dx}\right)_{1,2}\right] = \frac{\partial}{\partial T}\left[\left(\frac{dT}{dx}\right)_{2,1}\right] \quad \text{or} \quad \left(\frac{d^2 T}{dx^2}\right)_{1,2} = \left(\frac{d^2 T}{dx^2}\right)_{2,1}$$

$$\rho_\infty c_p S_l / \lambda_1 = \frac{Q_1}{\rho_\infty c_p S_l} w_1'(T_1)$$

The final expression is

$$S_l = \sqrt{\frac{\lambda_1 Q_1 w_{11}}{\rho_\infty^2 c_p^2 (T_1 - T_\infty)}} \qquad \frac{1}{(T_1 - T_\infty)} = \frac{w_1'(T_1)}{w_1(T_1)} \qquad (3.69)$$

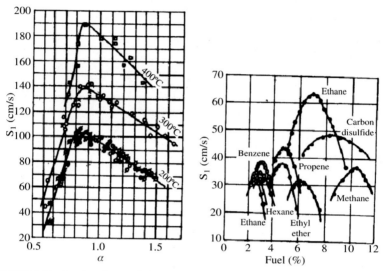

FIGURE 3.21 Laminar flame velocity.

The experimental results reported by Inozemtsev et al. [1] are shown in Fig. 3.21.

This shows the laminar flame velocity S as functions of fuel/air ratio and fuel types. S changes with the mixing ratio and reaches a maximum at the stoichiometric ratio. S is in the range of 0.2 m/s to 2 m/s, depending on the fuel type. S for acetylene and ethylene is much larger than that for methane. S is proportional to a $1.5 - 2$ power of the temperature. S is proportional to -0.15 to -0.33 power of the pressure. The flame velocity S is a mixture property, independent of flow velocity and sizes.

Using the experimental results of S versus temperature and pressure, the kinetics parameters can be found. For example, the reaction kinetics of CO−oxygen combustion are found to be:

$$\ln S_1 = (n/2 - 1)\ln p + c_1$$
$$\ln S_1^2 = -E/(RT_m) + c_2$$

$$\text{Exp: } w_{co} = (1.14 \sim 2.5) \times 10^6 Y_{CO} Y_{OX}^{0.25} \rho^{1.25} T^{-2.25} \exp\left(-\frac{46,000}{RT}\right)$$

3.4.4 Laminar Diffusion Flame

By simplifying the diffusion equation, the length of a gaseous-fuel−air coflowing laminar diffusion flame can be obtained as:

$$L = V/(2\pi D) \tag{3.70}$$

where V is the gaseous fuel and air equal flow rate, and D is the molecular diffusivity. It can be seen that the length of the laminar diffusion flame is independent of the nozzle diameter and the injection pressure. Assuming that the flame is an infinitively thin surface and the fuel and oxygen complete the reaction instantaneously at the location of the stoichiometric ratio, Burke and Schumann [6] derived a flame surface equation

$$\sum \frac{1}{\varphi} \frac{J_1\left(\varphi\frac{d}{2}\right) \cdot J_0(\varphi y_f)}{J_0^2\left(\varphi\frac{d'}{2}\right)} \exp\left(-\frac{D\varphi^2 x}{U}\right) = \frac{d'^2 C_2}{4d_t C_0} - \frac{d}{4} \tag{3.71}$$

where d and d' are the inner and outer diameters of the tube, respectively; J_0 and J_1 are zero-order and first-order Bessel functions, respectively; and C_1 and C_2 are gas fuel and oxygen concentrations at the nozzle exit, respectively. $C_0 = C_1 + \dfrac{C_2}{i}$ is the initial gas fuel concentration, i is the stoichiometric oxygen/fuel ratio. When the oxygen supply is less than that needed for the stoichiometric ratio, the flame is open-shaped; in contrast the flame is close-shaped. This result is in agreement with that observed in experiments. Hottel and Hawthorne [7] give a modified expression as

$$L = \frac{Ud^2\theta_f}{4D} = \frac{V\theta_f}{\pi D} = \frac{V}{\pi D}\frac{1}{4}\frac{1}{\ln\dfrac{1+a_t}{a_t - a_0}} \tag{3.72}$$

Hottel and Hawthorne also give an empirical expression as

$$L = A \lg V\theta_f + B \tag{3.73}$$

where A and B are empirical constants.

3.5 DROPLET EVAPORATION AND COMBUSTION

3.5.1 Background

Liquid-fuel combustion has different regimes: surface combustion; spray combustion; prevaporized combustion; but most often it takes the form of spray combustion. Liquid-fuel combustion always takes place in the gas phase, since the temperature at the liquid surface (lower than the boiling point) is much lower than the ignition temperature. Liquid-fuel combustion is always diffusion-controlled combustion, since the fuel vapor and oxygen come from two sides that are opposite to each other. It is found that the combustion rate is inversely proportional to the liquid density. Spray combustion takes different forms: prevaporized combustion; droplet diffusion combustion; and hybrid combustion. In all these cases droplet evaporation and combustion are basic processes in spray combustion.

3.5.2 Droplet Evaporation in Stagnant Air

Droplet evaporation in stagnant air under lower temperature is conventional heat and mass transfer. It can be described by the following relationships

$$Q = \pi d_p^2 q_w = \pi d_p^2 \lambda_w \left(\frac{dT}{dr}\right)_w = 2\pi d_p \lambda(T_g - T_w)$$

$$Q = G q_e \qquad G = 2\pi d_p \lambda(T_g - T_w)/q_e = 2\pi d_p \frac{\lambda}{c_p} B, \qquad B = \frac{c_p(T_g - T_w)}{q_e}$$

(3.74)

Hence, we have

$$G \sim d_p \quad G \sim \lambda \quad G \sim BG = \pi d_p^2 g_w = 2\pi d_p D\rho Y_w \quad D\rho Y_w = \frac{\lambda}{c_p} B$$

$$G \sim D\rho \quad G \sim Y_w \quad Y_w = B_w \exp(-E_w/RT_w) \quad B = \frac{c_p(T_g - T_w)}{q_e} = Le Y_w$$

(3.75)

It is seen that the evaporation rate is proportional to the droplet diameter (but not the droplet surface!) and the gas thermoconductivity, and is inversely proportional to the latent heat and gas specific heat.

3.5.3 Basic Equations for Droplet Evaporation and Combustion

To analyze more general cases of droplet evaporation under high temperature and combustion, a one-dimensional model, in spherical coordinate, as shown in Fig. 3.22, has been used by many investigators since the 1950s. The theoretical analysis is based on the following assumptions: quasisteady state

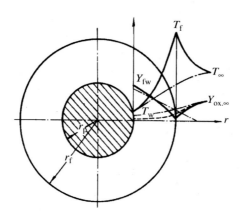

FIGURE 3.22 An evaporating and combusting droplet.

(neglect droplet diameter shrinking); no external convection; no radiation and body forces; spherically symmetric flow; and constant thermal properties in the gas layer surrounding the droplet. The basic equations for the one-dimensional laminar reacting gas flows surrounding the evaporating and combusting droplet and their boundary conditions are:

$$4\pi r^2 \rho v = 4\pi r^2 \rho_w v_w = G = \text{const} \quad p \approx \text{const}$$

$$\rho v \frac{dY_s}{dr} = \frac{1}{r^2}\frac{d}{dr}\left(r^2 D\rho \frac{dY_s}{dr}\right) - w_s$$

$$\rho v c_p \frac{dT}{dr} = \frac{1}{r^2}\frac{d}{dr}\left(r^2 \lambda \frac{dT}{dr}\right) + w_s Q_s$$

$$r = r_p : v = v_w \quad - D\rho \left(\frac{dY_s}{dr}\right)_w + Y_{sw}\rho_w v_w = \alpha \rho_w v_w \tag{3.76}$$

$$(s = F, \alpha = 1; s \neq F; \alpha = 0)$$

$$\lambda \left(\frac{dT}{dr}\right)_w = \rho_w v_w q_e = G q_e /(4\pi r_p^2) \quad Y_{Fw} = B_w \exp(-E_w/RT_w)$$

$$r = \infty$$

$$T = T_g; Y_F = Y_{pr} = 0; Y_{ox} = Y_{ox\infty}; Y_{iner} = Y_{iner\infty}$$

The integration from the droplet surface r_p to the arbitrary radial distance r gives

$$\frac{d}{dr}(GY_s) = \frac{d}{dr}\left(4\pi r^2 D\rho \frac{dY_s}{dr}\right) - 4\pi r^2 w_s \quad \frac{d}{dr}(Gc_p T) = \frac{d}{dr}\left(4\pi r^2 \lambda \frac{dT}{dr}\right) + 4\pi r^2 w_s Q_s$$

$$G\left[c_p(T - T_w) + q_e\right] = 4\pi r^2 \lambda \frac{dT}{dr} + \int_{r_p}^{r} 4\pi r^2 w_s Q_s dr \quad G[Y_F - 1] = 4\pi r^2 D\rho \frac{dY_F}{dr} - \int_{r_p}^{r} 4\pi r^2 w_F dr$$

$$GY_{ox} = 4\pi r^2 D\rho \frac{dY_{ox}}{dr} - \int_{r_p}^{r} 4\pi r^2 w_{ox} dr \quad w_{ox} = \beta w_F$$

3.5.4 Droplet Evaporation With and Without Combustion

In the case of pure evaporation, the reaction terms in the energy and species equations are eliminated, and the integration from the droplet surface to infinity gives

$$G = 2\pi d_p \frac{\lambda}{c_p} \ln\left[1 + \frac{c_p(T_g - T_w)}{q_e}\right] \quad G = 2\pi d_p D\rho \ln\left[1 + \frac{Y_{Fw}}{1 - Y_{Fw}}\right] \tag{3.77}$$

When Le $= 1$, we have

$$G = 2\pi d_p \frac{\lambda}{c_p} \ln[1 + B] \qquad B = \frac{c_p(T_g - T_w)}{q_e} = \frac{Y_{Fw}}{1 - Y_{Fw}}, \qquad Y_{Fw} = B_w \exp(- E_w/RT_w)$$

$$G \sim d_p \qquad G \sim \lambda \qquad G \sim \ln[1 + B]$$

when $B \ll 1$ $\ln[1 + B] \sim B$

$$(3.78)$$

It can be seen that in this case, that the evaporation rate is proportional to $\ln(1 + B)$, but not simply B. Since $\ln(1 + B)$ is smaller than B, it implies that the Stefan flux weakens the heat and mass transfer during evaporation.

In the case of evaporation with diffusion combustion, under the assumption of fast chemistry, for Le $= 1$, using the Zeldovich transformation to combine the fuel-vapor species and energy equations to a function Z, integration from the droplet surface to infinity gives

$$Z = c_p T + Y_{ox} Q_{ox} \qquad \text{Le} = 1, \qquad G = [(Z - Z_w) + q_e + Y_{oxw} Q_{ox}] = 4\pi r^2 \frac{\lambda}{c_p} \frac{dZ}{dr}$$

$$G = 2\pi d_p \frac{\lambda}{c_p} \ln\left[1 + \frac{Z_\infty - Z_w}{q_e + Y_{oxw} Q_{ox}}\right]$$

$$G = 2\pi d_p \frac{\lambda}{c_p} \ln\left[1 + \frac{c_p[T_g - T_w] + (Y_{ox\infty} - Y_{oxw})Q_{ox}}{q_e + Y_{oxw} Q_{ox}}\right]$$

$$(3.79)$$

for diffusion combustion, $Y_{oxw} \approx 0 \qquad G = 2\pi d_p \frac{\lambda}{c_p} \ln\left[1 + \frac{c_p[T_m - T_w]}{q_e}\right]$

$$= 2\pi d_p \frac{\lambda}{c_p} \ln\left[1 + B_f\right]$$

$$B_f = c_p(T_m - T_w)/q_e \qquad T_m = T_g + Y_{ox\infty} Q_{ox}/c_p$$

$$(3.80)$$

In this case the droplet evaporation/combustion rate is independent of reaction kinetics.

3.5.5   Droplet Evaporation and Combustion under Forced Convection

For the cases of droplet evaporation/combustion under forced convection, in the simplified one-dimensional analysis, a concept of "stagnant film" is introduces, as shown in Fig. 3.23. Use two hypothetical steps instead of the

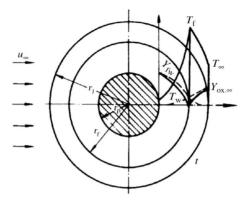

FIGURE 3.23   Droplet evaporation/combustion under forced convection.

real convection cases. First consider the hypothetical heat conduction equivalent to the real heat convection in a stagnant film without evaporation and
combustion. Then, consider evaporation and combustion in the stagnant film
without convection. The stagnant-film radius is defined by the following
relationships

$$Q = \pi d_p^2 h_*(T_g - T_w) = \frac{\pi d_1 d_p}{r_1 - r_p}\lambda(T_g - T_w) \quad Nu_* = h_* d_p/\lambda \quad Nu_* = d_1/(r_1 - r_p)$$

$$d_1 = d_p Nu_*/(Nu_* - 2)$$

$$(3.81)$$

Then, we have

$$Nu_* = 2 \quad d_1 = \infty; \quad Nu_* = \infty \quad d_1 = d_p$$

Ranze-Marshell   Law

$$Nu_* = 2 + 0.6 Re_p^{0.5} Pr^{0.33} \quad Re_p = |\vec{v}_g - \vec{v}_p| d_p/\nu$$

Therefore, by the integration of basic equations from the droplet surface
to the stagnant-film radius, using d_1 instead of infinity for the outer boundary, the evaporation with and without combustion under forced convection
can be obtained as

$$G = \pi d_p Nu_* \frac{\lambda}{c_p} \ln\left[1 + \frac{c_p(T_g - T_w)}{q_e}\right]$$

$$G = \pi d_p Nu_* \frac{\lambda}{c_p} \ln\left[1 + \frac{c_p(T_m - T_w)}{q_e}\right]$$

Define $Nu = \dfrac{h d_p}{\lambda} = G q_e d_p/\left[\pi d_p^2 \lambda(T_g - T_w)\right]$

The Nusselt numbers for the cases of high-temperature evaporation without combustion, evaporation with combustion and low-temperature evaporation are:

$$Nu = Nu_e = \frac{\pi d_p Nu_* \dfrac{\lambda}{c_p} \ln [1 + B] q_e d_p}{\pi d_p^2 \lambda (T_g - T_w)} = Nu_* \ln[1 + B]/B \qquad (3.82)$$

$$Nu = Nu_f = Nu_* \ln[1 + B_f]/B$$
$$Nu = Nu_*$$

Obviously, Stefan flux weakens heat and mass transfer and combustion enhances heat and mass transfer.

3.5.6 The d^2 Law

This law is called d^2 Law by Western scientists or Sreznevsky law by Russian scientists. In the case of the same fuel, unchanged gas temperature, and oxygen concentration and negligible change of Nu, we have

$$G = \pi d_p^2 \rho_l \frac{d(d_p/2)}{dt} = \frac{\pi d_p}{4} \rho_l \frac{d(d_p^2)}{dt}$$

$$K \equiv \frac{d(d_p^2)}{dt} = \frac{4G}{\pi d_p \rho_l}$$

$$K = 4 \frac{Nu_* \lambda}{\rho_l c_p} \ln(1 + B); \quad K_f = 4 \frac{Nu_* \lambda}{\rho_l c_p} \ln\left(1 + B_f\right) \qquad (3.83)$$

$$K = \text{const}; \quad K_f = \text{const} \quad K, K_f = \text{Evaporation constants}$$
$$d_{p0}^2 - d_p^2 = Kt \quad d_{p0}^2 - d_p^2 = K_f t$$
$$\text{Droplet Lifetime} \quad \tau_s = d_{p0}^2/K \quad \tau_s = d_{p0}^2/K_f$$

For light-oil droplet evaporation/combustion in high-temperature air, $K \approx 1$ mm^2/s, the lifetime of a 200-μm droplet is approximately 0.04 s.

From the above-stated theoretical analysis, the following conclusions can be drawn: (1) the evaporation rate $\propto$ droplet diameter and gas heat conductivity; (2) the evaporation rate $\propto ln(1 + B)$ or $ln(1 + B_f)$; (3) the evaporation rate $\propto$ square root of the gas relative velocity; (4) the evaporation rate with or without combustion is independent of the chemical kinetics; and (5) the d^2 law holds only for unchanged gas temperature, oxygen concentration, and Nu.

3.5.7 Experimental Results

The experimental results for droplet evaporation/combustion demonstrate the following phenomena.

(1) There are four regimes for droplet evaporation: fully enveloped flame; partially enveloped flame; wake flame (parachute flame); and pure evaporation, where the evaporation rate at first increases and then sharply drops, when the gas relative velocity increases (see Fig. 3.24); (2) d^2 law approximately holds; (3) the evaporation constants for kerosene, isooctane, diesel oil, paraffin oil, and heavy oil are 0.96, 0.95, 0.79, 0.7, and 0.5, respectively; (4) the evaporation constant $K \propto p^{0.25}$ for $p > 1$ atm, it is a result of natural convection, but $k \approx \text{inv}(p)$ for $p < 1$ atm (see Fig. 3.25); (5) $K \propto v^{0.5}$ for fully enveloped flame and pure evaporation; (6) $K \propto B^2$, but not $\ln(1 + B)$.

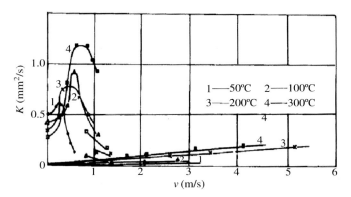

FIGURE 3.24   Evaporation rate versus gas velocity.

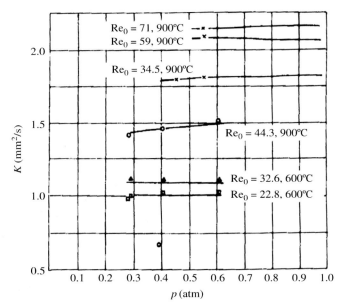

FIGURE 3.25   Evaporation rate versus pressure.

3.5.8 Droplet Ignition and Extinction

Assuming that ignition occurs at the hot boundary, using the thermal ignition theory, we get

$$\left(\frac{dT}{dr}\right)_1 = \sqrt{\frac{2Q_s}{\lambda}\int_{T_w}^{T_g} w_s dT} \qquad \left(\frac{dT}{dr}\right)_2 = G\left[c_p(T_g - T_w) + q_e\right]/(4\pi r_1^2 \lambda)$$

$$\left(\frac{dT}{dr}\right)_1 = \left(\frac{dT}{dr}\right)_2$$

$$G\left[c_p(T_g - T_w) + q_e\right]/(4\pi r_1^2 \lambda) = \sqrt{\frac{2Q_s}{\lambda}\int_{T_w}^{T_g} w_s dT}$$

$$G = \pi d_p \frac{\lambda}{c_p} \text{Nu}_* \ln\left[1 + c_p(T_g - T_w)/q_e\right] \qquad d_1 = d_p \text{Nu}_*/(\text{Nu}_* - 2) \tag{3.84}$$

$$\frac{1}{d_p^2}\left[\frac{(\text{Nu}_* - 2)^2}{\text{Nu}_*}\right]^2 = AT_g^n \exp(-E/RT_g)$$

when $\quad \text{Nu}_* >> 2, \quad \text{Nu}_* \sim (u_\infty d_p)^{1/2}$
for unchanged $T_g \quad u_\infty \sim d_p$
for unchanged $u_\infty \quad d_p \sim T_g^{-n} \exp(E/RT_g)$

The extinction condition can be obtained in a similar way as

$$(\text{Nu}_* - 2)^2/\text{Nu}_* = AT_m^n (S_l d_p c_p \rho/\lambda) \tag{3.85}$$

Therefore, it can be seen that the ambient gas temperature T_g has a greater effect on ignition than the oxygen concentration $Y_{ox\infty}$, and $Y_{ox\infty}$ has a greater effect on extinction than T_g.

3.6 SOLID-FUEL: COAL-PARTICLE COMBUSTION

3.6.1 Background

The solid fuel may be metals, nonmetals, solid propellants, and the widely used fossil fuel—coal. Coal is classified into anthracite, lean coal, bituminous coal, brown coal, and lignite, according to its age and degree of carbonization. Coal analysis may be ultimate analysis (carbon, hydrogen, oxygen, nitrogen, sulfur) or proximate analysis (volatile, fixed carbon, moisture, and ash). Coal is the cheapest fuel on the earth, but has disadvantages of pollution and erosion. The coal combustion processes include: heating, moisture evaporation, devolatilization, volatile ignition and combustion, simultaneous volatile and char combustion, and char combustion only (Fig. 3.26). The final stage takes the longest time and releases most of the heat.

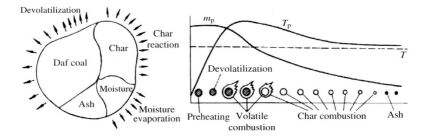

FIGURE 3.26   Coal-particle combustion.

3.6.2   Coal Pyrolyzation (Devolatilization)

When coal is heated to $500°C-600°C$, CO, C_mH_n, N_2, and H_2O are released, in a process called pyrolyzation or devolatilization. Coal particles consist of dry-and-ash-free coal (daf coal), char, ash, and moisture. Daf-coal pyrolyzation releases volatile. Heterogeneous char oxidation releases carbon monoxide and carbon dioxide. The coal-particle mass loss rate is the sum of the moisture evaporation rate, devolatilization rate, and char combustion rate. Coal devolatilization plays an important role in ignition and flame stabilization. Coal devolatilization also affects coal porosity, and hence indirectly affects char combustion. There are one-equation, two-equation, and multiequation models of devolatilization. The two-equation model is the most widely used, and is expressed by two reactions: One reaction dominates under conventional combustion temperature; the other dominates under very high temperatures. The devolatilization rate is proportional to the daf-coal mass and depends on the particle temperature in an exponential relationship. We have the following expressions

$$m = m_c + m_h + m_a + m_w$$
$$\dot{m} = \dot{m}_c + \dot{m}_h + \dot{m}_w = \dot{m}_v + \dot{m}_{hr} + \dot{m}_w$$

Daf coal $\rightarrow \alpha_1 \cdot$ Volatile 1 $+ (1 - \alpha_1) \cdot$ Char 1

$\rightarrow \alpha_2 \cdot$ Volatile 2 $+ (1 - \alpha_2) \cdot$ Char 2

$$\dot{m}_v = k\alpha m_c, \quad k = B_v \exp(-E_v/RT_p)$$
$$\dot{m}_v = m_c[\alpha_1 B_{v1} \exp(-E_{v1}/RT_p) + \alpha_2 B_{v2} \exp(-E_{v2}/RT_p)]$$
$$\dot{m}_c = \frac{dm_c}{dt} = -\frac{\dot{m}_{v1}}{\alpha_1} - \frac{\dot{m}_{v2}}{\alpha_2} = -m_c\left[B_{v1} \exp(-E_{v1}/RT_p) + B_{v2} \exp(-E_{v2}/RT_p)\right]$$

$\alpha_1 \approx \alpha_{\text{prox}} \quad \alpha_2 \approx 0.8 \quad B_{v2} > B_{v1} \quad E_{v2} > E_{v1}$

Low T_p: $\quad k_1 > k_2 \quad$ High T_p: $\quad k_2 > k_1$

$$(3.86)$$

3.6.3 Carbon Oxidation

Carbon oxidation is a heterogeneous reaction at the solid surface, since $T_b > T_i$. The carbon surface absorbs oxygen molecules and desorbs CO or CO_2. Surface oxidation reactions produce CO or CO_2, a reduction reaction produces CO, and a CO volume reaction produces CO_2.

$$
\begin{aligned}
&\text{Premary-surface}-1 \quad C+O_2 \rightarrow CO_2 + 94,200\ \text{kcal/mol} \\
&\text{Premary-surface}-2 \quad 2C+O_2 \rightarrow 2CO + 52,300\ \text{kcal/mol} \\
&\text{Secondary-surface} \quad\ \ C+CO_2 \rightarrow 2CO - 41,950\ \text{kcal/mol} \\
&\text{Secondary-volume} \quad 2CO+O_2 \rightarrow 2CO_2 + 136,200\ \text{kcal/mol}
\end{aligned}
\tag{3.87}
$$

For carbon oxidation kinetics, assume global kinetics of Arrhenius-type; first order (actually 0.5 order). It produces mainly CO_2 when $T < 1273K$, and mainly CO when T is in the range of $1273K - 2273K$. Therefore, we have

$$
w_{ox1} = B_1 \rho_w Y_{oxw} \exp(-E_1/RT_w) = k_{ox1} \rho_w Y_{oxw}
$$

$$
w_{ox2} = B_2 \rho_w Y_{oxw} \exp(-E_2/RT_w) = k_{ox2} \rho_w Y_{oxw}
$$

$$
w_{co_2,3} = B_3 \rho_w Y_{co_2 w} \exp(-E_3/RT_w) = k_{co_2,3} \rho_w Y_{co_2 w}
$$

$$
w_{ox4} = 6.6 \times 10^4 \rho Y_{co} \exp(-15,000/RT)
$$

$$
k_r = B_r \exp(-E_r/RT) = k_* \exp\left[-\frac{E_r}{RT}\left(1-\frac{T}{T_*}\right)\right] = k_* \exp\left(\frac{E_r}{RT_*}\right)\exp(-E_r/RT)
$$

$$
B_r = k_* \exp\left(\frac{E_r}{RT_*}\right) \quad k_* = 10\,m/s \quad T_* = 2000K
$$

$$
E_1 = (21 \sim 23) \times 10^3\ \text{kcal/mol} \quad E_2/E_1 = 1.2 \quad E_3/E_2 = 2.2
$$
$$
E_3 > E_2 > E_1 \quad B_3 > B_2 > B_1
$$
$$
\text{Low}\,T\,(E-\text{impor.})\ k_1 > k_2 > k_3 \quad \text{High}\,T\,(B-\text{import.})\ k_1 > k_2 > k_3
$$

$$
\tag{3.88}
$$

3.6.4 Carbon Oxidation—Basic Equations

$$
4\pi r^2 \rho v = 4\pi r_p^2 \rho_w v_w = G = \text{const} \quad p \approx \text{const}
$$

$$
\rho v \frac{dY_s}{dr} = \frac{1}{r^2}\frac{d}{dr}\left(r^2 D\rho \frac{dY_s}{dr}\right) - w_s \quad \rho v c_p \frac{dT}{dr} = \frac{1}{r^2}\frac{d}{dr}\left(r^2 \lambda \frac{dT}{dr}\right) + w_s Q_s
$$

$$
r = r_p \quad g_{sw} = -D\rho\left(\frac{dY_s}{dr}\right)_w + Y_{sw}\rho_w v_w = \sum_r B_{sr}\rho_w Y_{sw}\exp(-E_r/RT_p)
$$

$$\sum_r Q_{sr} B_{sr} \rho_w Y_{sw} \exp(-E_r/RT_p) = \varepsilon\sigma\left(T_g^4 - T_p^4\right) - \lambda\left(\frac{dT}{dr}\right)_w$$

$$\sum_s g_{sw} = \rho_w v_w = g_{cow} + g_{co_2w} + g_{o_2w} \qquad (3.89)$$

$$r = r_1 \quad T = T_g \quad Y_s = Y_{s\infty} \quad \sum_s Y_s = 1 \quad g_{iner} = 0$$

The stoichiometric relations at the particle surface are

$$g_{o_2w} = g_{o_2w}^{(1)} + g_{o_2w}^{(2)} \quad g_{cow} = g_{cow}^{(2)} + g_{cow}^{(3)} \quad g_{co_2w} = g_{co_2w}^{(1)} + g_{co_2w}^{(3)}$$

$$g_{o_2w}^{(1)} = -32/44 g_{co_2w}^{(1)} \quad g_{o_2w}^{(2)} = -32/56 g_{cow}^{(2)} \quad g_{co_2w}^{(3)} = -44/56 g_{cow}^{(3)}$$

$$56/44 g_{cow} + 56/32 g_{o_2w} + g_{cow} = 0 \quad g_c = g_{c_1} + g_{c_2} + g_{c_3} = \sum_s g_{sw} = \rho_w v_w = g$$

$$(3.90)$$

3.6.5 Carbon Oxidation—Single-Flame-Surface Model-Only Reaction 1 or 2 at the Surface

This is the case shown in Fig. 3.27B. There are only carbon oxidation reactions to produce CO and CO_2 at the surface. We get

$$G(Y_{o_2} - Y_{o_2w}) = 4\pi r^2 D\rho\left(\frac{dY_{o_2}}{dr}\right) - 4\pi r_p^2 D\rho\left(\frac{dY_{o_2}}{dr}\right)_w$$

$$Gc_p(T - T_p) = 4\pi r^2 \lambda\left(\frac{dT}{dr}\right) - 4\pi r_p^2 \lambda\left(\frac{dT}{dr}\right)_w$$

$$-4\pi r^2 D\rho\left(\frac{dY_{o_2}}{dr}\right) + GY_{o_2} = G_{o_2} = 4\pi r_p^2 g_{o_2w}$$

$$4\pi r^2 \lambda\left(\frac{dT}{dr}\right) = G\left[c_p(T - T_p) + q_w\right] \quad q_w = 4\pi r_p^2 \lambda\left(\frac{dT}{dr}\right)_w / G$$

$$G = G_{o_2} + G_{co_2} = -\frac{3}{8} G_{o_2}(\text{Reaction 1}) \quad \text{or} \quad G = -\frac{3}{4} G_{o_2}(\text{Reaction 2})$$

$$G = -G_{o_2}/\beta \quad \beta = 8/3 -- 4/3$$

$$4\pi r^2 D\rho\left(\frac{dY_{o_2}}{dr}\right) = G(Y_{o_2} + \beta) \quad 4\pi r^2 \lambda\left(\frac{dT}{dr}\right) = G\left[c_p(T - T_p) + q_w\right]$$

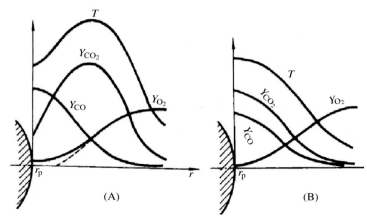

FIGURE 3.27   Char particle oxidation.

$$G = G_c = \pi d_p \mathrm{Nu}_* D\rho \ln\left[1 + \frac{Y_{o_2\,\infty} - Y_{o_2 w}}{\beta + Y_{o_2 w}}\right] = \pi d_p \mathrm{Nu}_* \frac{\lambda}{c_p} \ln\left[1 + \frac{c_p(T_g - T_p)}{q_w}\right]$$

$$G_{o_2} = -4\pi r_p^2 B_r Y_{o_2 w}\rho \exp(-E_r/\mathrm{RT}_p) \quad r = 1, 2$$

$$\frac{\pi d_p^3}{6} \rho_c c_c \frac{dT_p}{dt} = 4\pi r_p^2 \varepsilon\sigma(T_g^4 - T_p^4) - Gq_w + GQ_c$$

$$\text{(3.91)}$$

If there is steady state and radiation equilibrium, the wall heat flux will be equal to the combustion heat release, and we have

$$G = G_c = \pi d_p \mathrm{Nu}_* D\rho \ln\left[1 + \frac{Y_{o_2\,\infty} - Y_{o_2 w}}{\beta + Y_{o_2 w}}\right]$$

$$\text{(3.92)}$$

$$\frac{c_p(T_p - T_g)}{Q_c} = \frac{Y_{o_2\,\infty} - Y_{o_2 w}}{\beta + Y_{o_2 w}} \quad G_{o2} = -\pi d_p^2 B_r Y_{o_2 w}\rho \exp(-E_r/\mathrm{RT}_p)$$

The carbon combustion rate G depends on both reaction kinetics and heat and mass transfer, and G is larger for reaction 2 than that for reaction 1. From these results the regimes of diffusion combustion and kinetic combustion can be analyzed, as shown in Fig. 3.28.

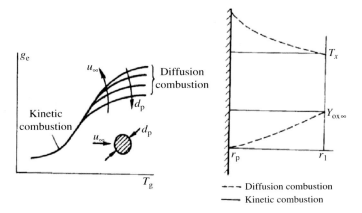

FIGURE 3.28   Diffusion combustion and kinetic combustion.

frequently $(Y_{O_2 \infty} - Y_{O_2 w})/(\beta + Y_{O_2 w}) \ll 1$

$\ln[1 + (Y_{O_2 \infty} - Y_{O_2 w})/(\beta + Y_{O_2 w})] \approx (Y_{O_2 \infty} - Y_{O_2 w})/\beta$

$g_{O_2 w} = G_{O_2}/(\pi d_p^2) = -\dfrac{\text{Nu}_* D\rho}{d_p}(Y_{O_2 \infty} - Y_{O_2 w}) = h_d^* \rho(Y_{O_2 \infty} - Y_{O_2 w})$

$g_{O_2 w} = -k_{O_2}\rho Y_{O_2 w} \quad Y_{O_2 w} = Y_{O_2 \infty} h_d^*/(h_d^* + k_{O_2})$

$g_{O_2 w} = -K Y_{O_2 \infty} \quad K = 1/\left(\dfrac{1}{h_d^*} + \dfrac{1}{k_{O_2}}\right)$

$k_{O_2} \gg h_d^* \quad D = \tau_d/\tau_c = k_{O_2}/h_d^* \gg 1 \quad K \approx h_d^* \quad Y_{O_2 w} \approx 0$ $\qquad$ (3.93)

$g_{O_2 w} \approx -h_d^* \rho Y_{O_2 \infty} = -\text{Nu}_* D\rho Y_{O_2 \infty}/d_p$

$g_{O_2 w} = \text{inv}(T_g) \sim \sqrt{\dfrac{u_\infty}{d_p}}$ Diffusion combustion

$T_p = T_g + Y_{O_2 \infty} Q_c/(\beta c_p)$, maximal

$k_{O_2} \ll h_d^* \quad D \ll 1 \quad K \approx k_{O_2} \quad Y_{O_2 w} \approx Y_{O_2 \infty} \quad T_p \approx T_g (T_p \text{ minimum})$

$g_{O_2 w} \approx -k_{O_2} Y_{O_2 \infty} \quad g_{O_2 w} \sim T_g = \text{inv}(u_\infty, d_p)$ Kinetic combustion

It can be seen that carbon combustion depends on both kinetics and heat and mass transfer. Carbon combustion may be diffusion combustion or kinetic combustion. The diffusion combustion takes place at higher temperatures, lower gas relative velocities, and larger particle sizes. In contrast, kinetic combustion happens at lower temperatures, higher gas relative velocities, and smaller particle sizes. For kinetic combustion, G increases with an increase in T and is independent of d_p and v. For diffusion combustion G increases with an increase in v and a decrease in d_p, and is independent

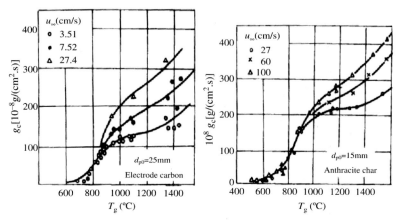

FIGURE 3.29 Carbon combustion rate.

of T. For carbon combustion the surface temperature is higher than the gas temperature and the surface oxygen concentration is lower than the surrounding temperature. The surface temperature is highest and the oxygen concentration is lowest for diffusion combustion. Conversely, the opposite results exist for kinetic combustion. As shown in Fig. 3.29, it was found that the theoretical results are in qualitative agreement with experimental results. When $T < 1173K$, there is kinetic combustion; when $T = 1173-1273K$, there is kinetic-diffusion combustion, and when $T = 1273-1473K$, there is diffusion combustion. However, when $T > 1473K$, there is again kinetic-diffusion combustion. A blue transparent CO flame around the carbon particle and local higher temperature (1753K) in the gas layer are observed. At higher temperatures the surface reduction reaction of CO_2 and the volume reaction of CO must be considered.

3.6.6 Carbon Oxidation—Two-Flame-Surface Model

Assuming that there are only a CO reduction reaction at the surface and a CO diffusion combustion in the stagnant film, the basic equations and boundary conditions are:

$$G\frac{dY_s}{dr} = \frac{d}{dr}\left(4\pi r^2 D\rho \frac{dY_s}{dr}\right) - w_s$$

$$Gc_p\frac{dT}{dr} = \frac{d}{dr}\left(4\pi r^2 \lambda \frac{dT}{dr}\right) + w_s Q_s$$

$$r = r_p \quad u = 0 \quad v = v_w \neq 0$$

$$g_{sw} = -D\rho \left(\frac{dY_s}{dr}\right)_w + Y_{sw}\rho_w v_w = B_{c_3}\rho_w Y_{sw} \exp(-E_3/RT_p)$$

$$Y_{O_2w} = 0 \quad \left(\frac{dY_{o2}}{dr}\right)_w = 0 \quad T_p - \text{is determined by Energy Eq.}$$

$$g_c = g_w = \sum g_{sw} = g_{cow} + g_{co_2w} = \rho_w v_w$$

$$g_{cow} = -\frac{11}{14} g_{co_2w} \quad g_c = g_w = -\frac{3}{11} g_{co_2w}$$

(3.94)

Using Zeldovich transformation, the following results are obtained:

Zeldovich Trans. $Y = Y_{O_2} + 4Y_{co_2}/11 \quad Z = c_p T + Y_{o_2} Q_{o_2,4}$

$$G\frac{dY}{dr} = \frac{d}{dr}\left(4\pi r^2 D\rho \frac{dY}{dr}\right) \quad G\frac{dZ}{dr} = \frac{d}{dr}\left(4\pi r^2 \frac{\lambda}{c_p}\frac{dZ}{dr}\right)$$

$$G(Y - Y_w) = 4\pi r^2 D\rho \left(\frac{dY}{dr}\right) - 4\pi r_p^2 D\rho \left(\frac{dY}{dr}\right)_w$$

$$G(Z - Z_w) = 4\pi r^2 \frac{\lambda}{c_p}\left(\frac{dZ}{dr}\right) - 4\pi r_p^2 \frac{\lambda}{c_p}\left(\frac{dZ}{dr}\right)_w$$

$$-D\rho \left(\frac{dY}{dr}\right)_w = -D\rho \left[\left(\frac{dY_{o2}}{dr}\right)_w + \frac{4}{11}\left(\frac{dY_{co2}}{dr}\right)_w\right] = -\frac{4}{11}D\rho \left(\frac{dY_{co2}}{dr}\right)_w =$$

$$= \frac{4}{11}(g_{co_2w} - Y_{co_2w}g_w) = -g_w\left(\frac{4}{3} + \frac{4}{11}Y_{co_2w}\right)$$

$$(Y - Y_w) = Y - \left(Y_{o_2w} + \frac{4}{11}Y_{co_2w}\right) = Y - \frac{4}{11}Y_{co_2w}$$

$$G(Y + 4/3) = 4\pi r^2 D\rho \left(\frac{dY}{dr}\right) \quad G = \pi d_p D\rho \text{Nu}_* \ln\left[1 + \frac{Y_{o_2\infty} - 4Y_{co_2w}/11}{\frac{4}{3} + 4Y_{co_2w}/11}\right]$$

(3.94)

$$G = \pi d_p \frac{\lambda}{c_p}\text{Nu}_* \ln\left[1 + \frac{c_p(T_g - T_p)}{q_w}\right]$$

$$G = G_c = \pi d_p^2 B_{c3}\rho_w Y_{co_2w} \exp(-E_3/RT_p)$$

$$T_P: (\pi d_p^3/6)\rho_c c_c\frac{dT_p}{dt} = \pi d_p^2 \varepsilon\sigma(T_g^4 - T_p^4) + Gq_w - GQ_{c3}$$

(3.95)

$4Y_{co_2w}/11$ instead of Y_{o_2w}; $T_p - Q_{c3}$ and $Q_{o_2,4}$, not Q_{c_1} or Q_{c_2}

In this case the carbon combustion rate G depends on diffusion and surface kinetics, but is independent of volume reaction kinetics. The difference between the one-flame-surface model and the two-flame-surface model is that the reduction kinetics of reaction 3 replaces the oxidation kinetics of reaction 1 or 2.

3.6.7   Coal-Particle Combustion

When analyzing the coal-particle combustion, all of the simultaneous moisture evaporation, devolatilization and char oxidation, and the three surface reactions should be considered. The basic equations for the gas layer surrounding the particle and the boundary conditions are:

$$G_s = -4\pi r^2 D\rho \frac{dY_s}{dr} + GY_s = \text{const}$$

$$4\pi r^2 \lambda \frac{dT}{dr} = G\left[c_p(T - T_p) + q_w\right]$$

(3.96)

$$G_s = G_{sw} = \pi d_p^2 \rho Y_{sw} \sum B_r \exp(-E_r/RT_p)$$

$$G = G_c + G_v + G_w \quad G_c = -3G_{o2}/4 - 3G_{co_2}/11$$

$$m_p c_c \frac{dT_p}{dt} = \pi d_p^2 \sigma \varepsilon(T - T) + Gq_w - G_w L_w - G_v \Delta h_v + \sum G_{cr} Q_{cr}$$

$$G_w = \pi d_p \text{Nu}_* D\rho \ln\left(\frac{G_s/G - Y_{s\infty}}{G_s/G - Y_{sw}}\right) \quad G = \pi d_p \text{Nu}_* \frac{\lambda}{c_p} \ln\left[1 + \frac{c_p(T_g - T_p)}{q_w}\right]$$

$$q_w = c_p(T_g - T_p)\left[\exp\left(\frac{Gc_p}{\pi d_p \text{Nu}_* \lambda}\right) - 1\right]^{-1}$$

$$G_w = \pi d_p \text{Nu}_* D\rho \ln\left(\frac{1 - Y_{w\infty}}{1 - Y_{ww}}\right) \quad Y_{ww} = B_w \exp(-E_w/RT_p)$$

$$G_v = m_c[\alpha_1 B_{v1} \exp(-E_{v1}/RT_p) + \alpha_2 B_{v2} \exp(-E_{v2}/RT_p)]$$

$$\frac{dm_c}{dt} = -m_c\left[B_{v1} \exp(-E_{v1}/RT_p) + B_{v2} \exp(-E_{v2}/RT_p)\right]$$

$$G = \pi d_p \text{Nu}_* D\rho \ln\left(\frac{G_{co_2}/G - Y_{co_2 \infty}}{G_{co_2}/G - Y_{co_2 w}}\right) = \pi d_p \text{Nu}_* D\rho \ln\left(\frac{G_{o_2}/G - Y_{o_2 \infty}}{G_{o_2}/G - Y_{o_2 w}}\right)$$

$$G_{o_2} = \pi d_p^2 \rho Y_{o_2 w}[B_1 \exp(-E_1/RT_p) + B_2 \exp(-E_2/RT_p)]$$

$$G_{co_2} = \pi d_p^2 \rho\left[\frac{8}{11}Y_{o_2 w} B_1 \exp(-E_1/RT_p) - Y_{co_2 w} B_3 \exp(-E_3/RT_p)\right]$$

$$G_c = -3G_{O_2}/4 - 3G_{CO_2}/11 \quad G_c = G - G_w - G_v$$

$$m_c c_c \frac{dT_p}{dt} = \pi d_p^2 \varepsilon \sigma (T_g^4 - T_p^4) + G c_p (T_g - T_p) \left[\exp\left(\frac{G c_p}{\pi d_p \mathrm{Nu}_* \lambda} \right) - 1 \right]^{-1}$$

$$- G_w L_w + G_{O_2}^{(1)} Q_{O_2}^{(1)} + G_{O_2}^{(2)} Q_{O_2}^{(2)} - G_{CO_2}^{(3)} Q_{CO_2}^{(3)}$$

$$G_{O_2}^{(1)} = \pi d_p^2 \rho Y_{O_2 w} B_1 \exp(-E_1/RT_p)$$

$$G_{O_2}^{(2)} = \pi d_p^2 \rho Y_{O_2 w} B_2 \exp(-E_2/RT_p)$$

$$G_{CO_2}^{(3)} = \pi d_p^2 \rho Y_{CO_2 w} B_3 \exp(-E_3/RT_p)$$

These equations have to be solved numerically. The simulation results for the combustion of a 1-cm Huainan bituminous coal particle and their comparison with experimental results are given in Figs. 3.30. It can be seen that the predictions are in fairly good agreement with the experimental results. The particle mass decreases sharply in the pyrolyzation period, then becomes almost unchanged when volatiles ignite and burn, and subsequently decreases again when char combustion takes place. The volatile combustion time is almost one-sixth (20 s) of the total combustion time (120 s). Hence char combustion plays a dominant role in coal combustion. The particle temperature remains almost unchanged during the time of char combustion.

Some other experimental results give the volatile ignition time, volatile combustion time, and char burnout time as

$$\tau_i = K_1 T_g^{-4} d_{p0} \text{ or } \tau_i = K_1 Y_{O_2 \infty}^{-0.15} T_g^{-2.5} d_{p0}^{1.2}$$

$$\tau_v = K_2 d_{p0}^2$$

$$\tau_{ch} = K_3 T_g^{-0.9} Y_{O_2 \infty}^{-1} d_{p0}^2 \tag{3.97}$$

$$\tau_{ch} = K_3 T_g^{-0.5} Y_{O_2 \infty}^{-1} d_{p0}^2$$

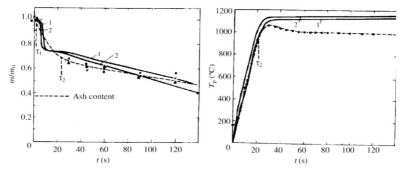

FIGURE 3.30 Particle mass (left) and temperature (right) change during combustion.

3.7 TURBULENT COMBUSTION AND FLAME STABILIZATION

3.7.1 Background

Practical flames are always turbulent. There are two problems with practical turbulent combustion: combustion intensity (flame length) and flame stabilization. The purpose of engineering is to increase the combustion intensity and flame stabilization. The well-known phenomena include: the length of turbulent diffusion flames is independent of inlet velocity; the length of turbulent premixed flames depends on inlet velocity and is shorter than that of laminar flames; the expanding angle of turbulent premixed flames behind a bluff body decreases along the axial distance and is independent of the mixing ratio and inlet velocity at high velocities; and increasing the inlet turbulence intensity reduces the flame length.

3.7.2 Turbulent Jet Diffusion Flame

The basic equations of a turbulent jet diffusion flame are:

$$\frac{\partial}{\partial x}(\rho u r) + \frac{\partial}{\partial r}(\rho v r) = 0$$

$$\rho u \frac{\partial u}{\partial x} + \rho v \frac{\partial u}{\partial r} = \frac{1}{r}\frac{\partial}{\partial r}\left(r\nu_T \rho \frac{\partial u}{\partial r}\right)$$

$$\rho u \frac{\partial Y_s}{\partial x} + \rho v \frac{\partial Y_s}{\partial r} = \frac{1}{r}\frac{\partial}{\partial r}\left(r D_T \rho \frac{\partial Y_s}{\partial r}\right) - w_s \tag{3.98}$$

$$\rho u c_p \frac{\partial T}{\partial x} + \rho v c_p \frac{\partial T}{\partial r} = \frac{1}{r}\frac{\partial}{\partial r}\left(r\lambda_T \frac{\partial T}{\partial r}\right) + w_s Q_s$$

Using the concept of Prandtl's mixing-length model, we have

$$\nu_T = D_T = \lambda_T/(c_p\rho) = cx^2\left|\frac{\partial u}{\partial r}\right|$$

The boundary conditions are:

$$r = r_1 \quad u = u_1; T = T_1; Y_{ox} = 0; Y_F = 1$$
$$r = r_2 \quad u = u_2; T = T_2; Y_{ox} = Y_{ox\,\infty}; Y_F = 0$$
$$r = r_f \quad Y_{ox} = Y_F = 0$$

Using the Zeldovich transformation, we have

$$Y = Y_{ox} - \beta Y_F \quad Z = c_p T + Y_{ox} Q_{ox}$$

$$\rho u \frac{\partial Y}{\partial x} + \rho v \frac{\partial Y}{\partial r} = \frac{1}{r} \frac{\partial}{\partial r} \left(r \frac{\lambda_T}{c_p} \frac{\partial Y}{\partial r} \right)$$

$$\rho u \frac{\partial Z}{\partial x} + \rho v \frac{\partial Z}{\partial r} = \frac{1}{r} \frac{\partial}{\partial r} \left(r \frac{\lambda_T}{c_p} \frac{\partial Z}{\partial r} \right)$$

Without solving these equations, it can be seen that the solution should be

$$\frac{Y - Y_2}{Y_1 - Y_2} = \frac{Z - Z_2}{Z_1 - Z_2} = \theta(x/R, r/R) \tag{3.99}$$

At the flame surface, there will be

$$\theta_f = \frac{Y_{ox \infty}}{Y_{ox \infty} + \beta} = \frac{c_p(T_f - T_2) - Y_{ox \infty} Q_{ox}}{- Y_{ox \infty} Q_{ox}} \tag{3.100}$$

Therefore, the adiabatic jet diffusion flame temperature can be obtained as

$$T_f = T_2 + \frac{Y_{ox \infty} Q_{ox}}{c_p(1 + Y_{ox \infty}/\beta)} \tag{3.101}$$

The flame shape is determined by

$$(x/R)_f = (x/R)_f[\theta_f, (r/R)_f] \tag{3.102}$$

The flame length is determined by

$$\begin{aligned} (x/R)_f &= L_f/R \quad r_f/R = 0 \\ L_f/R &= f(\theta_f) \quad L_f/R = a\theta_f + b \\ L_f &= R\left(a\frac{Y_{ox \infty} + \beta}{Y_{ox \infty}} + b \right) \end{aligned} \tag{3.103}$$

Experimental results give

$$L_f = R\left(10\frac{Y_{ox \infty} + \beta}{Y_{ox \infty}} + 4 \right) \tag{3.104}$$

From here there are the following relationships

$$\because Y_{ox \infty} \ll \beta; \quad 10\frac{Y_{ox \infty} + \beta}{Y_{ox \infty}} \gg 4 \quad \therefore L_f \sim R \quad L_f \sim 1/Y_{ox \infty}$$

It can be seen that the diffusion flame length depends on the tube radius and oxygen concentration, but is independent of the inlet velocity and other factors. The flame length is proportional to the tube radius and inversely proportional to the oxygen concentration.

3.7.3 Turbulent Premixed Flame—Damkohler–Shelkin's Wrinkled-Flame Model

In this model, it is assumed that the turbulent flame is a wrinkled laminar flame, consisting of elementary laminar flames. These elementary flames propagate at the speed of laminar flames S_l. Turbulence increases the flame surface area, hence increasing its burning velocity. This model is schematically shown in Fig. 3.31

In these cases, the dimensional analysis gives the propagation velocity of a turbulent flame

$$S_T F = S_l F_l \quad S_T = S_l F_l / F \quad F_l / F = f(\tau_{comb}/\tau_T)$$
$$\tau_{comb}/\tau_T = (l/S_l)/(l/u') = u'/S_l \quad F_l/F = f(u'/S_l) = A(u'/S_l)^n \quad (3.105)$$
$$S_T = A S_l^{1-n} u'^n$$

Experiments give A = 5.3, n = 0.67. Hence the flame length is

$$L = R u_\infty / S_T = R u_\infty / (A S_l^{0.33} u'^{0.67}) = R u_\infty^{0.33} \varepsilon^{-0.67} / (A S_l^{0.33}) \quad (3.106)$$

It can be seen that the turbulent combustion rate is affected by both turbulence and chemical kinetics, and turbulence has a stronger effect. The lengths of turbulent flames are affected by the inlet velocity, turbulence intensity, and reaction kinetics.

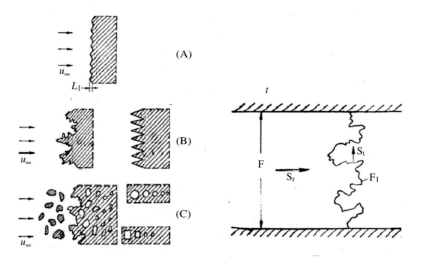

FIGURE 3.31 Wrinkled-flame model.

3.7.4 Turbulent Premixed Flame—Summerfield–Shetinkov's Volume Combustion Model

According to the wrinkled-flame model there are either fresh mixture or combustion products in eddies, but no intermediate states. However, according to the volume combustion model, turbulence affects combustion via mixing. Mixing and reaction occur at the same time. Different eddies have different temperatures, concentrations, and reactiveness. Turbulence intensifies combustion by increasing the mixing rate. The difference between the two models is schematically shown in Fig. 3.32

3.7.5 Flame Stabilization

In high-velocity flows in after-burners of aircraft engines and ramjet combustors, the recirculation of combustion products is used to stabilize the flame. Recirculating flows can be induced by bluff bodies, sudden expansion, recess, abrupt turning, opposed jets, and co-flow jets. The length of the recirculation zone is proportional to the bluff-body's diameter. Experiments have shown that combustion enlarges the central recirculation zone and shortens the corner recirculation zone. Experimental results reported by De-Zubay give the blow-off velocity as a function of pressure and the mixing ratio:

$$u_\infty/(D^{0.85}p^{0.95}) = f(\alpha) \text{ or approximately } u_\infty/(Dp) = f(\alpha) \qquad (3.107)$$

The recirculation zone behind a bluff body is shown in Fig. 3.33.

The mechanism of flame stabilization is shown in Fig. 3.34.

The flame stabilization condition can be given by an analytical model—Zukosky–Marble's ignition model—that is, ignition by hot combustion

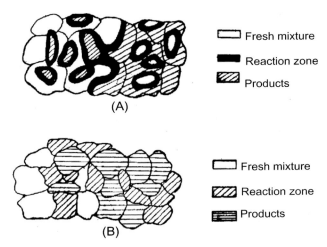

(A)

☐ Fresh mixture

■ Reaction zone

▨ Products

(B)

☐ Fresh mixture

▨ Reaction zone

▤ Products

FIGURE 3.32 Comparison of two turbulent combustion models: (A) the wrinkled-flame model; (B) the volume combustion model.

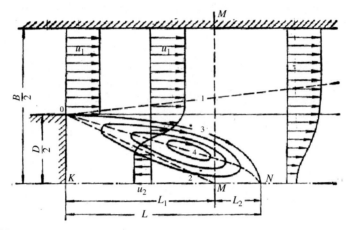

FIGURE 3.33   The recirculation zone behind a bluff body.

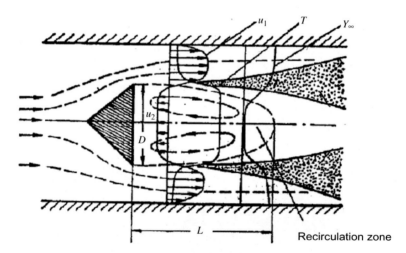

FIGURE 3.34   The mechanism of flame stabilization.

products. If the ignition distance x_i is greater than the length of the recirculation zone L ignition failure, the flame will be blown out.

At $x = x_i$, we have

$$\left(\frac{\partial T}{\partial y}\right)_1 = \left(\frac{\partial T}{\partial y}\right)_2$$

$$\left(\frac{\partial T}{\partial y}\right)_2 = \sqrt{\frac{2Q_s}{\lambda_T}\int_{T_\infty}^{T_m} w_s dT} \qquad \left(\frac{\partial T}{\partial y}\right)_1 = \frac{T_m - T_\infty}{b} = \frac{T_m - T_\infty}{c_1 x}$$

Furthermore, the derivation gives

$$\lambda_T = \rho_m c_p \nu_T = \rho_m c_p u' l = \rho_m c_p l^2 \left| \frac{u_m - u_\infty}{b} \right| = c_2 \rho_m c_p x u_\infty$$

$$\frac{(T_m - T_\infty)^2}{c_1^2 x^2} = \frac{2Q_s}{c_2 \rho_m c_p x u_\infty} \int_{T_\infty}^{T_m} w_s dT$$

$$x_i = c_3 u_\infty \rho_\infty (T_\infty / T_m) c_p (T_m - T_\infty)^2 / \left(Q_s \int_{T_\infty}^{T_m} w_s dT \right)$$

$$x_i = c_4 u_\infty (T_\infty / T_m) \overline{a} / (S_l)^2$$

The flame stabilization condition should be

$$x_i = L = c_5 D$$
$$D/u_\infty = c_6 (T_\infty / T_m) \overline{a} / S_l^2 \qquad (3.108)$$
$$\text{where} \quad \overline{a} = \overline{\lambda} / (c_p \overline{\rho})$$

Defining $\mathrm{Pe}_d = u_\infty D / \overline{a}$ $\mathrm{Pe}_f = S_l D / \overline{a}$, finally we have

$$\mathrm{Pe}_d = c_7 \mathrm{Pe}_f^2 \quad \text{or} \quad \mathrm{Re}_d = c_8 \mathrm{Pe}_f^2 \quad \text{where} \quad \mathrm{Re}_d = u_\infty D / \nu_\infty \qquad (3.109)$$

A comparison with experimental results gives

$$\mathrm{Pe}_d = 1.05 \mathrm{Pe}_f^2 \quad \text{or} \quad \mathrm{Re}_d = 1.45 \mathrm{Pe}_f^2 \qquad (3.110)$$

This is in agreement with the expression $u_\infty / (Dp) = f(\alpha)$

3.8 CONCLUSION ON COMBUSTION FUNDAMENTALS

Fundamentals of combustion theory are based on simplified analytical solutions. In the past, it was used only for qualitative analysis of experimental results for empirical engineering design. Now, it has become the basis of submodels in the CFD simulation of combustion processes, and is used indirectly to solve practical problems. Many new phenomena remain to be studied, such as microgravitational combustion, microscale combustion, supersonic combustion, pulsating combustion, combustion in electric and magnetic fields, self-propagating combustion in solid materials, etc.

REFERENCES

[1] L.X. Zhou, Theory and Numerical Modeling of Turbulent Gas-Particle Flows and Combustion, CRC Press, Florida, 1993.
[2] D.A. Frank-Kamenetsky, Diffusion and Heat Exchange in Chemical Kinetics., Princeton University Press, 1955 (Russian Edition: Диффузия и Теплопердача в Химической Кинетике, Изд. АНСССР, 1947).

[3]   N.N. Semenov, Thermal Theory of Combustion and Explosion., Progress in Physical Science 23 (1940) 251−292 (Russian Edition: Семённов,Н.Н., Тепловая теория горения и взрывов, УФН., 23(1940) 251-292).

[4]   L.N. Khitrin, Physics of Combustion and Explosion., National Science Foundation, 1962 (Russian Edition: Хитрин, Л.Н.,Физика горения и взрыва, Изд. МГУ, 1957).

[5]   L.A. Vulis, Thermal Regimes of Combustion., McGraw-Hill, N.Y, 1961 (Russian Edition, Energoizdat, 1952).

[6]   S.P. Burke, T.E.W. Schumann, Diffusion flames., Industrial and Engineering Chemistry 20 (1928) 998.

[7]   H.C. Hottel, W.R.Hawthorne, Third International Symposium on Combustion, 1949, p. 255.

Chapter 4

Basic Equations of Multiphase Turbulent Reacting Flows

In this chapter the basic equations of laminar or instantaneous multiphase flows are first considered and then the basic equations of turbulent multiphase flows are discussed. The laminar or instantaneous multiphase-flow equations are based on the concept of volume averaging, given separately by Zhou [1,2], S.L. Soo [3,4], Drew [5], Nigmatulin [6], Jackson [7], and Fan and Zhu [8].

4.1 THE CONTROL VOLUME IN A MULTIPHASE-FLOW SYSTEM

In a multifluid or two-fluid framework, the system is considered as a mixture of multiple/two fluids, in which the dispersed phase (particles/droplets/bubbles) and the continuous phase (liquid/gas) occupy the same space in a macroscopic sense (but occupying different spaces in a microscopic sense). Different phases interpenetrate each other and have their own sizes, velocities, concentrations, and temperatures. Now consider a control volume dV (corresponding to a computational cell) in a multiphase mixture with an external surface dA (see Fig. 4.1). If the phase k in this control volume occupies a volume of dV_k with a surface of dA_k (interface between phases), then the ratio $dv_k/dV = \alpha_k$ is just the local volume fraction of the phase k. To describe the dispersed phase using the multifluid concept in Eulerian coordinates, the condition to be satisfied is that the size of the control volume l should be much smaller than the geometrical size of the flow field L, but at the same time it should be much greater than the particle/bubble/droplet size d_k. This requires $d_k << l << L$, which means that the control volume contains sufficient numbers of particles. Only under this condition does the control volume behave like an elementary volume of the macroscopic flow field, and the local values of each variable in the control volume express the macroscopic properties of the flow field, but not the detailed flow field around each particle/droplet/bubble. For example, in a cyclone separator, L is the order of $0.5-1$ m, d_k is the order of $100\,\mu m$, and l may be taken as 0.01 m. For a large-size fluidized-bed furnace, if L is the order of 10 m, and d_k is the order of 1 mm, then l can be taken as 0.1 m.

Theory and Modeling of Dispersed Multiphase Turbulent Reacting Flows.
DOI: https://doi.org/10.1016/B978-0-12-813465-8.00004-1

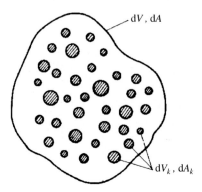

FIGURE 4.1   The control volume.

4.2   THE CONCEPT OF VOLUME AVERAGING

As already indicated, if the "microscopic" flow field inside a control volume is to be considered, then different phases will occupy different volumes. However, if the "macroscopic" flow field is to be considered, then different phases are treated as if they occupy the same space and interpenetrate each other, each having its own velocity, temperature, and volume fraction. The basic equations derived for single-phase flows are feasible only inside each phase to describe the "microscopic flows."

Therefore, to describe the "macroscopic" flow field, we have to use the concept of volume averaging. That is, each variable expressing the behavior of the "macroscopic" flow field is the volume-averaged value in the control volume dV (Fig. 4.1). Denoting the "microscopic" or real value inside a phase by the symbol "$\sim$" and the volume-averaged value in the control volume dV by the symbol $< >$, and using φ_k to express the generalized variable of the k-th phase ($k = f$; 1, 2, 3; 1, 2, 3... denotes the dispersed phases and f denotes the fluid phase), the volume-averaged value can be defined as

$$\phi_k = <\tilde{\phi}_k> = \frac{1}{dV} \int_{dV_k} \tilde{\phi}_k \delta V \quad dV_k/dV = \alpha_k \quad dV = \sum dV_k$$

$$\rho_k = <\tilde{\rho}_k> = \frac{dV_k}{dV} \frac{1}{dV_k} \int_{dV_k} \tilde{\rho}_k \delta V = \alpha_k \overline{\rho}_k$$

where $\overline{\rho}_k$ is the material density of the k-th particle/droplet/bubble phase. The relationships between the volume-averaged value of the derivatives of any scalar φ_k or vector/tensor φ_{kj} and the derivatives of volume-averaged values are [3,4]:

$$<\frac{\partial \tilde{\phi}_k}{\partial t}> = \frac{\partial <\tilde{\phi}_k>}{\partial t} - \frac{1}{dV} \int_{dA_k} \tilde{\phi}_k \tilde{v}_{sj} \cdot n_{kj} \cdot dA \qquad (4.1)$$

$$\left< \frac{\partial \tilde{\phi}_k}{\partial x_i} \right> = \frac{\partial <\tilde{\phi}_k>}{\partial x_i} + \frac{1}{dV} \int\limits_{dA_k} \tilde{\phi}_k n_{ki} \cdot dA \qquad (4.2)$$

$$\left< \frac{\partial \tilde{\phi}_{kj}}{\partial x_j} \right> = \frac{\partial <\tilde{\phi}_{kj}>}{\partial x_j} + \frac{1}{dV} \int\limits_{dA_k} \tilde{\phi}_{kj} \cdot n_{kj} \cdot dA \qquad (4.3)$$

where v_{sj} is the interface velocity due to phase change, and n_{ki} and n_{kj} are unit vectors in the direction of external normal to the interface.

4.3 "MICROSCOPIC" CONSERVATION EQUATIONS INSIDE EACH PHASE

If the dispersed phase is considered as a pseudo-fluid, and the basic conservation equations for single-phase flows are applied to an elementary control volume dV_k of the k-th dispersed phase, then the continuity, momentum, energy, and species conservation equations inside each phase for low Mach number and constant specific-heat flows can be given as

$$\frac{\partial \tilde{\rho}_k}{\partial t} + \frac{\partial}{\partial x_j}(\tilde{\rho}_k \tilde{v}_{kj}) = 0 \qquad (4.4)$$

$$\frac{\partial}{\partial t}(\tilde{\rho}_k \tilde{v}_{ki}) + \frac{\partial}{\partial x_j}(\tilde{\rho}_k \tilde{v}_{kj} \tilde{v}_{ki}) = -\frac{\partial \tilde{\rho}_k}{\partial x_i} + \frac{\partial}{\partial x_j}(\tilde{\tau}_{kji}) + \tilde{\rho}_k g_i \qquad (4.5)$$

$$\frac{\partial}{\partial t}(\tilde{\rho}_k c_k \tilde{T}_k) + \frac{\partial}{\partial x_j}(\tilde{\rho}_k \tilde{v}_{kj} c_k \tilde{T}_k) = \frac{\partial}{\partial x_j}\left(\tilde{\lambda}_k \frac{\partial \tilde{T}_k}{\partial x_j} \right) + \tilde{q}_c - \tilde{q}_r \qquad (4.6)$$

$$\frac{\partial}{\partial t}(\tilde{\rho}_k \tilde{Y}_{ks}) + \frac{\partial}{\partial x_j}(\tilde{\rho}_k \tilde{v}_{kj} \tilde{Y}_{ks}) = \frac{\partial}{\partial x_j}\left(\tilde{D}_k \tilde{\rho}_k \frac{\partial \tilde{Y}_{ks}}{\partial x_j} \right) - \tilde{w}_{ks} \qquad (4.7)$$

where the indices k and s denote the k-th phase and species s, respectively. The second and third terms on the right-hand side of Eq. (4.6) express the reaction heat release and radiative heat transfer, and the second term on the right-hand side of Eq. (4.7) is the reaction rate of species s in the k-th phase.

4.4 THE VOLUME-AVERAGED CONSERVATION EQUATIONS FOR LAMINAR/INSTANTANEOUS MULTIPHASE FLOWS

For practical multiphase flows, it is needed to understand only the macroscopic flow behavior, but not the detailed "microscopic" flow behavior (the flow behavior around each particle/droplet/bubble), hence what we need is the volume-averaged conservation equations. As examples, based on Eqs. (4.1−4.3), let $\varphi_k = \rho_k$ and taking the volume-averaged values of each term in Eq. (4.4), we have

$$< \frac{\partial \tilde{\rho}_k}{\partial t} > = \frac{\partial < \tilde{\rho}_k >}{\partial t} - \frac{1}{dV} \int_{dA_k} \tilde{\rho}_k \tilde{v}_{sj} \cdot n_{kj} \cdot dA$$

$$< \frac{\partial}{\partial x_j}(\tilde{\rho}_k \tilde{v}_{kj}) > = \frac{\partial}{\partial x_j} < \tilde{\rho}_k \tilde{v}_{kj} > + \frac{1}{dV} \int_{dA_k} \tilde{\rho}_k \tilde{v}_{kj} \cdot n_{kj} \cdot dA$$

$$\frac{\partial \rho_k}{\partial t} + \frac{\partial}{\partial x_j}(\rho_k v_{kj}) = \frac{\partial < \tilde{\rho}_k >}{\partial t} + \frac{\partial}{\partial x_j} < \tilde{\rho}_k \tilde{v}_{kj} > = -\frac{1}{dV} \int_{dA_k} \tilde{\rho}_k (\tilde{v}_{kj} - \tilde{v}_{sj}) \cdot n_{kj} \cdot dA$$

It should be noted that $\tilde{v}_{kj} - \tilde{v}_{sj}$ is the velocity of the k-th phase relative to the interface, it is called the Stefan-flux velocity (see Ref. [2]). Hence we have

$$-\frac{1}{dV} \int_{dA_k} \tilde{\rho}_k (\tilde{v}_{kj} - \tilde{v}_{sj}) \cdot n_{kj} \cdot dA = -\frac{1}{dV} \int_{dV_k} \tilde{S}_k dV = S_k$$

where S_k is the volume-averaged value of the source term due to mass change per unit volume of multiphase flows. For simplification of notation by dropping the superscripts and volume-averaging notation, the obtained volume-averaged continuity equations for the dispersed phase (particle/droplet/bubble) and continuous phase (gas/liquid) are:

$$\begin{cases} \dfrac{\partial \rho_k}{\partial t} + \dfrac{\partial}{\partial x_j}(\rho_k v_{kj}) = S_k \\[2mm] \dfrac{\partial \rho}{\partial t} + \dfrac{\partial}{\partial x_j}(\rho v_j) = S \end{cases} \qquad (k = 1, 2, 3, 4) \qquad (4.8)$$

where

$$S = -\sum S_k = -\sum n_k \dot{m}_k \qquad \dot{m}_k = \frac{dm_k}{dt}$$

$$\rho_k = n_k m_k = n_k \pi d_k^3 \bar{\rho}_k / 6$$

n_k and m_k are the number density and mass of the k-th particle phase. Similarly, taking $\varphi_k = \rho_k v_{ki}$ and doing the volume averaging procedure for each term of Eq. (4.5) as

$$< \frac{\partial}{\partial t} \tilde{\rho}_k \tilde{v}_{ki} > = \frac{\partial}{\partial t} < \tilde{\rho}_k \tilde{v}_{ki} > - \frac{1}{dV} \int_{dA_k} \tilde{\rho}_k \tilde{v}_{ki} \tilde{v}_{sj} \cdot n_{kj} \cdot dA$$

$$< \frac{\partial}{\partial x_j}(\tilde{\rho}_k \tilde{v}_{kj} \tilde{v}_{ki}) > = \frac{\partial}{\partial x_j} < \tilde{\rho}_k \tilde{v}_{kj} \tilde{v}_{ki} > + \frac{1}{dV} \int_{dA_k} \tilde{\rho}_k \tilde{v}_{ki} \tilde{v}_{kj} \cdot n_{kj} \cdot dA$$

$$- \left< \frac{\partial \tilde{p}_k}{\partial x_i} \right> = - \frac{\partial <\tilde{p}_k>}{\partial x_i} - \frac{1}{dV} \int_{dA_k} \tilde{p}_k \delta_{ij} \cdot n_{kj} \cdot dA$$

$$\left< \frac{\partial \tilde{\tau}_{kij}}{\partial x_j} \right> = \frac{\partial}{\partial x_j} <\tilde{\tau}_{kij}> + \frac{1}{dV} \int_{dA_k} \tilde{\tau}_{kij} \cdot n_{kj} \cdot dA$$

$$<\tilde{\rho}_k g_i> = \rho_k g_i \qquad <\tilde{p}_k> = \alpha_k p$$

the volume-averaged momentum equations for the k-th phase can be obtained as

$$\frac{\partial}{\partial t} <\tilde{\rho}_k \tilde{v}_{ki}> + \frac{\partial}{\partial x_j} <\tilde{\rho}_k \tilde{v}_{kj} \tilde{v}_{ki}> = \frac{\partial}{\partial x_i} (\alpha_k p) + \frac{\partial}{\partial x_j} <\tilde{\tau}_{kij}>$$

$$+ \rho_k g_i + \frac{1}{dV} \int_{dA_k} (-\tilde{p}_k \delta_{ij} + \tilde{\tau}_{kij}) \cdot n_{kj} \cdot dA - \frac{1}{dV} \int_{dA_k} \tilde{\rho}_k \tilde{v}_{ki} (\tilde{v}_{kj} - \tilde{v}_{sj}) \cdot n_{kj} \cdot dA$$

The fourth term on the right-hand side of the above equation is the summation of pressure and viscous forces at the phase interface due to two-phase interactions, i.e., the summation of drag force, lift force, buoyancy force, virtual-mass force, etc., between a gas and particle/droplet or bubble and liquid, namely:

$$\frac{1}{dV} \int_{dA_k} (-\tilde{p}_k \delta_{ij} + \tilde{\tau}_{kij}) \cdot n_{kj} \cdot dA = F_{di} + F_{Mi} + F_{si} + F_{vmi}$$

where F_{di} is the drag force, F_{Mi} is the Magnus force, F_{si} is the Saffman force, and F_{vmi} is the virtual-mass force. The drag force can be expressed by

$$F_{di} = \sum_k n_k (\pi d_k^2/4) \rho |\vec{v} - \vec{v}_k| (v_i - v_{ki}) = \sum_k \frac{\rho_k}{\tau_{rk}} (v_i - v_{ki})$$

$$\tau_{rk} = d_k^2 \bar{\rho}_p (1 + \mathrm{Re}_k^{2/3}/6)^{-1}/(18\mu) \qquad \mathrm{Re}_k = |\vec{v} - \vec{v}_k| d_k/\nu$$

The last term on the right-hand side of the above equation is the momentum source term due to mass change

$$-\frac{1}{dV} \int_{dA_k} \tilde{\rho}_k \tilde{v}_{ki} (\tilde{v}_{kj} - \tilde{v}_{sj}) \cdot n_{kj} \cdot dA = <\tilde{v}_i \tilde{S}> = v_i S$$

Dropping the notations "$\sim$" and "$< >$," the volume-averaged gas/fluid momentum equation for dilute multiphase flows can be obtained as

$$\frac{\partial}{\partial t} (\rho v_i) + \frac{\partial}{\partial x_j} (\rho v_j v_i) = -\frac{\partial p}{\partial x_i} + \frac{\partial \tau_{ji}}{\partial x_j} + \Delta \rho g_i + \sum_k \frac{\rho_k}{\tau_{rk}} (v_{ki} - v_i)$$

$$+ F_{Mi} + F_{si} + F_{vmi} + v_i S \tag{4.9}$$

The third term on the RHS of Eq. (4.9) is the gravitational term accounting for the buoyancy effect, and the last five terms are the phase interaction terms. For dilute multiphase flows there are no interparticle collisions, and hence no particle pressure and particle viscosity, i.e.,

$$<\tilde{p}_k> = 0 \qquad <\tilde{\tau}_{kij}> = 0$$

Therefore, the volume-averaged particle momentum equation becomes

$$\frac{\partial}{\partial t}(\rho_k v_{ki}) + \frac{\partial}{\partial x_j}(\rho_k v_{kj} v_{ki}) = \rho_p g_i + \frac{\rho_k}{\tau_{rk}}(v_i - v_{ki}) + F_{k,Mi} + F_{k,si} + F_{k,vmi} + v_i S_k$$

(4.10)

Finally, in a similar way, for low Mach number, constant specific heat, dilute multiphase flows, neglecting interparticle collision and forces other than the drag force and the gravitational force, the instantaneous fluid and particle continuity, momentum, energy, and species equations can be obtained as

$$\frac{\partial \rho}{\partial t} + \frac{\partial}{\partial x_j}(\rho v_j) = S \tag{4.11}$$

$$\frac{\partial \rho_k}{\partial t} + \frac{\partial}{\partial x_j}(\rho_k v_{kj}) = S_k \tag{4.12}$$

$$\frac{\partial}{\partial t}(\rho v_i) + \frac{\partial}{\partial x_j}(\rho v_j v_i) = -\frac{\partial p}{\partial x_i} + \frac{\partial \tau_{ji}}{\partial x_j} + \Delta \rho g_i +$$
$$\sum_k \frac{\rho_k}{\tau_{rk}}(v_{ki} - v_i) + v_i S \tag{4.13}$$

$$\frac{\partial}{\partial t}(\rho_k v_{ki}) + \frac{\partial}{\partial x_j}(\rho_k v_{kj} v_{ki}) = \rho_k g_i + \frac{\rho_k}{\tau_{rk}}(v_i - v_{ki}) + v_i S_k \tag{4.14}$$

$$\frac{\partial}{\partial t}(\rho c_p T) + \frac{\partial}{\partial x_j}(\rho v_j c_p T) - \frac{\partial p}{\partial t} = \frac{\partial}{\partial x_j}\left(\lambda \frac{\partial T}{\partial x_j}\right) + w_s Q_s - q_r + \sum n_k Q_k + c_p T S$$

(4.15)

$$\frac{\partial}{\partial t}(\rho_k c_k T_k) + \frac{\partial}{\partial x_j}(\rho_k v_{kj} c_k T_k) = n_k(Q_h - Q_k - Q_{rk}) + c_p T S_k \tag{4.16}$$

$$\frac{\partial}{\partial t}(\rho Y_s) + \frac{\partial}{\partial x_j}(\rho v_j Y_s) = \frac{\partial}{\partial x_j}\left(D\rho \frac{\partial Y_s}{\partial x_j}\right) - w_s + \alpha_s S \tag{4.17}$$

where S is the mass source term due to particle/droplet evaporation, devolatilization, and heterogeneous reactions

$$S = -\sum_k S_k = -\sum_k n_k \dot{m}_k \qquad \dot{m}_k = \frac{dm_k}{dt}$$

where Q_k is the convective heat transfer between each particle and the fluid, q_r is the fluid radiative heat transfer, Q_{rk} is the particle radiative heat transfer, w_s and $w_s Q_s$ are the reaction rate and reaction heat release for species s of the fluid phase, respectively, Q_h is the heat release or absorption due to particle evaporation (or condensation), devolatilization and heterogeneous reactions, and α_s is the fraction of contribution to species s due to phase change.

In establishing the above equations, attention should be paid to two problems. One is the source term in the fluid momentum equation due to particle mass change. The result of volume averaging gives $v_i S$, as pointed out by the present author [9], but not $v_{ki} S_k$, as proposed by some investigators [10], and also not the product of particle mass change times the velocity slip between gas and particles. Another problem is the distribution of reaction heat release between two phases in the gas and particle energy equations. Some investigators assume that all of the reaction heat is at first imposed to the gas phase, and then to the particles by convective heat transfer between two phases [10], the result being that the predicted gas temperature is unbelievably high. Other investigators assume that all of the reaction heat is at first imposed to the particles, and then to the gas by convective heat transfer between two phases [11], the result again being that the predicted particle temperature is unbelievably high. Experiments show that in the reacting two-phase flows, such as the pulverized-coal flame, in the fully developed-flow region, the particle temperature is somewhat higher than the gas temperature by about 10%. Therefore, the approach taken by the present author is that the gas-phase reaction heat is imposed to the gas phase, and the particle heating effect, including particle evaporation, devolatilization, and heterogeneous combustion, is imposed to the particle phase.

The above-derived are conservation equations in Eulerian coordinates. Alternatively, the particle conservation equations can be expressed in Lagrangian coordinates. Putting $\rho_p = n_p m_p$ into the particle continuity equation, we have

$$\frac{\partial}{\partial t}(n_k m_k) + \frac{\partial}{\partial x_j}(n_k m_k v_{kj}) = \left[\frac{\partial n_k}{\partial t} + \frac{\partial}{\partial x_j}(n_k v_{kj})\right] m_k + n_k \dot{m}_k = n_k \dot{m}_k$$

$$\frac{\partial n_k}{\partial t} + \frac{\partial}{\partial x_j}(n_k v_{kj}) = 0$$

For steady flows, we have

$$\frac{\partial}{\partial x_j}(n_k v_{kj}) = 0 \tag{4.18}$$

For a stream tube with cross-sections of A_1 and A_2 the use of the Gauss theorem gives

$$
\begin{aligned}
N_k &= \int_{A_1} n_k v_{kn} dA = \int_{A_2} n_k v_{kn} dA \\
&= \int_A n_k v_{kn} dA = const
\end{aligned}
\tag{4.19}
$$

where v_{kn} is the particle velocity component perpendicular to the cross-sections of the stream tube, and N_k is the total particle number flux. Eq. (4.19) implies that total particle number flux remains unchanged along particle trajectories. Using the particle continuity equation (Eq. 4.12), the particle momentum equation (Eq. 4.14) can be changed to

$$
\rho_k \left(\frac{\partial v_{ki}}{\partial t} + v_{kj} \frac{\partial v_{ki}}{\partial x_j} \right) = \rho_k \frac{dv_{ki}}{dt_k} = \rho_k g_i + \left(\frac{\rho_k}{\tau_{rk}} + S_k \right) (v_i - v_{ki})
$$

Dividing each term by ρ_k gives the particle momentum equation in Lagrangian coordinates

$$
\frac{dv_{ki}}{dt_k} = \left(\frac{1}{\tau_{rk}} + \frac{\dot{m}_k}{m_k} \right) (v_i - v_{ki}) + g_i
\tag{4.20}
$$

In a similar way the particle energy equation in Lagrangian coordinates can be derived. Therefore, the particle conservation equations in Lagrangian coordinates can be obtained as

$$
\left\{
\begin{aligned}
&\int_A n_k v_{kn} dA = N_k = const \\
&\frac{dv_{ki}}{dt_k} = \left(\frac{1}{\tau_{rk}} + \frac{\dot{m}_k}{m_k} \right) (v_i - v_{ki}) + g_i \\
&\frac{dT_k}{dt_k} = \left[Q_h - Q_k - Q_{rk} + \dot{m}_k (c_p T - c_k T_k) \right] / (m_k c_k)
\end{aligned}
\right.
\tag{4.21}
$$

4.5 THE REYNOLDS-AVERAGED EQUATIONS FOR DILUTE MULTIPHASE TURBULENT REACTING FLOWS

Although the conservation equations for turbulent multiphase flows were mentioned by S.L. Soo in Reference [4] more than 40 years ago, the so-called "turbulent diffusion coefficient" instead of the laminar viscosity is a unclear concept. Therefore, actually what are used are still equations for laminar flows. Starting in the 1980s, Elghobashi [12], Zhou [13], and other investigators separately proposed the Reynolds-averaged equations for turbulent multiphase flows. For multiphase turbulent flows, the method of Reynolds averaging for single-phase flows, namely the decomposition of instantaneous variables into time-averaged and fluctuation components and then once again taking their time-averaged values, can be adopted. Using the generalized variable ϕ_k to denote the instantaneous value of velocity, temperature, concentration, or volume fraction for each phase and species concentration, the time-averaged value is defined by

$$\overline{\phi_k} = \frac{1}{T} \int_t^{t+T} \phi_k dt \ (T \text{ approaches a large value}) \tag{4.22}$$

where T should be much larger than the turbulence integral time scale and simultaneously should be much smaller than the time period of macroscopic flows (e.g., the period of wave motion). The relationships among the instantaneous, time-averaged, and fluctuation values and their operation rules are

$$\phi = \overline{\phi} + \phi', \overline{(\overline{\phi})} = \overline{\phi}, \overline{\phi'} = \overline{\overline{\phi}\phi'} = 0, \overline{\phi'\psi'} \neq 0$$

For dilute multiphase turbulent flows, neglecting the interparticle collision, fluid-density fluctuation, particle mass and its changing-rate fluctuation, particle number density–fluid velocity correlation and some unsteady correlation terms, taking Reynolds averaging for the above-obtained instantaneous equations, the time-averaged continuity, momentum, energy, and species conservation equations for multiphase turbulent flows can be obtained as (for simplicity omit the superscript bar for time-averaged values)

$$\frac{\partial \rho}{\partial t} + \frac{\partial}{\partial x_j}(\rho V_j) = S \tag{4.23}$$

$$\frac{\partial \rho_k}{\partial t} + \frac{\partial}{\partial x_j}(\rho_k V_{kj}) = S_k - \frac{\partial}{\partial x_j}(m_k \overline{n_k v_k}) \tag{4.24}$$

$$\frac{\partial N_k}{\partial t} + \frac{\partial}{\partial x_j}(N_k V_{kj}) = -\frac{\partial}{\partial x_j}(\overline{n_k v_{kj}}) \tag{4.25}$$

$$\frac{\partial}{\partial t}(\rho V_i) + \frac{\partial}{\partial x_j}(\rho V_j V_i) = -\frac{\partial p}{\partial x_i} + \Delta \rho g_i + \sum \rho_k (V_{ki} - V_i)/\tau_{rk} + V_i S$$
$$- \frac{\partial}{\partial x_j}(\overline{v_j v_i}) + \sum m_k \overline{n_k v_{ki}} \tag{4.26}$$

$$\frac{\partial}{\partial t}(N_p V_{pj}) + \frac{\partial}{\partial x_j}(N_p V_{pi} V_{pj}) = N_p g_i - \frac{\partial}{\partial x_j}(N_p \overline{v_{pi} v_{pj}} + V_{pj} \overline{n_p v_{pi}} + V_{pi} \overline{n_p v_{pj}})$$
$$+ \frac{1}{\tau_{rp}}\left[N_p(V_i - V_{pi}) - \overline{n_p v_{pi}}\right] + \frac{N_p \dot{m}_p}{m_p}(V_i - V_{pi})$$

$$\tag{4.27}$$

$$\frac{\partial}{\partial t}(\rho h) + \frac{\partial}{\partial x_j}(\rho V_j h) = \frac{\partial}{\partial x_j}\left(\lambda\frac{\partial T}{\partial x_j}\right) - \frac{\partial q_{rj}}{\partial x_j} + \sum n_k Q_k + hS - \frac{\partial}{\partial x_j}(\overline{\rho v_j' h'})$$

(4.28)

$$\frac{\partial}{\partial t}(N_p h_p) + \frac{\partial}{\partial x_j}(N_p V_{pj} h_p) = \frac{N_p}{m_p}(Q_h - Q_p - Q_{rp}) - \frac{\partial}{\partial x_j}(N_p \overline{v_{pj}' h_p'} + V_{pj}\overline{n_p h_p'} + h_p \overline{n_p v_{pj}})$$

$$+ \frac{N_p \dot{m}_p}{m_p}(h - h_p)$$

(4.29)

$$\frac{\partial}{\partial t}(\rho Y_s) + \frac{\partial}{\partial x_j}(\rho V_j Y_s) = \frac{\partial}{\partial x_j}\left(D\rho\frac{\partial Y_s}{\partial x_j}\right) - w_s - \alpha_s S_k - \frac{\partial}{\partial x_j}(\overline{\rho v_j' Y_s'})$$ (4.30)

Eqs. (4.23)–(4.30) give the general description of multiphase turbulent flows. The unknown correlation terms, i.e., the Reynolds stress, Reynolds heat flux, and Reynolds mass flux, need to be closed by multiphase turbulence models.

4.6 THE PDF EQUATIONS FOR TURBULENT TWO-PHASE FLOWS AND STATISTICALLY AVERAGED EQUATIONS

The above-stated conservation equations for turbulent multiphase flows are obtained by Reynolds averaging for instantaneous volume-averaged equations, based on the concept of pseudo-fluids of the dispersed phase. A question is raised: Is the concept of pseudo-fluids feasible, if the number of particles is too small in the control volume? Alternatively, some investigators consider the particles as a randomly moved statistical group and use the kinetic theory to derive the PDF (probability density function) transport equation for particles. Then the statistical averaging gives the conservation equations for two-fluid modeling. These statistically averaged equations are close to those obtained by Reynolds averaging or mass-weighed averaging. Hence, the two-fluid approach can cover a wide range of problems and is not limited by the number of particles in the control volume. Still in the 1960s, F.A. Williams [14] and L.X. Zhou [15] independently proposed the statistical conservation equations for a particle group, however at that time it was not related to turbulence. Starting in the 1980s, S.B. Pope [16] proposed the PDF transport equation for single-phase reacting flows, using the Monte-Carlo method to solve the PDF equation for turbulent combustion. From the beginning of the 1990s, Zaichik [17], Reeks [18], and Simonin [19] independently derived the particle PDF equations for turbulent two-phase flows, using these PDF equations to obtain the particle conservation equations in two-fluid modeling. Afterwards, L.X. Zhou et al. [20] derived the particle joint PDF equation in gas-particle velocity space, using this PDF equation to obtain the two-fluid model equations. The PDF distribution for complex swirling and nonswirling turbulent gas-particle flows is directly solved by numerical methods [21,22], and is used to directly obtain the particle Reynolds stress and particle turbulent kinetic energy. The PDF equation is used not only to obtain the two-fluid model equations, but also to close the two-phase turbulence model, which will be discussed later.

Let us at first consider the statistical conservation equations for a particle group in laminar two-phase flows. In a most general case for a dispersed system, like a particle/bubble/droplet group, each particle has its own size, velocity, temperature, and material density; particles of the same size may have different velocities, temperatures, and material densities. The PDF for the particle number density in the range of

$$V_p \to V_p + dV_p, d_p \to d_p + d(d_p), T_p \to T_p + dT_p, \qquad \bar{\rho}_p \to \bar{\rho}_p + d\bar{\rho}_p$$

can be defined as

$$\phi(V_p, d_p, T_p, \bar{\rho}_p, x, t)dv_p d(d_p)dT_p d\bar{\rho}_p dxdt$$

Based on the Louville theorem in statistical mechanics for a collisionless particle group, the volumetric changing rate in the phase space should be zero, so the PDF conservation equation in the phase space is

$$\frac{\partial \phi}{\partial t} + \frac{\partial}{\partial v_p} \left(\phi \frac{dV_p}{dt} \right) + \frac{\partial}{\partial d_p} \left(\phi \frac{d(d_p)}{dt} \right) + \frac{\partial}{\partial T_p} \left(\phi \frac{dT_p}{dt} \right) + \frac{\partial}{\partial \bar{\rho}_p} \left(\phi \frac{d\bar{\rho}_p}{dt} \right) + \frac{\partial}{\partial x} \left(\phi \frac{dx}{dt} \right) = 0$$

$$(4.31)$$

Now, let us consider turbulent two-phase flows. For simplicity, using capital letters to denote instantaneous values, lower-case letters to denote fluctuation values, and the symbol $< \, >$ to denote statistically averaged values, the PDF for an instantaneous variable at the time instant t and the location x_j can be defined by

$$p_s \equiv p_s(V_i, V_{pi}; x_j, t)$$

where p_s is the joint PDF in the two-phase velocity space, and V_i, V_{pi} are the gas and particle velocity coordinates, respectively. Based on the property of the δ function, the one possible realization of p_s can be defined by

$$p'_s(V_i, V_{pi}; x_j, t) = \prod_{j=1}^{3} \delta(U_i(x_j, t) - V_i) \prod_{j=1}^{3} \delta(U_{pi}(x_j, t) - V_{pi})$$

Its statistically averaged value in the phase space is

$$p_s(V_i, V_{pi}; x_j, t) = <p'_s(V_i, V_{pi}; x_j, t)>$$

where the symbol $<\ >$ expresses the statistically averaged value after tests of infinite times (namely the mathematical expectation of p'_s). Similarly, the PDF for a fluctuation variable in the turbulent two-phase flow field can be defined by

$$p_f(v_i, v_{pi}; x_j, t) = <p'_f(v_i, v_{pi}; x_j, t)>$$

where v_i, v_{pi} are the gas and particle fluctuation velocity coordinates, respectively, and we also have

$$p'_f(v_i, v_{pi}; x_j, t) = \prod_{j=1}^{3} \delta(u_i(x_j, t) - v_i) \prod_{j=1}^{3} \delta(u_{pi}(x_j, t) - v_{pi})$$

It can be seen that from the property of PDF for any physical variable $Q = Q(V_i, V_{pi})$ or $q = q(v_i, v_{pi})$, the statistically averaged value in the geometrical space should be

$$<Q(V_i, V_{pi})> = \iint p_s(V_i, V_{pi}) Q(V_i, V_{pi}) dV_i dV_{pi}$$

$$<q(v_i, v_{pi})> = \iint p_f(v_i, v_{pi}) q(v_i, v_{pi}) dv_i dv_{pi}$$

For dilute two-phase flows, neglecting the forces acting on the particles other than the drag force and using the instantaneous two-phase flow equations, the transport equations for p_s and p_f can be obtained as

$$\frac{\partial p_{si}}{\partial t} + (U_j + U_{pj}) \frac{\partial p_{si}}{\partial x_j} = \frac{\partial}{\partial V_i} < \frac{p'_{si}}{\rho} \frac{\partial P}{\partial x_i} > - < \frac{p'_{si}}{\rho} \frac{\partial \tau_{ij}}{\partial x_j} \delta_{ij} >$$

$$- \left[g_i + \sum \frac{\rho_p}{\rho \tau_{rp}} (<U_{pi}> - <U_i>) + U_{pj} \frac{\partial U_i}{\partial x_j} \right] \frac{\partial p_{si}}{\partial V_i}$$

$$- \left[g_i + \left(\frac{1}{\tau_{rp}} + \frac{\dot{m}_p}{m_p} \right) (<U_{pi}> - <U_i>) + U_j \frac{\partial U_{pi}}{\partial x_j} \right] \frac{\partial p_{si}}{\partial V_{pi}} \tag{4.32}$$

$$- \frac{\partial}{\partial V_i} <p'_{si} (\sum f_{rpi})> - \frac{\partial}{\partial V_i} <p'_{si} f_{ri}>$$

$$\frac{\partial p_{fi}}{\partial t} + (<U_j> + <U_{pj}>) \frac{\partial p_{fi}}{\partial x_j} + (u_j + u_{pj}) \frac{\partial p_{fi}}{\partial x_j}$$

$$= \frac{\partial}{\partial v_i} < \frac{p'_{fi}}{\rho} \frac{\partial P'}{\partial x_i} > - < \frac{p'_{fi}}{\rho} \frac{\partial \tau'_{ij}}{\partial x_j} \delta_{ij} > + \frac{\partial p_{fi}}{\partial v_i} \left(u_j \frac{\partial <U_i>}{\partial x_j} \right)$$

$$+ \frac{\partial p_{fi}}{\partial v_{pi}} \left(u_{pj} \frac{\partial <U_{pi}>}{\partial x_j} \right) - <p'_{fi} \left(\frac{\rho'}{\rho} g_i \right) > - (<U_{pj}> + u_{pj})$$

$$\times \frac{\partial u_i}{\partial x_j}\frac{\partial p_{fi}}{\partial v_i} - (<U_j> + u_j)\frac{\partial u_{pi}}{\partial x_j}\frac{\partial p_{fi}}{\partial v_{pi}} - \frac{\partial}{\partial v_i} <p'_{fi}\left(\sum f_{rpi}\right)>$$

$$- \frac{\partial}{\partial v_{pi}} <p'_{fi}\left(\sum f_{ri}\right)>$$

(4.33)

where $f_{rpi} = \frac{\rho_p}{\rho \tau_{rp}}(u_{pi} - u_i), f_{ri} = \left(\frac{1}{\tau_{rp}} + \frac{\dot{m}_p}{m_p}\right)(u_i - u_{pi})$ express the gas and particle fluctuation drag, respectively, U_j, U_{pj} are gas and particle instantaneous velocities, respectively, $U_j = <U_j> + u_j$, $U_{pj} = <U_{pj}> + u_{pj}$, and u_j and u_{pj} are gas and particle fluctuation velocities, respectively. When neglecting the fluctuations of gas and particle densities and particle mass changing rate, by multiplying any macroscopic variable with the PDF transport equation and doing the integral in the velocity space, the statistically averaged gas and particle continuity and momentum equations for turbulent two-phase flows can be obtained as

$$\frac{\partial \rho}{\partial t} + \frac{\partial}{\partial x_j}(\rho <U_j>) = S$$

(4.34)

$$\frac{\partial \rho_p}{\partial t} + \frac{\partial}{\partial x_j}(\rho_p <U_{pj}>) = S_p = n_p \dot{m}_p$$

(4.35)

$$\frac{\partial <U_i>}{\partial t} + <U_j>\frac{\partial <U_i>}{\partial x_j} + \frac{\partial <u_i u_j>}{\partial x_j} = - <\frac{1}{\rho}\frac{\partial P}{\partial x_i}>$$

$$+ <\frac{1}{\rho}\frac{\partial \tau_{ij}}{\partial x_j}> + g_i + \sum \frac{\rho_p}{\rho \tau_{rp}}(<U_{pi}> - <U_i>) + <\sum f_{rpi}>$$

(4.36)

$$\frac{\partial <U_{pi}>}{\partial t} + <U_{pj}>\frac{\partial <U_{pi}>}{\partial x_j} + \frac{\partial <u_{pi}u_{pj}>}{\partial x_j}$$

$$= g_i + \left(\frac{1}{\tau_{rp}} + \frac{\dot{m}_p}{m_p}\right)(<U_i> - <U_{pi}>) + <f_{ri}>$$

(4.37)

It can be seen that these statistically averaged equations are almost the same as those obtianed from the Reynolds time-averaging or mass-weighed averaging, with only some minor differences.

4.7 THE TWO-PHASE REYNOLDS STRESS AND SCALAR TRANSPORT EQUATIONS

The Reynolds time-averaged, mass-weighed averaged, or statistically averaged equations are an unclosed system, since the correlation terms in these equations, such as the fluid/gas and particle Reynolds stress, heat flux, mass flux

$$\overline{v_i v_j}, \overline{v_{pi} v_{pj}}, \overline{v_j h'}, \overline{v_j h'_p}, \overline{v_j Y'_s}, \overline{n_p v_{pi}}, \overline{n_p v_{pj}}$$

are unknown terms. To create closure of these equations, many years ago, it was decided to further derive the transport equations for these correlation terms. Still in the 1940s, P.Y. Chou [23] proposed the idea of deriving the Reynolds stress transport equation for single-phase flows. In the 1990s, L.X. Zhou proposed deriving the two-phase Reynolds stress equations [24]. There are different methods to derive the Reynolds stress transport equation. One is: (1) Write the N-S equations separately for the instantaneous velocity components V_i and V_j; (2) Multiply the N-S equation for V_i by V_j and multiply the N-S equation for V_j by V_i and take their summation to obtain the equation for $V_i V_j$; (3) Make the time averaging for the above-obtained equation to get the equation for the time-averaged value of the product of instantaneous velocities $\overline{V_i V_j}$; (4) Do a similar procedure for the time-averaged velocity components to obtain the equation for the product $\overline{V}_i \overline{V}_j$; (5) Subtract the equation for $\overline{V}_i \overline{V}_j$ from the equation for $\overline{V_i V_j}$ to obtain the Reynolds stress equation for $\overline{v_i v_j}$.

For two-phase flows we have

$$\frac{\partial}{\partial t}(\rho V_i) + \frac{\partial}{\partial x_m}(\rho V_m V_i) = -\frac{\partial p}{\partial x_i} + \frac{\partial \tau_{mi}}{\partial x_m} + \rho g_i \beta \Delta T + \sum_k \frac{\rho_k}{\tau_{rk}}(V_{ki} - V_i) + V_i S$$

$$\frac{\partial}{\partial t}(\rho V_j) + \frac{\partial}{\partial x_m}(\rho V_m V_j) = -\frac{\partial p}{\partial x_j} + \frac{\partial \tau_{mj}}{\partial x_m} + \rho g_j \beta \Delta T + \sum_k \frac{\rho_k}{\tau_{rk}}(V_{kj} - V_j) + V_j S$$

$$\frac{\partial}{\partial t}(N_k V_{ki}) + \frac{\partial}{\partial x_m}(N_k V_{km} V_{ki}) = N_k g_i + \left(\frac{1}{\tau_{rk}} + \frac{\dot{m}_k}{m_k}\right) N_k (V_i - V_{ki})$$

$$\frac{\partial}{\partial t}(N_k V_{kj}) + \frac{\partial}{\partial x_m}(N_k V_{km} V_{kj}) = N_k g_j + \left(\frac{1}{\tau_{rk}} + \frac{\dot{m}_k}{m_k}\right) N_k (V_j - V_{kj})$$

Doing the above-stated procedure for these four equations gives the exact transport equations of the fluid/gas and particle Reynolds stresses for turbulent two-phase flows as

$$\frac{\partial}{\partial t}(\rho \overline{v_i v_j}) + \frac{\partial}{\partial x_m}(\rho V_m \overline{v_i v_j}) = D_{ij} + P_{ij} + G_{ij} + \Pi_{ij} - \varepsilon_{ij} + G_{p,ij} + G_{R,ij} \quad (4.38)$$

$$\frac{\partial}{\partial t}(N_k \overline{v_{ki} v_{kj}}) + \frac{\partial}{\partial x_m}(N_k V_{km} \overline{v_{ki} v_{kj}}) = D_{k,ij} + P_{k,ij} + G_{k,ij} + \varepsilon_{k,ij} \quad (4.39)$$

where D_{ij}, P_{ij}, G_{ij}, Π_{ij}, and ε_{ij} are the diffusion, shear production, buoyancy production, pressure-strain, and dissipation terms of the fluid/gas stresses, and they have the same meanings as those for single-phase flows [1]:

$$D_{ij} = -\frac{\partial}{\partial x_m}\left[\rho\overline{v_i v_j v_m} + \overline{p'v_j}\delta_{im} + \overline{p'v_i}\delta_{jm} - \mu(\frac{\partial}{\partial x_m}\overline{v_i v_j})\right];$$

$$P_{ij} = -\rho\left(\overline{v_i v_m}\frac{\partial V_j}{\partial x_m} + \overline{v_j v_m}\frac{\partial V_i}{\partial x_m}\right); G_{ij} = \beta\rho(g_i\overline{v_j T'} + g_j\overline{v_i T'});$$

$$\Pi_{ij} = p'\left(\frac{\partial v_i}{\partial x_j} + \frac{\partial v_j}{\partial x_i}\right); \varepsilon_{ij} = -2\mu\left(\frac{\partial v_i}{\partial x_m}\frac{\partial v_j}{\partial x_m}\right)$$

The last two terms on the right-hand side of Eq. (4.38)

$$G_{p,ij} = \sum_k\frac{\rho_k}{\tau_{rk}}(\overline{v_{ki}v_j} + \overline{v_{kj}v_i} - 2\overline{v_i v_j}) \text{ and } G_{R,ij} = \overline{v_i v_j}S \text{ are the source terms}$$

of fluid stress due to particle drag and particle reaction (including evaporation, devolatilization, and heterogeneous reaction), respectively. The three terms on the right-hand side of Eq. (4.39) are the particle stress diffusion, shear production, and the production/dissipation terms due to particle drag. These are

$$D_{k,ij} = -\frac{\partial}{\partial x_m}\left[N_k\overline{v_{km}v_{ki}v_{kj}}\right]$$

$$P_{k,ij} = -(V_{km}\overline{n_k v_{kj}} + N_k\overline{v_{km}v_{kj}})\frac{\partial V_{ki}}{\partial x_m} V_{km}\overline{n_k v_{ki}} + N_k\overline{v_{km}v_{ki}})\frac{\partial V_{kj}}{\partial x_m}$$

$$G_{k,ij} = +\overline{n_k v_{kj}}g_i + \overline{n_k v_{ki}}g_j$$

$$\varepsilon_{k,ij} = \left(\frac{1}{\tau_{rk}} + \frac{\dot{m}_k}{m_k}\right)\left[N_k(\overline{v_{ki}v_j} + \overline{v_{kj}v_i} - 2\overline{v_{ki}v_{kj}}) + (V_i - V_{ki})\overline{n_k v_{kj}} + (V_j - V_{kj})\overline{n_k v_{ki}}\right]$$

The transport equations for $\overline{v_{ki}v_j}, \overline{v_{kj}v_i}, \overline{n_k v_{ki}}, \overline{n_k v_{kj}}$, and $\overline{n_k n_k}$ can also be derived. For example, the transport equation of two-phase velocity correlation is [24]:

$$\frac{\partial}{\partial t}(\overline{v_{ki}v_j}) + (V_{km} + V_m)\frac{\partial}{\partial x_m}(\overline{v_{ki}v_j}) = -\frac{\partial}{\partial x_m}(\overline{v_m v_{ki}v_j} + \overline{v_{km}v_{ki}v_j})$$

$$+ \frac{1}{\rho\tau_{rk}}\left[\rho_k\overline{v_{ki}v_{kj}} + \rho\overline{v_i v_j} - (\rho_k + \rho)\overline{v_{ki}v_j}\right] - \left(\overline{v_{km}v_j}\frac{\partial V_{ki}}{\partial x_m} + \overline{v_{ki}v_m}\frac{\partial V_j}{\partial x_m}\right) - \frac{k}{\varepsilon}\overline{v_{ki}v_i}\delta_{ij}$$

$$(4.40)$$

Taking $i = j$, the two-phase turbulent kinetic energy equations can be derived from Eqs. (4.38) and (4.39) as:

$$\frac{\partial}{\partial t}(\rho k) + \frac{\partial}{\partial x_m}(\rho V_m k) = -\frac{\partial}{\partial x_m}\left(\overline{\rho v_m v_i^2}/2 + \overline{p' v_m} - \mu \frac{\partial k}{\partial x_m}\right) - \overline{\rho v_m v_i}\frac{\partial V_i}{\partial x_m}$$

$$+ \beta \rho g_i \overline{v_i T'} - \mu \overline{\left(\frac{\partial v_i}{\partial x_m}\right)^2} + \sum_k \frac{N_k m_k}{\tau_{rk}}(\overline{v_{ki} v_i} - 2k) + kS$$

$$\frac{\partial}{\partial t}(N_k k_k) + \frac{\partial}{\partial x_m}(N_k V_{km} k_k) = -\frac{\partial}{\partial x_m}(N_k \overline{v_{km} v_{ki}^2}/2) - \left(N_k \overline{v_{km} v_{ki}}\frac{\partial V_{ki}}{\partial x_m}\right)$$

$$+ \left(\frac{1}{\tau_{rk}} + \frac{\dot{m}_k}{m_k}\right)[N_k(\overline{v_{ki} v_i} - 2k_k) + (V_i - V_{ki})\overline{n_k v_{ki}}]$$

(4.41)

Similarly, the transport equation of the fluid heat flux for turbulent two-phase flows can be obtained as

$$\frac{\partial}{\partial t}(\rho \overline{v_i T'}) + \frac{\partial}{\partial x_m}(\rho V_m \overline{v_i T'}) = -\frac{\partial}{\partial x_m}\left(\rho \overline{v_m v_i T'} + \overline{p' T'}\delta_{im} - \frac{\lambda}{c_p}\overline{v_i \frac{\partial T'}{\partial x_m}} - \mu \overline{T'\frac{\partial v_i}{\partial x_m}}\right)$$

$$- \rho\left(\overline{v_i v_m}\frac{\partial T}{\partial x_m} + \overline{v_m T'}\frac{\partial V_i}{\partial x_m}\right) - \beta \rho g_i \overline{T'^2} - \left(\frac{\lambda}{c_p} + \mu\right)\overline{\frac{\partial v_i}{\partial x_m}\frac{\partial T'}{\partial x_m}} + \overline{p'\frac{\partial T'}{\partial x_m}} + \overline{v_i w_s'}Q_s$$

$$+ \sum \overline{n_k v_{ki}}Q_k + c_p \overline{S v_i T'}$$

(4.42)

and also the transport equation for the mean square value of the fluid temperature fluctuation is obtained as

$$\frac{\partial}{\partial t}(\rho \overline{T'^2}) + \frac{\partial}{\partial x_m}(\rho V_m \overline{T'^2}) = -\frac{\partial}{\partial x_m}\left(\rho \overline{v_m T'^2} - \frac{\lambda}{c_p}\frac{\partial \overline{T'^2}}{\partial x_m}\right) - 2\overline{v_m T'}\frac{\partial T}{\partial x_m} - 2\frac{\lambda}{c_p}\overline{\left(\frac{\partial T'}{\partial x_m}\right)^2}$$

$$+ \overline{T' w_s'}Q_s + \sum \overline{n_k T'}Q_k + c_p \overline{S T'^2}$$

(4.43)

Other scalars, such as the species mass flux and its RMS value of fluctuation, can also be derived. Starting with the PDF transport equation, by integrating over the PDF, the statistically averaged two-phase Reynolds stress equations can also be derived [20]. Their form is similar to that for the equations derived by Reynolds time averaging or mass-weighed averaging with only some minor differences. There are still many unknown correlation terms in Eqs. (4.38)–(4.43). The closure models for these terms are called two-phase turbulence models, which will be discussed in the following chapters of this book.

REFERENCES

[1] L.X. Zhou, Theory and Numerical Modeling of Turbulent Gas-Particle Flows and Combustion., CRC Press, Florida, 1993.

[2] L.X. Zhou, Combustion Theory and Reacting Fluid Dynamics [in Chinese]., Science Press, Beijing, 1986.

[3] S.L. Soo, Multiphase Fluid Dynamics., Science Press, Beijing and Hong Kong, 1990.

[4] S.L. Soo, Fluid Dynamics of Multiphase Systems. Blaisdell (Ginn), New York, 1967.

[5] D.A. Drew, Averaged field equations for two-phase media., Stud. Appl. Mech. 50 (1971) 133−166.

[6] R.I. Nigmatulin, Spatial averaging in the mechanics of heterogeneous and dispersed systems. Inter, J. Multiphase Flow 5 (1979) 353−385.

[7] B. Jackson, Locally averaged equations of motion for a mixture of identical spherical particles and a Newtonian fluid., Chem. Eng. Sci. 52 (1997) 2457−2469.

[8] L.S. Fan, C. Zhu, Principles of Gas-Solid Flows., Cambridge University Press, Cambridge, 1998.

[9] L.X. Zhou, Multiphase fluid dynamics of gas-particle systems with phase change., Progress in Mechanics [in Chinese] 12 (1982) 141−150.

[10] L.D. Smoot, P.J. Smith, Coal Combustion and Gasification., Plenum Press, London, 1985.

[11] D.B. Stickler. AIAA 17th Aerospace Science Meeting, Paper 79-0298, 1979.

[12] S.E. Elghobashi, T.W. Abou-Arab, A two-equation turbulence model of two-phase flows., Physics of Fluids 26 (1983) 931−940.

[13] L.X. Zhou, X.Q. Huang, Prediction of confined turbulent gas-particle jets by an energy equation model of particle turbulence., Science in China A33 (1990) 52−59.

[14] F.A. Williams, Combustion Theory., Addison-Wesley, N.Y, 1965.

[15] L.X. Zhou. Modeling of liquid-spray combustors, Proceedings of the Conference on Experimental Techniques of Gas Turbine Combustors [in Chinese]. China Industry Press, Beijing, 1965.

[16] S.B. Pope, PDF methods for turbulent reactive flows., Progress Energy Combustion Sci. 11 (1985) 119−192.

[17] I.V. Derevich, L.I. Zaichik, The equations for the probability density of the particle velocity and temperature in a turbulent flow simulated by the Gauss stochastic field., Prikl. Mat. Mekh. [in Russian] 54 (1990) 767.

[18] M.W. Reeks, On a kinetic equation for the transport of particles in turbulent flows, Phys. Fluids A3 (1991) 446−456.

[19] O. Simonin. Continuum modeling of dispersed turbulent two-phase flows. VKI Lectures: Combustion in Two-phase Flows, 1996.

[20] Zhou, Lixing, Lin, Wenyi, Li, Yong, New statistical theory and a k-ε-PDF model for simulating gas-particle flows., Tsinghua Sci. Technol. 2 (1997) 628−632.

[21] Y. Li, L.X. Zhou, A k-ε-PDF two-phase turbulence model for simulating sudden-expansion particle-laden flows., J. Thermal Sci. 5 (1996) 34−38.

[22] L.X. Zhou, Simulation of strongly swirling gas-particle flows using a DSM-PDF two-phase turbulence model., Powder Technol. 113 (2000) 70−79.

[23] P.Y. Chou, On velocity correlations and the solution of the equation of the fluctuation., Q. Appl. Math. 3 (1945) 38−54.

[24] L.X. Zhou, C.M. Liao, T. Chen, A unified second-order moment two-phase turbulence model for simulating gas-particle flows., ASME-FED 185 (1994) 307−313.

Chapter 5

Modeling of Single-Phase Turbulence

5.1 INTRODUCTION

The governing equations given in Chapter 4 are not a closed system, since there are unknown correlations in these equations. In this chapter we will discuss the closure models for single-phase turbulent flows, that is, so-called turbulence models for single-phase flows. This is the basis for developing the two-phase turbulence models, which will be discussed in Chapter 6. It is well known that turbulence is one of the unsolved complex theoretical problems in fluid mechanics. However, in treating practical engineering problems, different methods were used in the past. The simplest method is taking a constant "turbulent viscosity" or "turbulent diffusivity" as a fluid property, which was assumed to be several hundred times the molecular viscosity. This was used in early hydrodynamic simulation and combustor flow predictions. For many years in jet flows the integral methods based on semi-empirical velocity and temperature profiles were used. In these approaches there is no knowledge of turbulence. Obviously, when predicting more complicated engineering turbulent flows, for example, three-dimensional recirculating or swirling flows, the methods of constant viscosity or of jet flows are not appropriate, because these types of flows are far from simple jet flows and the turbulent viscosity is not a constant. Nowadays, one of the fundamental studies on turbulence is direct numerical simulation (DNS). This is used to directly solve the instantaneous 3-D Navier–Stokes equation in Kolmogorov dissipation scales. DNS can give the information of all scales without using any closure models. However, even for small-sized computational domains and low Reynolds number flows, DNS still needs huge computational requirement, and at present it cannot be used in solving practical engineering problems. However, its database can be used for validating and improving the following discussed turbulence models for so-called Reynolds-averaged Navier–Stokes (RANS) modeling. An alternative method is called large-eddy simulation (LES). This is used directly to solve the N-S equation in large-eddy grid sizes using white-noise filtering. For unresolved small scales so-called subgrid scale (SGS) stress models are still needed. Compared with RANS modeling, LES can give more detailed information

Theory and Modeling of Dispersed Multiphase Turbulent Reacting Flows.
DOI: https://doi.org/10.1016/B978-0-12-813465-8.00005-3

about the production and development of turbulence structures and its statistical results are more accurate than those given by RANS modeling, hence it is now considered as a new generation of CFD modeling. Currently, the most widely used method in solving complex engineering turbulent flows is still RANS modeling, based on solving a Reynolds-averaged N-S equation, which was first proposed by P.-Y. Chou more than 50 years ago and later realized by Launder et al. It is called the method of turbulence models. The basic idea of turbulence models is to simulate unknown higher-order correlations using lower-order correlations or averaged-flow properties by physical knowledge, in order to close the time-averaged-flow equations or transport equations of correlation moments. Frequently, in engineering applications our major concern is the time-averaged velocity, temperature, species concentration, and turbulence properties, but not details on the turbulence structure and its development. It can be seen in the following text that although the turbulence models are the simplest in theory, they are sufficiently complex in engineering application. In general, when dealing with engineering problems, the method of turbulence models based on solving Reynolds time-averaged N-S equations is still a most effective and economic approach. There are two categories of turbulence models. The first is eddy-viscosity models, and the most widely used and tested within the last 40 years is the k-ε two-equation model. The second is the Reynolds stress equation model, which has obtained more and more applications during the last 20 years.

5.2 THE CLOSURE OF SINGLE-PHASE TURBULENT KINETIC ENERGY EQUATION

Excluding the drag-force term and the particle-mass change term from the fluid turbulent kinetic energy equation in two-phase flows, Eq. (4.53), given in Chapter 4, Basic Equations of Multiphase Turbulent Reacting Flows, the exact transport equation of gas/fluid turbulent kinetic energy can be obtained as

$$
\frac{\partial}{\partial t}(\rho k) + \frac{\partial}{\partial x_m}(\rho V_m k) = -\frac{\partial}{\partial x_m}\left(\overline{\rho v_m v_i^2}/2 + \overline{p' v_m} - \mu\frac{\partial k}{\partial x_m}\right) - \rho\overline{v_m v_i}\frac{\partial V_i}{\partial x_m}
$$
$$
+ \beta\rho g_i\overline{v_i T'} - \mu\overline{\left(\frac{\partial v_i}{\partial x_m}\right)^2}
$$

(5.1)

The physical meaning of this equation is that the transport of the turbulent kinetic energy depends on its convection, diffusion, production due to mean velocity gradient and buoyancy, and viscous dissipation. The method of obtaining turbulent viscosity by solving a differential equation was first proposed by Kolmogorov (1942) and Prandtl (1945) [1]. They defined

$$
\mu_T = c_\mu\rho k^{1/2}l, \quad \nu_T = c_\mu k^{1/2}l
$$

Historically, the one-equation model, or the energy-equation model, has been developed, in which a closed turbulent kinetic energy equation and a presumed algebraic expression of the turbulence scale l was used. In order to close Eq. (5.1), imitating the Newton law, Fourier law, and Fick law for viscous force, heat conductivity, and molecular diffusion, the method of gradient modeling is used. The third-order correlations are simulated using the second-order correlations and the second-order correlations are simulated using the mean-flow gradients. Hence the diffusion, shear production, and buoyancy production terms in Eq. (5.1) are simulated by

$$-\overline{\rho v'_k\left(p'/\rho + v'^2_i/2\right)} + \mu\frac{\partial k}{\partial x_k} = \left(\frac{\mu_T + \mu}{\sigma_k}\right)\frac{\partial k}{\partial x_k} = \frac{\mu_e}{\sigma_k}\frac{\partial k}{\partial x_k}$$

$$-\overline{\rho v'_i v'_k}\frac{\partial v_i}{\partial x_k} = \mu_T\left(\frac{\partial v_k}{\partial x_i} + \frac{\partial v_i}{\partial x_k}\right)\frac{\partial v_i}{\partial x_k}$$

$$\beta\rho g_i\overline{v'_i T'} = -\beta g_k\frac{\mu_T}{\sigma_T}\frac{\partial T}{\partial x_k}$$

In simulating the dissipation term, the difficulty arises due to lack of a simple concept like gradient modeling. In this case the Kolmogorov concept and dimensional analysis are used. The last term on the right-hand side of Eq. (4.1), namely the dissipation term, can be defined as

$$\mu\overline{\left(\frac{\partial v'_i}{\partial x_k}\right)^2} = c_D\rho\varepsilon$$

where ε is the dissipation rate of turbulent kinetic energy, which should have the dimension of $\mu_T k/l^2$, i.e.,

$$\rho\varepsilon \sim \mu_T k/l^2 \sim \rho k^{1/2}l\cdot k/l^2 \sim \rho k^{3/2}/l$$

Hence the dissipation term can be simulated by

$$\mu\overline{\left(\frac{\partial v'_i}{\partial x_k}\right)^2} = c_D\rho\varepsilon = c_D\rho k^{3/2}/l$$

Finally, the closed transport equation of turbulent kinetic energy is

$$\frac{\partial}{\partial t}(\rho k) + \frac{\partial}{\partial x_k}(\rho v_k k) = \frac{\partial}{\partial x_k}\left(\frac{\mu_e}{\sigma_k}\frac{\partial k}{\partial x_k}\right) + G_k + G_b - c_D\rho k^{3/2}/l \qquad (5.2)$$

where $\mu_e = \mu + \mu_T, \mu_T = c_\mu\rho k^2/\varepsilon, G_k = \mu_T\left(\frac{\partial v_i}{\partial x_k} + \frac{\partial v_k}{\partial x_i}\right), G_b = -\beta g_k\frac{\mu_T}{\sigma_T}\frac{\partial T}{\partial x_k}.$

When neglecting the convection and diffusion terms, i.e., there is a local equilibrium between the production and diffusion, the k-equation model will reduce to the mixing-length model, or a zero-equation or an algebraic model as

$$\mu_T = \rho l_m^2 \left| \frac{\partial v_i}{\partial x_j} + \frac{\partial v_j}{\partial x_i} \right| \qquad\qquad -\rho \overline{v_i' v_j'} = \rho l_m^2 \left| \frac{\partial v_i}{\partial x_j} + \frac{\partial v_j}{\partial x_i} \right| \left(\frac{\partial v_i}{\partial x_j} + \frac{\partial v_j}{\partial x_i} \right) \quad (5.3)$$

where l_m is the "mixing length," determined by empirical or intuitive judgment, for example, $l_m = cx$ for a jet, x is the distance in the flow direction, and c is an empirical constant. The advantage of the mixing-length model is its simplicity, intuition, and ability to be used without using differential equations. It was successfully used in jets, boundary layers, pipe flows, and nozzle flows. However, this model gives unreasonable zero turbulent viscosity at the axis of pipe flows and in the convergent channel flows behind a grid, where the velocity gradient is zero. The turbulence produced near the wall in pipe flows can diffuse to the near-axis region, and the turbulence produced behind the grid can be transported by convection to the downstream region. The shortcoming of the mixing-length model is ignoring the convection and diffusion of turbulence. Besides, for complex flows, such as recirculating and swirling flows, it is difficult to give the algebraic expression of the mixing length. Obviously, the k-equation or one-equation model is more reasonable than the mixing-length model. However, for very simple flows, the mixing-length model is sufficient and it is not necessary to use the k-equation model, and for complex flows, there is the same difficulty in specifying the generalized expression for the turbulence scale l. Therefore, the k-equation model still cannot be widely used and is regarded as an intermediate step for developing two-equation models.

5.3   THE K-ε TWO-EQUATION MODEL AND ITS APPLICATION

In turbulent flows there is not only the transport of turbulent kinetic energy, but also the transport of other properties. For example, the turbulence scales have convection, diffusion, production, and dissipation. The stretch of energy-containing large eddies leads to the formation of small eddies, and conversely, the dissipation of small eddies (Kolmogorov eddies) may form large eddies. Spalding and Launder summarized the turbulence parameter $Z = k^m l^n$ [1], proposed by different investigators and different forms of two-equation models as k-f (Kolmogorov), k-ε (P.Y. Zhou, Harlow, Nukayama), k-l (Rodi, Spalding), k-kl (Ng, Spalding), and k-w (Spalding) models, where

$$f = k^{1/2}/l, \ \varepsilon = k^{3/2}/l, \ w = k/l^2$$

The simulation results and experimental validation within the last nearly 40 years indicate that various two-equation models give almost the same results and among them the well-known k-ε equation model is the most widely used and well tested in different applications, such as in power engineering, aeronautical and astronautical engineering, chemical and metallurgical engineering, and hydraulic engineering. Using a method similar to that used in derivation of the k-equation, the exact transport equation of the dissipation rate of turbulent kinetic energy can be derived. Furthermore, using the gradient modeling of the diffusion term for the ε-equation, the assumption of the proportionality of the production and diffusion terms for the ε-equation to those of the k-equation, and using the dimensional analysis, the closed form of the ε-equation can be obtained as

$$\frac{\partial}{\partial t}(\rho\varepsilon) + \frac{\partial}{\partial x_k}(\rho v_k \varepsilon) = \frac{\partial}{\partial x_k}\left(\frac{\mu_e}{\sigma_\varepsilon}\frac{\partial\varepsilon}{\partial x_k}\right) + \frac{\varepsilon}{k}(c_1 G_k - c_2 \rho\varepsilon) \qquad (5.4)$$

Eqs. (5.2) and (5.4) constitute the k-ε two-equation turbulence model. The application and experimental validation of this model for nearly 40 years have shown that this model can successfully predict nonbuoyant plain jets, boundary layers over flat plates, 2-D and 3-D nonswirling recirculating flows, or weakly swirling flows. However, it fails to predict strongly swirling flows, buoyant flows, gravity-stratified flows, boundary layers over curved walls, low-Re flows, and round jets. In some cases it gives results that not only quantitatively but also qualitatively disagree with experiments.

Figs. 5.1−5.3 show the axial velocity and turbulent kinetic energy for nonswirling and weakly swirling flows, predicted using the k-ε model and their comparison with experimental results [2]. The predicted velocity profiles are in good agreement with the measurement results. The predicted shape, size, and location of the recirculation zone are in fair agreement with the measurement results. However, the length of the near-wall recirculation zone is underpredicted by about 5%−15%. This discrepancy may be caused

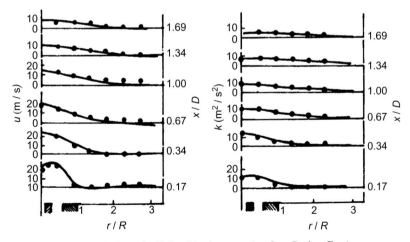

FIGURE 5.1   Axial velocity and turbulent kinetic energy ($s = 0$, − Pred., • Exp.).

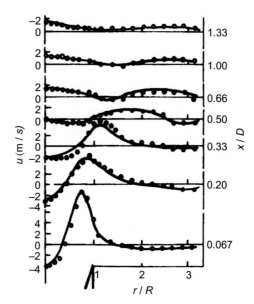

FIGURE 5.2 Axial velocity ($s = 0.3$, − Pred., • Exp.).

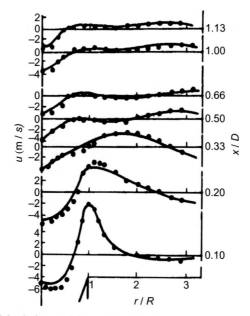

FIGURE 5.3 Axial velocity ($s = 0.5$, − Pred., • Exp.).

by the model deficiency, numerical diffusion, inaccuracy in specifying inlet conditions, and measurement errors. Figs. 5.4 and 5.5 give the predicted axial and tangential velocities for strongly swirling cyclonic flows [18] and their comparison with experimental results. It can be seen that there is a qualitative difference between the k-ε modeling results (dashed line) and experimental results. The k-ε model eliminates the measured central recirculation zone and the Rankine-vortex structure (solid-body rotation plus free vortex), which are crucial to the performance of cyclone separators and cyclone combustors.

The simulation of buoyant recirculating flows [2] shows that even the k-ε model with buoyancy terms cannot predict temperature stratification in gravity-stratified flows, induced by hot water discharged into a cold water environment (Fig. 5.6). Owing to the drawback of the k-ε model in simulating swirling and buoyant flows, different modified k-ε models have been proposed.

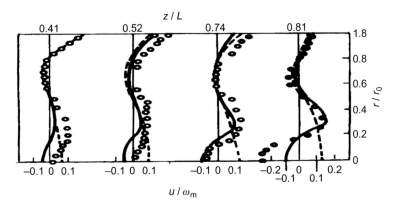

FIGURE 5.4   Axial velocity for strongly swirling flows (○ Exp., — ASM, --- k-ε).

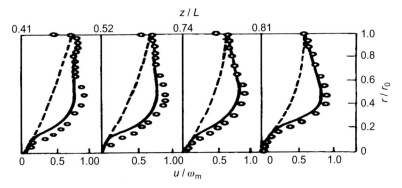

FIGURE 5.5   Tangential velocity for strongly swirling flows (○ Exp., — ASM, --- k-ε).

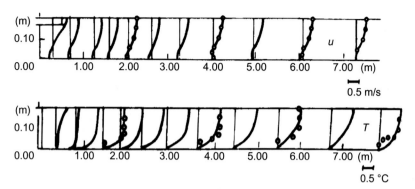

FIGURE 5.6 Velocity and temperature distributions in gravity-stratified flows (O Exp., — Modified k-ε).

The main idea is to modify the source term in the ε-equation, taking into account the low-Re, swirl, and buoyancy effects in the framework of the isotropic viscosity model. The low Reynolds number k-ε-f_μ model [3], two-scale k-ε model [4], fourth-order tensor viscosity model [5], improved closure of ε-equation [6], and RNG-k-ε model [7] have been proposed. The computation practice shows that these kinds of modification may give some improvement, but sometimes the results are not as good as expected. The reason for this is that swirling and buoyant flows are anisotropic turbulent flows. The concept of isotropic turbulent viscosity should be abandoned.

5.4 THE SECOND-ORDER MOMENT CLOSURE OF SINGLE-PHASE TURBULENCE

The turbulence in most practical flows is anisotropic. The swirl or buoyancy forces will reduce or enhance turbulence in the centrifugal or gravitational directions due to the effect of centrifugal or gravitational forces, but have no effect in other directions. Therefore, the turbulence is anisotropic. In such cases the turbulent viscosity is not a scalar, but a tensor. Hence, the Boussinesq expression, based on the concept of eddy viscosity, should be abandoned and the second-order moment closure to directly solve the Reynolds stress equations, accounting for the swirl or buoyancy effects, should be developed. This closure is the simplest in the theory of turbulence, but is the most complex in engineering predictions. The exact Reynolds stress equation for a gas or fluid, derived in Chapter 4, is

$$\frac{\partial}{\partial t}\left(\rho\overline{v_i v_j}\right) + \frac{\partial}{\partial x_k}\left(\rho V_k \overline{v_i v_j}\right) = D_{ij} + P_{ij} + G_{ij} + \Pi_{ij} - \varepsilon_{ij} + G_{p,ij} + G_{R,ij}$$

where D_{ij}, P_{ij}, G_{ij}, Π_{ij}, and ε_{ij} are the diffusion, shear production, buoyancy production, pressure$-$strain, and dissipation terms, respectively, they are

$$D_{ij} = -\frac{\partial}{\partial x_k}\left[\rho\overline{v_i v_j v_k} + \overline{p'v_j}\delta_{ik} + \overline{p'v_i}\delta_{jk} - \mu\left(\frac{\partial}{\partial x_k}\overline{v_i v_j}\right)\right]$$

$$P_{ij} = -\rho\left(\overline{v_i v_k}\frac{\partial V_j}{\partial x_k} + \overline{v_j v_k}\frac{\partial V_i}{\partial x_k}\right) \quad G_{ij} = \beta\rho\left(g_i\overline{v_j T'} + g_j\overline{v_i T'}\right)$$

$$\varepsilon_{ij} = -2\mu\overline{\left(\frac{\partial v_i'}{\partial x_k}\frac{\partial v_j'}{\partial x_k}\right)} \quad \Pi_{ij} = \overline{p'\left(\frac{\partial v_i'}{\partial x_j} + \frac{\partial v_j'}{\partial x_i}\right)}$$

The last two terms in the right-hand side of Eq. (4.50)

$$G_{p,ij} = \sum_k \frac{\rho_k}{\tau_{rk}}\left(\overline{v_{ki}v_j} + \overline{v_{kj}v_i} - 2\overline{v_i v_j}\right) \text{ and } G_{R,ij} = \overline{v_i v_j}S$$

are the fluid Reynolds stress source terms caused by the drag force and particle reactions (including evaporation, devolatilization, and char reaction), respectively. When omitting these two terms, the single-phase fluid Reynolds stress equation can be obtained as

$$\frac{\partial}{\partial t}\left(\rho\overline{v_i v_j}\right) + \frac{\partial}{\partial x_k}\left(\rho V_k\overline{v_i v_j}\right) = D_{ij} + P_{ij} + G_{ij} + \Pi_{ij} - \varepsilon_{ij} \qquad (5.5)$$

The second-order moment closure uses physical concepts to close Eq. (5.5), that is, not further deriving the transport equations for the unknown third-order correlations and other correlations except the second-order correlations themselves. This is the idea of turbulence modeling, proposed by Launder and Spalding, based on the original suggestion of P.-Y. Chou [8]. The basic principles of turbulence modeling are:

1. Consider the physical meanings of each term;
2. Use dimensional analysis;
3. Keep the modeled term with the same characteristics as those of the original term in coordinate transformation;
4. Use gradient modeling for the third-order terms;
5. Assume isotropic dissipation;
6. Take a return-to-isotropy assumption for the pressure$-$strain term.

Based on the above-stated principles, taking gradient modeling as the diffusion term, that is to simulate the third-order term by using the second-order term, the simplest Daly$-$Harlow (DH) model was proposed as

$$D_{T,ij} = -\frac{\partial}{\partial x_k}\left(\rho\overline{v_i v_j v_k} + \overline{p'v_j}\delta_{ik} + \overline{p'v_i}\delta_{jk}\right) = \frac{\partial}{\partial x_k}\left[c_s\rho\frac{k}{\varepsilon}\overline{v_k'v_l'}\frac{\partial}{\partial x_l}\left(\overline{v_i'v_j'}\right)\right]$$

Alternatively, a more complex model for the diffusion term was proposed by Launder and Hanjalic from the transport equation of the third-order moment [9]

$$D_{T,ij} = \frac{\partial}{\partial x_k}\left[c_s \rho \frac{k}{\varepsilon}\left(\overline{v_k' v_l'}\frac{\partial}{\partial x_l}\overline{v_i' v_j'} + \overline{v_i' v_l'}\frac{\partial}{\partial x_l}\overline{v_j' v_k'} + \overline{v_j' v_l'}\frac{\partial}{\partial x_l}\overline{v_i' v_k'} \right) \right]$$

Obi and Hara derived a more complex six-term expression [10]. However, different models for the third-order moment give the second-order moments, i.e., the Reynolds stress components and the third-order moments with only a slight difference. Hence the DH model is widely used. Assuming the dissipation is isotopic, and using the dimensional analysis, we have

$$\varepsilon_{ij} = -2\mu\left(\frac{\partial v_i'}{\partial x_m}\frac{\partial v_j'}{\partial x_m} \right) = \frac{2}{3}\rho \varepsilon \delta_{ij}$$

The pressure–strain term Π_{ij} is also called a turbulence redistribution term. It is divided into $\Pi_{ij,1}$, $\Pi_{ij,2}$, and $\Pi_{ij,3}$. The first term expresses the interaction between the stresses in different directions. The second term reflects the interaction between the stress and mean flows. The third term expresses the interaction between the turbulence and the buoyancy force. For the first term the Rotta model is proposed based on the concept of return-to-isotropy. The second term is called an IPM (isotropization of production) model. When adopting a similar closure model for the third term, the Launder–Rotta (LR) model is finally obtained as

$$\Pi_{ij} = \Pi_{ij,1} + \Pi_{ij,2} + \Pi_{ij,3} \quad \Pi_{ij,1} = -c_1\left(\varepsilon/k \right)\left(\overline{v_i' v_j'} - \frac{2}{3}\delta_{ij}k \right)$$

$$\Pi_{ij,2} = -c_2\left(P_{ij} - \frac{2}{3}\delta_{ij}G_k \right) \quad \Pi_{ij,3} = -c_3\left(G_{ij} - \frac{2}{3}\delta_{ij}G_b \right)$$

Subsequently, a linear IPCM model [11] and a GL model [12] were proposed. The so-called IPCM model is a "isotropization of production and convection model," in which the second term is modeled as

$$\Pi_{ij,2} = -c_2\left[(P_{ij} - c_{ij}) - \frac{1}{3}\delta_{ij}(G_k - c_{kk}) \right]$$

where c_{ij} and c_{kk} are convection terms in the Reynolds stress equation and turbulent kinetic energy equation, respectively. The GL model and more complex nonlinear SSG and FLT models can be found in Ref. [13] and [14].

5.5   THE CLOSED MODEL OF REYNOLDS STRESSES AND HEAT FLUXES

When neglecting the molecular stress diffusion, adopting the LR model for the pressure−strain term and taking the closure methods for the Reynolds heat flux equation and the equation of RMS temperature fluctuation, similar to those used in the Reynolds stress equation, the final closed form of transport equations for Reynolds stresses, heat fluxes, and RMS temperature fluctuation can be obtained as

$$
\frac{\partial}{\partial t}\left(\rho\overline{v_i'v_j'}\right) + \frac{\partial}{\partial x_k}\left(\rho v_k\overline{v_i'v_j'}\right) = \frac{\partial}{\partial x_k}\left[c_s\rho\frac{k}{\varepsilon}\overline{v_k'v_l'}\frac{\partial}{\partial x_l}\left(\overline{v_i'v_j'}\right)\right] - c_1\rho\frac{\varepsilon}{k}\left(\overline{v_i'v_j'} - \frac{2}{3}\delta_{ij}k\right)
$$

$$
- c_2\left(P_{ij} - \frac{2}{3}G_k\right) - c_3\left(G_{ij} - \frac{2}{3}G_b\right)
$$

$$
+ P_{ij} + G_{ij} - \frac{2}{3}\rho\varepsilon\delta_{ij}
$$

(5.6)

$$
\frac{\partial}{\partial t}\left(\rho\overline{v_i'T'}\right) + \frac{\partial}{\partial x_k}\left(\rho v_k\overline{v_i'T'}\right) = \frac{\partial}{\partial x_k}\left[c_{sT}\rho\frac{k}{\varepsilon}\overline{v_k'v_l'}\frac{\partial}{\partial x_l}\left(\overline{v_i'T'}\right)\right] - \rho\left(\overline{v_i'v_k'}\frac{\partial T}{\partial x_k} + \overline{v_k'T'}\frac{\partial v_i}{\partial x_k}\right)
$$

$$
- \beta\rho g_i\overline{T'^2} - c_{1T}\frac{\varepsilon}{k}\rho\overline{v_i'T'} + c_{2T}\rho\overline{v_k'T'}\frac{\partial v_i}{\partial x_k} + c_{3T}\beta\rho g_i\overline{T'^2}
$$

(5.7)

$$
\frac{\partial}{\partial t}\left(\rho\overline{T'^2}\right) + \frac{\partial}{\partial x_k}\left(\rho v_k\overline{T'^2}\right) = \frac{\partial}{\partial x_k}\left[c_T\rho\frac{k}{\varepsilon}\overline{v_k'v_l'}\frac{\partial}{\partial x_l}\left(\overline{T'^2}\right)\right] - 2\overline{v_k'T'}\frac{\partial T}{\partial x_k} - \frac{1}{R}\frac{\varepsilon}{k}\overline{T'^2}
$$

(5.8)

In addition, we have the anisotropic k and ε equations

$$
\frac{\partial}{\partial t}(\rho k) + \frac{\partial}{\partial x_k}(\rho v_k k) = \frac{\partial}{\partial x_k}\left(c_s\rho\frac{k}{\varepsilon}\overline{v_k'v_l'}\frac{\partial k}{\partial x_l}\right) + G_k + G_b - \rho\varepsilon \qquad (5.9)
$$

$$
\frac{\partial}{\partial t}(\rho\varepsilon) + \frac{\partial}{\partial x_k}(\rho v_k\varepsilon) = \frac{\partial}{\partial x_k}\left(c_\varepsilon\rho\frac{k}{\varepsilon}\overline{v_k'v_l'}\frac{\partial\varepsilon}{\partial x_l}\right) + \frac{\varepsilon}{k}\left[c_{\varepsilon1}(G_k + G_b)(1 + c_{\varepsilon2}R_f) - c_{\varepsilon3}\rho\varepsilon\right]
$$

(5.10)

Eqs. (5.6)−(5.10) express the transport equation model of Reynolds stresses and heat fluxes, where R_f on the right-hand side of Eq. (4.10) is called the "flux Richardson number," which is an empirical factor. For example, in the case of swirling flows, we have

$$
R_f = \frac{k^2}{\varepsilon^3}\left(\frac{w}{r^2}\right)\left[\frac{\partial}{\partial r}(wr)\right]
$$

Its physical meaning is: in the solid-body rotation zone, w increases with an increase in r, the Richardson number is greater than zero, the increase of swirl leads to an increase in ε and reduction in k, i.e., the centrifugal force will reduce the turbulent kinetic energy. In contrast, in the potential vortex zone, w decreases with an increase in r, the Richardson is smaller than zero, and the centrifugal force will enhance the turbulent kinetic energy. In the above equations the k equation and stress equations are not independent of each other, since the turbulent kinetic energy is the sum of three normal components of Reynolds stresses divided by a factor of 2. As for the empirical constants in the DSM (differential stress model), although their generality remains to be further studied, the computation tests of some investigators recommend the values as shown in Table 5.1. The merits of the DSM are being able to automatically account for the swirl, buoyancy, curvature, and near-wall effects, without using empirical modification. Hence, in recent years the DSM has gained more and more applications in predicting complex flows, such as swirling and buoyant flows.

Currently, the Reynolds stress model is still not as widely used as the k-ε model. This is because: (1) for 3-D problems the DSM needs to solve 11 equations, but the k-ε model needs to solve only two equations; (2) the DSM model needs 14 empirical constants, but the k-ε model needs only three constants; (3) for DSM, six Reynolds stress components and three thermal fluxes should be given at the inlet, but sometimes these details are lacking in experimental data. However, as computer hardware rapidly develops and computation experience is accumulated, these problems are gradually being solved. Some investigators proposed multiscale Reynolds stress equation models, but there has been no breakthrough in this aspect. On the other hand, the closure of the ε equation in the DSM is the same as that in the k-ε model, therefore, the closure of the ε equation should be improved. Sometimes the DSM gives results that are not as good as those obtained by the k-ε model for various reasons. Some practical problems in application of the DSM are discussed in Reference [15]. One of these is: which is better at solving six stress equations and solving five stress equations plus a k equation? The answer is that the latter method is better, because the turbulent kinetic energy obtained by solving the k equation is more exact than that obtained by the summation of

TABLE 5.1 Suggested Empirical Constants in DSM

c_s	c_1	c_2	c_3	c_{1T}	c_{2T}	c_{3T}
0.24	2.2	0.55	0.55	3.0	0.5	0.5
R	c_{sT}	c_T	c	$c_{\varepsilon 1}$	$c_{\varepsilon 2}$	$c_{\varepsilon 3}$
0.8	0.11	0.13	0.15	1.44	1.92	0.8

three normal Reynolds stress components. The second is the numerical scheme. Since the DSM includes more source terms, when using the first-order scheme, the numerical diffusion may be very large, it may obscure the merits of the DSM, and the DSM results may be worse than the k-ε model results. Therefore the schemes with higher orders than those of the second-order scheme should be used in DSM. The third is the diffusion term in the stress equation. In some commercial software an isotropic diffusion coefficient is used to substitute the anisotropic diffusion coefficient. Numerical studies show that the anisotropic diffusion coefficient gives better results. Besides, there are some solution technique problems with different point of views, such as directly solving DSM equations or gradual transition from the k-ε model solution to the DSM solution, etc.

5.6 THE ALGEBRAIC STRESS AND FLUX MODELS— EXTENDED K-ε MODEL

From the above discussion it can be seen that the standard k-ε model is rather simple, but not general. On the other hand, the Reynolds stress equation model (DSM) is more general, but rather complex. Launder and Rodi [16] proposed an algebraic stress and flux model (ASM), as a compromise between the k-ε model and the DSM, trying to combine their generality, simplicity, and economy. The ASM model includes algebraic expressions of stresses and heat fluxes and k and ε equations with anisotropic diffusion terms, hence it is also called an extended k-ε model. Its main purpose is reducing the stress and flux equations to algebraic expressions, whilst retaining the basic features of anisotropic turbulence.

The first approximation made by Rodi is an assumption of the difference between the stress convection and diffusion being proportional to the difference between the convection and diffusion of the turbulent kinetic energy (TKE). In other words, the source terms on the right-hand side of the stress equation are assumed to be proportional to those of the TKE equation, namely

$$\left(P_{ij} + G_{ij} + \Pi_{ij} + \varepsilon_{ij}\right) \sim (G_k + G_b - \rho\varepsilon)$$

The dimensional analysis gives

$$\left(P_{ij} + G_{ij} + \Pi_{ij} + \varepsilon_{ij}\right) = \frac{\overline{v_i' v_j'}}{k}(G_k + G_b - \rho\varepsilon)$$

Therefore, we have the following algebraic stress expressions

$$\overline{v_i' v_j'} = k\left[\frac{2}{3}\delta_{ij} + \frac{(1 - c_2)\left(P_{ij} - \frac{2}{3}\delta_{ij}G_k\right) + (1 - c_3)\left(G_{ij} - \frac{2}{3}\delta_{ij}G_b\right)}{G_k + G_b + (c_1 - 1)\rho\varepsilon}\right] \quad (5.11)$$

The second Rodi' approximation assumes that the difference between the stress convection and diffusion is zero, namely the stress production equals its dissipation. In that case we have

$$\left(P_{ij} + G_{ij} + \Pi_{ij} + \varepsilon_{ij}\right) = 0$$

When further introducing the approximation $c_2 = c_3$, $G_k + G_b = \rho\varepsilon$, the following algebraic stress expression can be obtained

$$\overline{v_i'v_j'} = \frac{2}{3}\lambda\delta_{ij}k - (1-\lambda)\frac{k}{\varepsilon}\left(\overline{v_i'v_k'}\frac{\partial v_j}{\partial x_k} + \overline{v_j'v_k'}\frac{\partial v_i}{\partial x_k} + \beta g_i\overline{v_j'T'} + \beta g_j\overline{v_i'T'}\right) \quad (5.12)$$

where $\lambda = (c_1 + c_2 - 1)/c_1$

Comparing this expression with the Boussinesq formula of isotropic scalar viscosity

$$\overline{v_i'v_j'} = \frac{2}{3}\delta_{ij}k - c_\mu\frac{k^2}{\varepsilon}\left(\frac{\partial v_j}{\partial x_i} + \frac{\partial v_j}{\partial x_i}\right)$$ it can be seen that beside the difference

in empirical constants and buoyancy terms, the main difference is using the anisotropic viscosity $(1-\lambda)\frac{k}{\varepsilon}\left(\overline{v_i'v_k'}\right)$ or $(1-\lambda)\frac{k}{\varepsilon}\left(\overline{v_j'v_k'}\right)$ instead of the isotropic viscosity $c_\mu k^2/\varepsilon$. That is, c_μ is no more a constant, but a tensor, like $c_\mu \sim \overline{v_i'v_k'}/k$. It depends on buoyancy or centrifugal forces. Using a similar method the algebraic expressions of heat fluxes and RMS temperature fluctuation can be obtained as

$$\overline{v_i'T'} = \frac{k}{c_{1T}\varepsilon}\left[\overline{v_i'v_k'}\frac{\partial T}{\partial x_k} + (1-c_{2T})\overline{v_k'T'}\frac{\partial v_i}{\partial x_k} + (1-c_{3T})\beta g_i\overline{T'^2}\right] \quad (5.13)$$

$$\overline{T'^2} = -2R\frac{k}{\varepsilon}\overline{v_k'T'}\frac{\partial T}{\partial x_k} \quad (5.14)$$

Therefore, finally, the algebraic stress model (ASM) or extended k-ε model (k-ε-a model) is given by

$$\frac{\partial}{\partial t}(\rho k) + \frac{\partial}{\partial x_k}(\rho v_k k) = \frac{\partial}{\partial x_k}\left(c_s\rho\frac{k}{\varepsilon}\overline{v_k'v_l'}\frac{\partial k}{\partial x_l}\right) + G_k + G_b - \rho\varepsilon$$

$$\frac{\partial}{\partial t}(\rho\varepsilon) + \frac{\partial}{\partial x_k}(\rho v_k\varepsilon) = \frac{\partial}{\partial x_k}\left(c_\varepsilon\rho\frac{k}{\varepsilon}\overline{v_k'v_l'}\frac{\partial\varepsilon}{\partial x_l}\right) + \frac{\varepsilon}{k}\left[c_{\varepsilon 1}(G_k + G_b)\left(1 + c_{\varepsilon 2}R_f\right) - c_{\varepsilon 3}\rho\varepsilon\right]$$

$$\overline{v_i'v_j'} = \frac{2}{3}\lambda\delta_{ij}k - (1-\lambda)\frac{k}{\varepsilon}\left(\overline{v_i'v_k'}\frac{\partial v_j}{\partial x_k} + \overline{v_j'v_k'}\frac{\partial v_i}{\partial x_k} + \beta g_i\overline{v_j'T'} + \beta g_j\overline{v_i'T'}\right)$$

$$\overline{v_i'T'} = \frac{k}{c_{1T}\varepsilon}\left[\overline{v_i'v_k'}\frac{\partial T}{\partial x_k} + (1-c_{2T})\overline{v_k'T'}\frac{\partial v_i}{\partial x_k} + (1-c_{3T})\beta g_i\overline{T'^2}\right]$$

$$\overline{T'^2} = -2R\frac{k}{\varepsilon}\overline{v'_k T'}\frac{\partial T}{\partial x_k} \tag{5.15}$$

Obviously, the ASM to a certain degree reflects the features of anisotropic turbulence related to buoyancy or swirl. In comparison with DSM, it can reduce the number of differential equations and number of empirical constants, and does not require use of the inlet stress and flux components. However, the present author and his colleagues found that in simulating very strongly swirling flows, such as cyclonic flow, the central recirculation zone and the Rankine-vortex structure cannot be properly predicted, even using the original ASM, because the ASM actually neglects or remarkably underpredicts the effect of stress convection, and hence leads to significant errors. Subsequently, an improved ASM was proposed [17]. Its basic purpose is to add a nongradient convection term to the algebraic stress expression, keeping the simplicity of algebraic expressions and partially taking the effect of stress convection into account. For example, the original ASM expression according to the first Rodi approximation is

$$P_{ij} + \Pi_{ij} - \frac{2}{3}\delta_{ij}\rho\varepsilon = \frac{\overline{v'_i v'_j}}{k}(G_k - \rho\varepsilon)$$

The improved ASM expression is

$$A_{ij} + P_{ij} + \Pi_{ij} - \frac{2}{3}\delta_{ij}\rho\varepsilon = \frac{\overline{v'_i v'_j}}{k}(G_k - \rho\varepsilon) \tag{5.15a}$$

where A_{ij} is the nongradient part of stress convection. For example,

$$A_{rr} = -2\rho\overline{v'w'}w/r$$

In very strongly swirling flows we have $w \gg u \gg v$, and this part of the convection term is much larger than other parts of the convection term, hence it is more reasonable to account for this term.

5.7 THE APPLICATION OF DSM AND ASM MODELS AND THEIR COMPARISON WITH OTHER MODELS

Fig. 5.7 gives the predicted temperature, velocity, turbulent kinetic energy, and its dissipation rate for the buoyancy-stratified flows formed in hot water discharged into a cold water region using the k-ε model with the buoyancy source term, the k-ε model with buoyancy-modified Prandtl number σ_T and the ASM, and their comparison with experimental results [2]. It can be seen that the ASM can better predict the temperature and velocity stratification than other models. Fig. 5.8 shows the velocity distribution of nonswirling coaxial sudden-expansion flows using k-ε, ASM, and DSM models and its comparison with measurement results, reported in Reference [2]. Obviously,

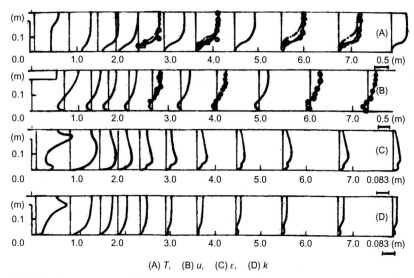

(A) T,   (B) u,   (C) ε,   (D) k

FIGURE 5.7   Buoyancy-stratified flows (○ Exp, — ASM, -●- Modified k-ε, --- Modified k-ε with modified turbulent Pr number).

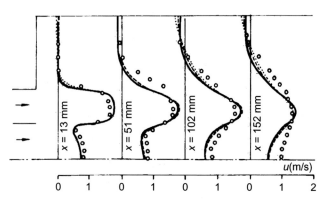

FIGURE 5.8   Axial velocity of sudden-expansion flows (○ Exp., — DSM, --- ASM, ... k-ε).

there is only a slight difference in the results obtained using different models and all of the prediction results are in agreement with experiments. However, for the axial velocity (Fig. 5.9) and tangential velocity (Fig. 5.10) of swirling coaxial flows, only the ASM and DSM give better results. For axial and tangential velocities in the above-discussed strongly swirling flows (Figs. 5.4 and 5.5), the ASM can properly predict the central recirculation zone in axial velocity profiles and Rankine-vortex structure in tangential velocity profiles (solid lines), whereas the k-ε model cannot predict these flow structures and its results are qualitatively wrong.

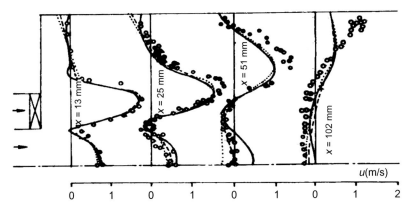

FIGURE 5.9 Axial velocity of swirling sudden-expansion flows (○ Exp, — DSM, --- ASM, ... k-ε).

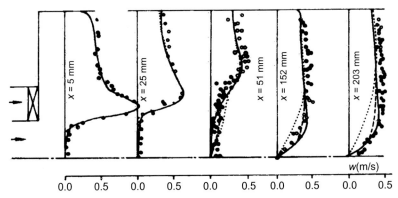

FIGURE 5.10 Tangential velocity of swirling sudden-expansion flows (○ Exp., — DSM, --- ASM, ... k-ε).

Fig. 5.11 gives the predicted axial velocity of swirling sudden-expansion flows using DSM with different pressure—strain models [18]. Except in the near-axis region at $x/D = 1.25$ the SSG model shows larger errors, at other regions different models give prediction results with slight differences that are close to measurement results.

Fig. 5.12 shows the predicted tangential velocity of swirling sudden-expansion flows using these models. The results are similar to those of axial velocity. Figs. 5.13 and 5.14 give the predicted RMS axial and tangential fluctuation velocities using different pressure—strain models. In most regions the difference between different models is not obvious, and both of the IPCM and FLT models underpredict the RMS velocity.

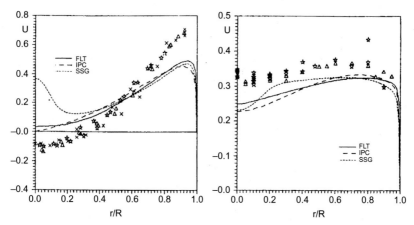

FIGURE 5.11 Axial velocity of swirling sudden-expansion flows (Left: $x/D = 1.25$; right: x/D = 2.5; Dots: Exp.).

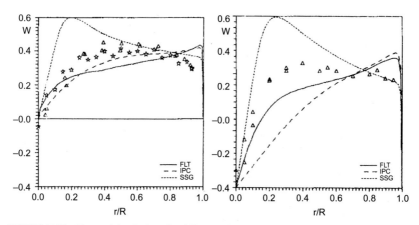

FIGURE 5.12 Tangential velocity of swirling sudden-expansion flows (Left: $x/D = 1.25$; right: x/D = 2.5; Dots: Exp.).

Simulation of swirling flows in a coaxial chamber (Fig. 5.15) with a swirl number of 0.53 was made by the author and his colleague, using DSM with both IPCM and GL pressure–strain models [19]. The predicted axial velocity by IPCM was better (Fig. 5.16), but the predicted tangential velocity by GL was better (Fig. 5.17).

Fig. 5.18 gives the axial (left) and tangential (right) velocities of strongly swirling flows in an annular duct predicted using the k-ε, original ASM, and

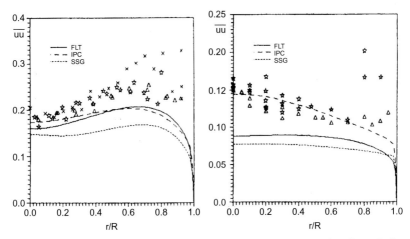

FIGURE 5.13 RMS axial fluctuation velocity of swirling sudden-expansion flows (Left: $x/D = 1.25$; right: $x/D = 2.5$; Dots: Exp.).

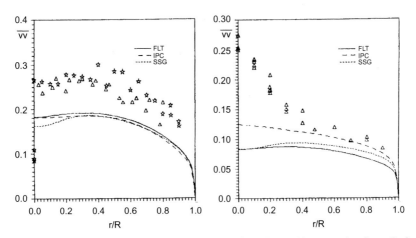

FIGURE 5.14 RMS tangential fluctuation velocity of swirling sudden-expansion flows (Left: $x/D = 1.25$; right: $x/D = 2.5$; Dots: Exp.).

improved ASM models [17]. It is seen that only the improved ASM can properly predict the central recirculation zone and the Rankine-vortex structure. Both the k-ε and original ASM fail to do that.

The validation of different turbulence models by experimental data was reported in 1997 [18]. A typical case is the prediction of a 2-D wall jet, as shown in Fig. 5.19. Figs. 5.20 and 5.21 are the predicted axial velocity, shear stress, longitudinal and lateral normal stresses given by different DSM and

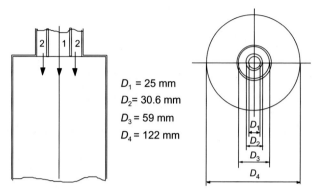

FIGURE 5.15   A coaxial swirl chamber.

D_1 = 25 mm
D_2= 30.6 mm
D_3 = 59 mm
D_4 = 122 mm

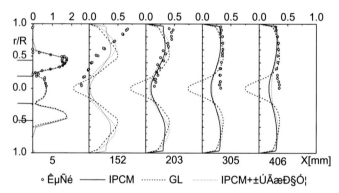

FIGURE 5.16   Axial velocity ($s = 0.53$, m/s).

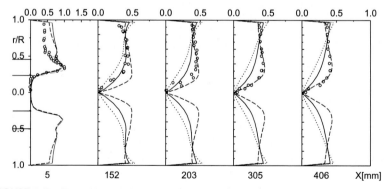

FIGURE 5.17   Tangential velocity ($s = 0.53$, m/s).

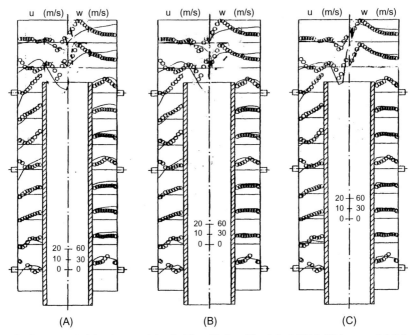

FIGURE 5.18 Axial and tangential velocities: (A) k-ε; (B) original ASM; (C) improved ASM.

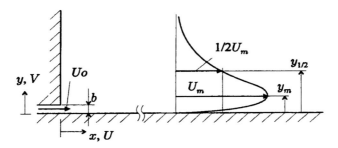

FIGURE 5.19 Two-dimensional wall jet.

k-ε models, respectively. It can be seen that there is a slight difference in the predicted axial velocity and shear stress by two kinds of models and the prediction results are close to the experimental results. However, different k-ε models underpredict the longitudinal normal stress and overpredict the lateral normal stress, showing the drawback of the isotropic models. The DSM models however give somewhat different results, but the prediction results are not far from the experimental results, and correctly give the longitudinal fluctuation as being stronger than the lateral fluctuation.

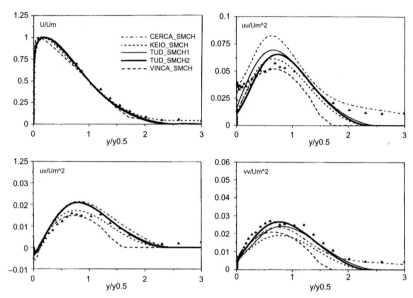

FIGURE 5.20 Predictions by different DSM (Dots-Exp).

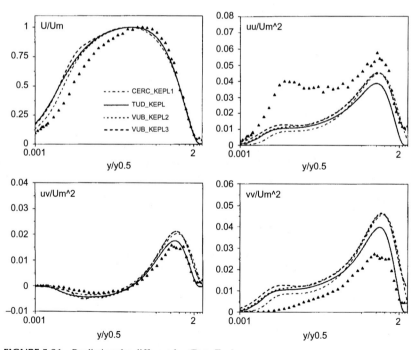

FIGURE 5.21 Predictions by different k-ε (Dots-Exp).

Fig. 5.22 show the predicted normal stresses of strongly swirling flows in a cyclone separator in tangential and axial directions by using different DSM models [20]. It can be seen that only the modified IPCM + wall pressure−strain model, proposed by the author and his colleague, gives reasonable results, and the original DSM model remarkably underpredicts the normal stresses.

In summary, there are no general reasonable and economic turbulence models applicable to all cases. Even the same model used in simulating the same case by different investigators gives different results. The more complex models, such as the multiscale models and third-order moment closures, are limited in their applications. It is suggested for engineering applications that the mixing-length model be used in boundary-layer, nozzle and jet flows; different k-ε models can be used in 2-D and 3-D nonbuoyant, non-swirling, or weakly swirling and recirculating flows; different DSM or ASM models can be used in buoyant and strongly swirling flows. As regards to many problems, such as the closure of the diffusion term, the pressure−strain term and the dissipation term in the stress transport equation, remain to be further studied.

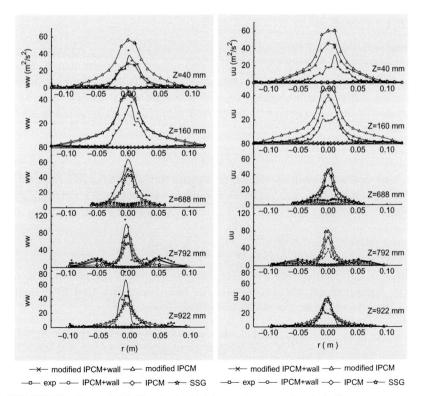

FIGURE 5.22 Tangential (left) and axial (right) normal stresses of cyclonic flows.

5.8   LARGE-EDDY SIMULATION

The large-eddy simulation (LES) is an approach between the above-discussed Reynolds-averaged Navier–Stokes (RANS) modeling and direct numerical simulation (DNS). In LES the dynamics of 3-D unsteady turbulent flows in large eddies is solved directly, since the turbulent energy and its anisotropy are contained mainly in large eddies, affected by the flow geometry, and are not universal, whereas the effect of motion in small dissipative eddies is less important and more universal, the motions in small eddies are unresolved and are simulated using very simple models. In comparison with RANS modeling, LES can give more detailed unsteady structures of turbulence and more accurate statistical results than RANS modeling. The DNS directly solves 3-D unsteady turbulent flows in all scales down to the Kolmogorov dissipation scale, without using any models. It has huge computational cost and cannot be used in simulating practical high-Reynolds-number flows. In comparison with DNS, LES can save a great deal more computational cost and can be applied in simulating high-Reynolds-number flows. Hence, in recent years, LES has attracted more and more attention and it has prospects to become a new generation of CFD tools.

5.8.1   Filtration

In order to carry out LES, the first step is to make a filtering operation, i.e., to decompose a variable into a filtered or resolved component and an unresolved residual component, i.e., subgrid scale (SGS) component. Actually, filtering is a volume averaging in a grid size scale of Δ. Define $f(x,t)$ as a variable of all scales, $\bar{f}(x,t)$ is its filtered value, and $f'(x,t)$ is the residual value, that is

$$f(x,t) = \bar{f}(x,t) + f'(x,t)$$

This decomposition, or volume averaging, formally looks like Reynolds' time averaging, but here $\bar{f}(x,t)$ is a random variable, not a deterministic variable, and $f'(x,t)$ is not zero. The filtered variable can be expressed by

$$\bar{f}(x) = \int_V G\left(x - x'\right) f\left(x'\right) dx'$$

where $G(x)$ is a filter function. The integral domain V is the whole flow field. For a 3-D space, $G = G_1 G_2 G_3$. The most commonly used filter is the box filter (top-hat filter) or white-noise filter, expressed by

$$G(x) = \begin{cases} 1/\Delta & \left|x - x'\right| < \Delta/2 \\ 0 & \left|x - x'\right| > \Delta/2 \end{cases}$$

This filter corresponds to a volume averaging of the variable in the interval $x - \Delta/2 < x' < x + \Delta/2$. Here Δ is the filter scale, which is equal to the grid size. For incompressible flows, the filtered continuity and momentum equations can be obtained as

$$\partial \bar{u}_i / \partial x_i = 0 \tag{5.16}$$

$$\frac{\partial \bar{u}_i}{\partial t} + \frac{\partial}{\partial x_j} \left(\overline{u_i u_j} \right) = -\frac{1}{\rho} \frac{\partial \bar{P}}{\partial x_i} + \upsilon \frac{\partial^2 \bar{u}_i}{\partial x_i \partial x_j} \tag{5.17}$$

Owing to $u_i = \bar{u}_i + u_i'$, the term $\overline{u_i u_j}$ can be expressed as

$$\overline{u_i u_j} = \bar{u}_i \bar{u}_j + \left(\overline{\bar{u}_i \bar{u}_j} - \bar{u}_i \bar{u}_j \right) + \overline{\bar{u}_i u_j'} + \overline{u_i' \bar{u}_j} + \overline{u_i' u_j'}$$

Therefore, we have

$$\frac{\partial \bar{u}_i}{\partial t} + \frac{\partial}{\partial x_j} \left(\bar{u}_i \bar{u}_j \right) = -\frac{1}{\rho} \frac{\partial \bar{p}}{\partial x_i} + \frac{\partial}{\partial x_j} \left[\nu \frac{\partial \bar{u}_i}{\partial x_i} - \tau_{ij} \right] \tag{5.18}$$

where the subgrid scale stress (SGS stress, also called pseudo-Reynolds stress) is

$$\tau_{ij} = \left(\overline{\bar{u}_i \bar{u}_j} - \bar{u}_i \bar{u}_j \right) + \overline{\bar{u}_i u_j'} + \overline{u_i' \bar{u}_j} + \overline{u_i' u_j'}.$$

Further, it is assumed that $\overline{\bar{u}_i \bar{u}_j} = \bar{u}_i \bar{u}_j$, $\overline{\bar{u}_i u_j'} = \overline{u_i' \bar{u}_j} = 0$, so this term can be simplified as

$$\tau_{ij} = \overline{u_i' u_j'}.$$

5.8.2 SGS Stress Models

The SGS stress is an unknown, reflecting the effect of small-scale fluctuations on large-scale fluctuations. Its importance and magnitude are much lower than those of the Reynolds stresses. To solve Eq. (5.18), it is necessary to give the SGS stress models. The first is the simplest and most widely used SGS stress model—the Smagorinsky eddy-viscosity model [21]

$$\tau_{ij} = -2\nu_T \bar{S}_{ij} + \frac{1}{3} \delta_{ij} \tau_{kk} \tag{5.19}$$

where $\nu_T = C_s^2 \Delta^2 |\bar{S}|$; $|\bar{S}| = (2\bar{S}_{ij}\bar{S}_{ij})^{1/2}$; $\bar{S}_{ij} = (\partial \bar{u}_i / \partial x_j + \partial \bar{u}_j / \partial x_i)/2$; δ_{ij} is Kronecker-Delta, C_s is an empirical constant, ranging from 0.1 to 0.566 for different types of flows, mostly taken as 0.16. The second is the so-called Smagorinsky—Lilly model [22], which is given by

$$\mu_t = \rho L_s^2 |\bar{S}| \tag{5.20}$$

where $L_s = \min(Kd, C_s V^{1/3})$; $K = 0.42$; $C_s = 0.1$, d is the distance to the wall and V is the volume of the computation cell. The third one is the Germano dynamic SGS stress model [23]. In many cases the Smagorinsky coefficient C_s is not a constant. Germano proposed a dynamics eddy-viscosity model for modification of the Smagorinsky model's constant by using two-time filtration, in which the coefficient C_s is not a constant, but is a function determined by

$$C_s = -\frac{1}{2}\frac{\langle L_{ij}\bar{S}_{ij}\rangle}{\hat{\bar{\Delta}}^2\left\langle|\hat{\bar{S}}|\hat{\bar{S}}_{ij}\bar{S}_{ij}\right\rangle - \overline{\Delta}^2\left\langle|\hat{\bar{S}}|\hat{\bar{S}}_{ij}\bar{S}_{ij}\right\rangle} \tag{5.21}$$

where $L_{ij} = \overline{\bar{u}_i\bar{u}_j} - \hat{\bar{u}}_i\hat{\bar{u}}_j$, $\hat{\overline{\Delta}} = 2\overline{\Delta}\times\overline{\Delta}$ is the filter size/grid size, and

$$\hat{\bar{S}}_{ij} = \frac{1}{2}\left(\frac{\partial\hat{\bar{u}}_i}{\partial x_j} + \frac{\partial\hat{\bar{u}}_j}{\partial x_i}\right)\quad|\hat{\bar{S}}| = \sqrt{2\hat{\bar{S}}_{ij}\hat{\bar{S}}_{ij}}$$

The fourth is Kim's SGS energy k-equation model [24], which is given by

$$\tau_{ij}^{sgs} = -2\bar{\rho}\nu_t\left(\bar{S}_{ij} - \frac{1}{3}\bar{S}_{kk}\delta_{ij}\right) + \frac{2}{3}\bar{\rho}k^{sgs}\delta_{ij} \tag{5.22}$$

where $\nu_t = C_\nu(k^{sgs})^{1/2}\overline{\Delta}$

$$\frac{\partial\bar{\rho}k^{sgs}}{\partial t} + \frac{\partial}{\partial x_i}(\bar{\rho}\,\bar{u}_ik^{sgs}) = P^{sgs} - D^{sgs} + \frac{\partial}{\partial x_i}\left(\frac{\bar{\rho}\nu_t}{\Pr_t}\frac{\partial k^{sgs}}{\partial x_i}\right) \tag{5.23}$$

where $k^{sgs} = \frac{1}{2}\left[\overline{u_k^2} - \bar{u}_k^2\right]$; $P^{sgs} = -\tau_{ij}^{sgs}(\partial\bar{u}_i/\partial x_j)$; $D^{sgs} = C_\varepsilon\bar{\rho}(k^{sgs})^{3/2}/\overline{\Delta}$

Coefficients C_ν and C_ε may be dynamically determined.

5.8.3 LES of Swirling Gas Flows

The swirling gas flows with a swirl number of $s = 0.53$ were studied by Hu and Zhou et al. [25] using both LES with a Smagorinsky SGS stress model and RANS modeling with an IPCM + wall version of the Reynolds stress equation model. Figs. 5.23 and 5.24 show the predicted tangential and axial fluctuation velocities, respectively. It can be seen that LES, even using the simplest SGS stress model, gives obviously better results than RANS modeling using the most complex turbulence model. LES can also simulate the peaks of $\overline{ww}$ and $\overline{uu}$ at the axis, observed in experiments, whereas RANS modeling fails to do that.

Figs. 5.25 and 5.26 give the LES-predicted instantaneous vorticity maps and velocity vectors, respectively.

The instantaneous flow behavior, showing detailed turbulence structures, given by LES is much more complex than those given by RANS modeling, using IPCM and GL versions of the Reynolds stress equation model, which can show only the near-axis and near-corner recirculating flows (Figs. 5.27 and 5.28).

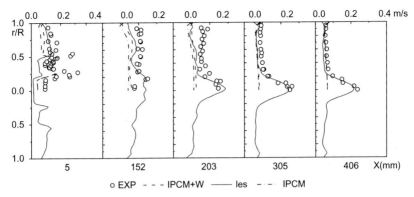

FIGURE 5.23 Tangential fluctuation velocity ($s = 0.53$).

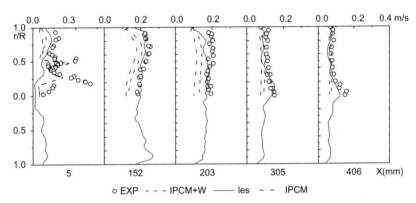

FIGURE 5.24 Axial fluctuation velocity.

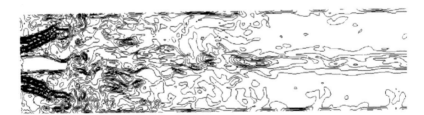

FIGURE 5.25 Instantaneous vorticity maps (LES).

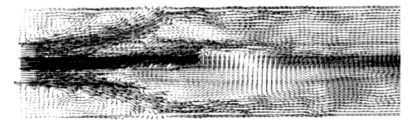

FIGURE 5.26 Instantaneous velocity vectors (LES).

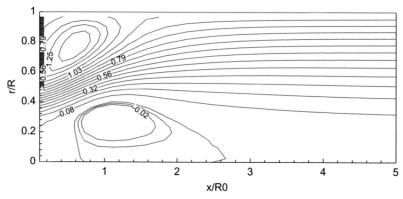

FIGURE 5.27 Streamlines (RSM-IPCM).

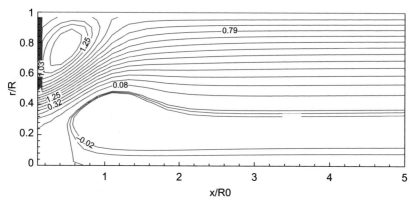

FIGURE 5.28 Streamlines (RSM-GL).

5.9 DIRECT NUMERICAL SIMULATION

Direct numerical simulation (DNS) is used to solve instantaneous Navier–Stokes equations, resolving all scales, down to Kolmogorov dissipation scales, without using any models. It can give the most accurate results, with high spatial and temporal resolution, which cannot even be given by measurements. However, its computer requirements are extremely high, and increase rapidly with Reynolds number. Therefore, its application is limited to low-Reynolds number flows and small-size computation domains. Therefore, currently, DNS cannot be used to simulate practical high-Reynolds-number complex flows. However, the DNS database can be used to validate the RANS and LES SGS models. Since there are no closure models in DNS, DNS only has highly accurate numerical methods and proper boundary conditions. The adopted numerical methods include spectral method, pseudo-spectral method, and finite difference method with a highly accurate compact difference scheme. DNS needs periodic boundary conditions at the inlet and walls. The Kolmogorov scale is determined by

$$\eta = \left(\nu^3/\varepsilon\right)^{1/4}$$

where ν is the kinematic viscosity and ε is the dissipation rate of turbulent kinetic energy. It is known that $\varepsilon \sim \left(u'\right)^3/l$ where u' is the RMS fluctuation velocity, and l is the integral scale. $l = k^{3/2}/\varepsilon$. The three-dimensional grid number should be at least $N = (N_x)^3 = (\mathrm{Re}_l)^{9/4}$. It was estimated by Pope [26] that for $\mathrm{Re}_l = 6000$, the required grid number should be 2×10^9, and the computation time 20 months.

As an example, the DNS of channel flows was made by Xu [27] using a spectral method with Galerkin–Tau spectral expansion; Fourier transformation in x, z directions; Chebyshev transformation in y direction, and third-order scheme in time. The predicted instantaneous isosurface of fluctuation velocity shows the coherent structures—the strip structures at the near-wall region, indicating the burst behavior of near-wall turbulence. The typical structures of wall turbulence, such as low-speed streaks, bursting events, and vortical structures, are presented. The DNS database was used to obtain the turbulence statistics, the budget of terms in the Reynolds stress equation, and to validate the closure models. Fig. 5.29 gives the budget of terms in the Reynolds stress equation for $\overline{u'u'}$ given by DNS database. It can be seen that in most regions the production and dissipation terms play dominant roles, and other terms play minor roles. Fig. 5.30 shows the comparison of the turbulent diffusion term—the third-order correlation $-\overline{u'u'v'}$, given by the

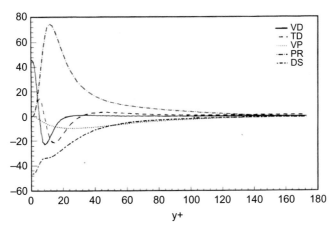

FIGURE 5.29 The fudget for $\overline{u'u'}$ given by DNS [27].
(VD, viscous diffusion; TD, turbulent diffusion; VP, velocity–pressure correlation or pressure–strain correlation; PR, production; DS, turbulent dissipation).

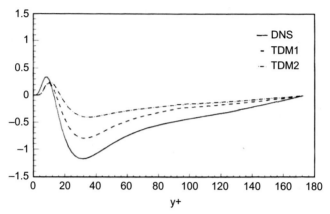

FIGURE 5.30   Modeled third-order correlation $-\overline{u'u'v'}$ and DNS results [27].

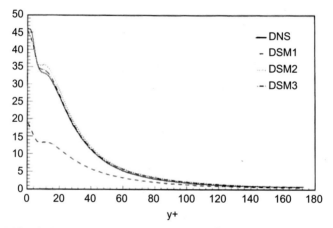

FIGURE 5.31   Modeled dissipation term ε_{11} and DNS results [27].

Daly—Harlow model (TDM1) and the Hanjalic—Launder model (TDM2) with the DNS results. Although the tendency is the same, both models over-predict the third-order correlation. Fig. 5.31 is the modeled dissipation term and DNS results, where DSM1 indicates the isotropic model, DSM2 indicates the anisotropic model, and DSM3 indicates the Hanjalic—Launder modified model. The isotropic model remarkably underpredicts the dissipation, whereas the anisotropic models are in good agreement with the DNS results.

Fig. 5.32 shows the modeled pressure—strain term and DNS results, where "Rotta" indicates the Launder—Rotta model, "QI" indicates the quasi-isotropic model, and "wall" indicates the near-wall model. It can be seen that in the near-wall region there is a discrepancy between all models and DNS results, and in other regions the agreement is fairly good.

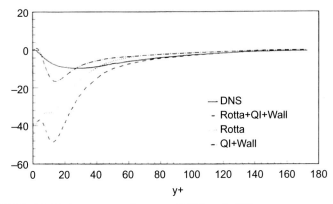

FIGURE 5.32   Modeled pressure—strain term and DNS results [27].

It should be pointed out that these results are obtained only for simple and low-Re flows. For complex high Re flows, the validation of turbulence models by DNS remains to be studied in the future.

REFERENCES

[1]   B.E. Launder, D.B. Spalding, Mathematical Models of Turbulence, Academic Press, New York, 1972.

[2]   L.X. Zhou, Theory and Numerical Modeling of Turbulent Gas-Particle Flows and Combustion, CRC Press Florida, USA, 1993.

[3]   H.J. Sung, T.S. Park, A new low-Reynolds-number k-ε-f_μ model for predictions involving multiple surfaces. Proc. International Symposium on Mathematical Modeling of Turbulent Flows, pp. 56—63, Tokyo, 1995.

[4]   S. Ko, Derivation of a two-scale k-ε turbulence model, in: Proceedings of International Symposium on Mathematical Modeling of Turbulent Flows, pp. 205—210, Tokyo, 1995.

[5]   Z.D. Wang, N.W. Jiang, S.Q. Wu, On the modeling of eddy viscosity tensor, in: Proceedings of International Symposium on Mathematical Modeling of Turbulent Flows, pp. 345—351, Tokyo, 1995.

[6]   Y. Nagano, M. Shimada, in: Z.S. Zhang, Y. Miyake (Eds.), Critical assessment and reconstruction of dissipation-rate equations using direct simulations, The Recent Developments in Turbulence Research, Proceedings of Sino-Japan Workshop of Turbulent Flows, International Academic Publishers, 1995, pp. 189—217.

[7]   V.S. Yakhot, A. Orszag, S. Thangam, Development of turbulence models for shear flows by a double expansion technique, Phys. Fluids A4 (1992) 1510—1520.

[8]   B.E. Launder, Turbulence Modeling. Lecture Notes at Tsinghua University, 1988.

[9]   B.E. Launder, Second-moment closure and its use in modelling turbulent industrial flows, Numerical Methods in Fluids, 9 (1989): 963—985

[10]   S. Obi, H. Hara. An algebraic model of turbulent diffusion. in: Proceedings of International Symposium on Mathematical Modeling of Turbulent Flows, pp. 274—279, Tokyo, 1995.

[11]   S. Fu, in: Z.S. Zhang, Y. Miyake (Eds.), Modeling of the strongly swirling flows with second-moment closures. Proceedings of Sino-Japan Workshop of Turbulent Flows, International Academic Publishers, 1995, pp. 22—41.

[12]   M.M. Gibson, B.E. Launder, On the calculation of horizontal turbulent free shear flows under gravitational influence, Trans. ASME, J. Heat Transfer, 98, 1976, pp. 81—87.

[13]   S. Yunemura, M. Yamamoto. Proposal of a multiple-time-scale Reynolds stress model for turbulent boundary layers. in: Proceedins of International Symposium on Mathematical Modeling of Turbulent Flows, pp. 381—386, Tokyo, 1995.

[14]   K. Hanjalic, S. Jakirlic, I. Hadzic, Expanding the limits of "equilibrium" second-moment turbulence models. in: Proceedings of International Symposium on Mathematical Modeling of Turbulent Flows, pp. 9—16, Tokyo, 1995.

[15]   K.-C. Chang, On application of the Reynolds-stress equation model, Proceedings of Sixth International Symposium on Flow Modeling and Turbulence Measurements, Florida, 1996.

[16]   W. Rodi, in: W. Kollman (Ed.), Turbulence Models in Environmental Problems. Prediction Methods for Turbulent Flows, McGraw Hill, 1980.

[17]   J. Zhang, S. Nieh, L.X. Zhou, A new version of algebraic stress model for simulating strongly swirling flows, Numer. Heat Transfer B22 (1992) 49—62.

[18]   L.X. Zhou, Dynamics of multiphase turbulent reacting fluid flows, Defense Industry Press, 2002.

[19]   L.X. Zhou, Y. Xu, Simulation of swirling gas-particle flows using an improved second-order moment two-phase turbulence model, Powder Technol. 116 (2001) 2—3.

[20]   L.Y. Hu, L.X. Zhou, J. Zhang, M.X. Shi, Studies on strongly swirling flows in the full space of a volute cyclone separator, AIChE J. 51 (2005) 740—749.

[21]   J. Smagorinsky, General circulation experiments with the primitive equations, 1. The basic experiment, Monthly Weather Rev. 91 (1963) 99—164.

[22]   D.G. Lilly, On the application of the eddy viscosity concept in the inertial subrange of turbulence, NACR Manuscript 123 (1966).

[23]   M. Germano, U. Piomelli, P. Moin, W.H. Cabot, A dynamic sub-grid scale eddy viscosity model, Phys. Fluids A3 (1991) 1760—1765.

[24]   W.W. Kim, S. Menon, H.C. Mongia, Large-eddy simulation of a gas turbine combustor flow, Combustion Sci. Technol. 143 (1999) 25—62.

[25]   L.Y. Hu, L.X. Zhou, J. Zhang, Large-eddy simulation of a swirling diffusion flame using a SOM SGS combustion model, Numer. Heat Transfer B50 (2006) 41—48.

[26]   S.B. Pope, Turbulent Flows, Cambridge University Press, Cambridge, UK, 2000.

[27]   C.X. Xu. Ph.D. Thesis, Department of Engineering Mechanics, Tsinghua University, 1995.

Chapter 6

Modeling of Dispersed Multiphase Turbulent Flows

6.1 INTRODUCTION

There are different approaches for modeling turbulent dispersed multiphase flows. The fundamental approach is the direct numerical simulation (DNS), including the fully resolved or full-scale DNS (FDNS) and the point-particle DNS (PDNS). In FDNS, particles/droplets/bubbles are simulated to occupy finite-size volumes, no models for gas-particle interaction forces are needed and the main task is the numerical methods to deal with the interface, such as the volume-of fluid (VOF), level-set, front-tracking, and immerse-boundary methods, etc. In PDNS, particles/droplets/bubbles are treated as point sources, not occupying finite-size volumes and models for the particle drag force, lift force, and particle mass change (evaporation/devolatilization/combustion) are needed. In the point-particle approaches, beside PDNS, there are large-eddy simulation (LES) and Reynolds-averaged (RANS) modeling. These approaches include Eulerian−Eulerian (E − E, two-fluid) modeling and Eulerian−Lagrangian (E − L) modeling. E − L modeling, when accounting for interparticle collision, is called discrete element modeling (DEM). The PDNS can give instantaneous velocities and detailed turbulence structures, but has too large a computational requirement, and hence it is used for fundamental studies and provides database for validating models in LES and RANS modeling. The FDNS can give the detailed shape of interface and instantaneous velocities inside each phase, but needs a still larger computational requirement and cannot be used to simulate practical dispersed multiphase flows with phase change and combustion. For engineering dispersed multiphase flows only Eulerian−Eulerian and Eulerian−Lagrangian approaches for RANS modeling or LES can be used. The former needs a particle turbulence model and particle stress model, whereas the latter needs a fluid instantaneous velocity model and interparticle collision model.

Many engineering applications, including the commercial computer codes, adopt E-L models, but E-E models have their specific features. It was pointed by Crowe in 1991 [1] that "the advantage of the two-fluid model is that the algorithm developed for the conveying phase can be easily modified for the particulate phase. Also the storage and computational time are not as

Theory and Modeling of Dispersed Multiphase Turbulent Reacting Flows.
DOI: https://doi.org/10.1016/B978-0-12-813465-8.00006-5

121

excessive as it may be for the trajectory models." The development of two-fluid models or multifluid models can be traced back to the 1960s. Their pioneer was S.L. Soo [2]. In his book "Fluid Dynamics of Multiphase Systems" published in 1967, particles are treated as a pseudo-fluid. However, in his concept, the velocity slip between the fluid and particles is considered to be much smaller than the velocity itself and this velocity slip forms a drift force for particle diffusion. In his equations the "particle diffusivity" is an empirical or semiempirical constant, not definitely related to turbulence properties. In fact, this concept of "small slip" is approximately true only for a few cases like pipe flows laden with small particles. For most cases, for example, in the case of liquid spray injected in reverse direction into high-velocity gas flows, the velocity slip between the gas and droplets may be very large, and is not related to the droplet diffusion. Therefore, in 1982 two-fluid model equations with the separate velocity slip and particle diffusion were given by Zhou [3].

The key problem of two-fluid modeling is the closure models of particle turbulence (particle turbulent fluctuation), leading to particle diffusion/dispersion. Early studies in this direction were made by Elghobashi et al. [4– 6], combining the gas $k - \varepsilon$ turbulence model with algebraic particle turbulence models (we called this a $k - \varepsilon - A_p$ model). In fact, similar approaches have been taken by Melville and Bray [7], Chen and Wood [8], Mostafa and Mongia [9], etc. All of these closure models for the particle turbulence are based on the idea of Hinze−Tchen's particle-tracking-fluid theory of particle fluctuation, originally proposed by Tchen [10], and finally developed by Hinze [11]. According to the Hinze−Tchen model, the particle fluctuation should always be weaker than the fluid fluctuation and, the larger the particle size, the weaker the particle fluctuation. Hence larger particles should diffuse slower than smaller particles. However, in the experiments of enclosed gas-particle jets done by Zhou et al. [12] it was found that 165-μm particles diffuse faster than 26-μm particles, and in experiments done by Borner and Durst [13] it was found that the particle RMS fluctuation velocity is larger than the gas RMS fluctuation velocity in enclosed gas-particle jets. Therefore, a transport equation theory of particle turbulence was proposed by Zhou and Huang [14], and a $k - \varepsilon - k_p$ two-phase turbulence model against the $k - \varepsilon - A_p$ model was proposed and used to simulate a gas-particle jet [15]. From that time on, the two-phase turbulence models, in particular the particle turbulence models, were systematically studied and developed by many investigators. There are two groups of two-fluid modelers: the first group includes Zhou [16] and Tu [17], who directly use the Reynolds expansion and averaging to derive and close the particle turbulent kinetic energy and particle Reynolds stress equations. The other group includes Zaichik, Reeks, and Simonin, who derive and close the particle Reynolds stress equation based on the probability density function (PDF) equations, called the "kinetic approach."

In Chapter 5, the single-phase turbulence models are discussed. These models can be used as a basis for developing two-phase or multiphase turbulence models. There are two sides to this method. On one hand, the dispersed-phase (i.e., particles/droplets/bubbles) turbulence (turbulent fluctuation) causes turbulent diffusion or dispersion. On the other hand, the dispersed phase affects the fluid turbulence, called turbulence modification/modulation. In the Eulerian–Eulerian or two-fluid approach, the dispersed phase is treated as a pseudo-fluid, having its own turbulent fluctuation, causing its turbulent transport of mass, momentum, and energy, and the dispersed-phase turbulence is determined not only by the effect of fluid turbulence, but also by its own convection, diffusion, and production due to the mean velocity gradient.

According to the original concept of trajectory model, only fluid-phase fluctuation is considered, and the dispersed phase does not have the same turbulence properties as the fluid phase. In the frame of the two-fluid model, the dispersed phase has turbulence properties, like those of the fluid phase, such as particle turbulent kinetic energy and particle Reynolds stress. Presently, the main two-fluid modelers, like Zaichik, Simonin, Reeks, and the present author agree with this point of view. Nowadays, the concepts of particle turbulence, particle turbulent kinetic energy, and particle Reynolds stress are accepted by all investigators in the multiphase flow community, including Lagrangian particle modelers.

The pioneering work on particle turbulence model was done by S.M. Tchen. Based on the concept of particle tracking local fluid turbulence, the relationship between the particle turbulent viscosity or diffusivity and the fluid turbulent viscosity was obtained. Subsequently, it was derived by Hinze as an algebraic expression of the particle turbulence model. Until the mid-1980s this model was widely used by two-fluid modelers; it was then discovered by the present author and his colleagues by experiments that larger particles diffuse faster than smaller particles and subsequently it was found that the particle fluctuation is greater than the gas fluctuation in experiments of enclosed gas-particle jets by Durst's research group. These results encouraged the author to study the two-phase turbulence model using the method of single-phase turbulence modeling. A transport equation model of particle turbulent kinetic energy was proposed and the $k - \varepsilon - k_p$ two-phase turbulence model was formulated. Later, when dealing with anisotropic two-phase turbulent flows, a unified second-order moment (USM) or two-phase Reynolds stress model was proposed. Almost at the same time some investigators introduced the concept of probability density function (PDF) to construct the particle Reynolds stress equation. Meanwhile the author also used the PDF concept to establish the two-phase turbulence models, such as the $k - \varepsilon - \text{PDF}$ and DSM-PDF models. In the following text we will discuss the USM, $k - \varepsilon - k_p$, $k - \varepsilon - \text{Ap}$, $k - \varepsilon - \text{PDF}$, and DSM-PDF two-phase turbulence models, and conclude with a brief discussion of LES and DNS of dispersed two-phase flows.

6.2 THE HINZE–TCHEN'S ALGEBRAIC MODEL OF PARTICLE TURBULENCE

S. M. Tchen [10] considered a single particle entering a fluctuating gas eddy and tracking the gas motion, using Taylor's statistical theory of gas turbulence. After Hinze's derivation, the ratio of the particle turbulent viscosity over the gas turbulent viscosity or the particle turbulent diffusivity over the gas turbulent diffusivity was obtained as

$$\nu_p/\nu_T = D_p/D_T = (k_p/k)^2 = (1 + \tau_{r1}/\tau_T)^{-1} \qquad (6.1)$$

where ν_p, ν_T are particle and gas turbulent viscosity, respectively, D_p, D_T are particle and gas turbulent diffusivity, respectively, and $\tau_{r1} = \rho_s d_p^2/(18\mu)$, $\tau_T = k/\varepsilon$ are the particle Stokes relaxation time and gas fluctuation time, respectively. Eq. (6.1) is called the Hinze–Tchen algebraic model (A_p model) of particle turbulence. According to this model, the particle turbulent fluctuation is always smaller than the gas turbulent fluctuation and the larger the particle size, the smaller the particle fluctuation. This model, like the algebraic mixing-length model of gas turbulence, is rather simple and intuitive, and was widely used by many two-fluid modelers. However, as indicated above, experiments show that in many cases or in some regions of the flow field the particle fluctuation is stronger than the gas fluctuation, and the fluctuation of larger particles is stronger than that of smaller particles. The reason is that the A_p model, based on the particle-tracking gas motion, considers only the effect of local gas turbulence, and neglects the effect of convection, diffusion, and mean-flow production of the particle turbulent kinetic energy. Its drawback is similar to that of the mixing-length model. Therefore, based on the development of the single-phase turbulence models, the transport equation models of particle turbulent kinetic energy and particle Reynolds stress were developed by the present author.

6.3 THE UNIFIED SECOND-ORDER MOMENT TWO-PHASE TURBULENCE MODEL

The unified second-order moment model (USM model), proposed by the author [16], has both two-phase turbulences closed using the second-order moment method, i.e., using the two-phase Reynolds stress transport equations. It is an extension of the single-phase Reynolds stress equation to two-phase flows; however, it is not a simple duplication. In the case of turbulent gas-particle flows there are interaction forces (drag force, lift force, etc.), leading to forming new source terms in both gas and particle Reynolds stress equations. These source terms need closure models. This is the special feature of two-phase turbulence models.

In Chapter 4, the exact form of two-phase Reynolds stress equations was derived as

$$\frac{\partial}{\partial t}(\rho\overline{v_i v_j}) + \frac{\partial}{\partial x_m}(\rho V_m \overline{v_i v_j}) = D_{ij} + P_{ij} + G_{ij} + \Pi_{ij} - \varepsilon_{ij} + G_{p,ij} + G_{R,ij} \quad (6.2)$$

$$\frac{\partial}{\partial t}(N_k \overline{v_{ki} v_{kj}}) + \frac{\partial}{\partial x_m}(N_k V_{km} \overline{v_{ki} v_{kj}}) = D_{k,ij} + P_{k,ij} + G_{k,ij} + \varepsilon_{k,ij} \quad (6.3)$$

where the terms on the right-hand side of Eq. (6.2) D_{ij}, P_{ij}, G_{ij}, Π_{ij}, and ε_{ij} are diffusion, shear production, buoyancy production, pressure−strain, and dissipation terms of the gas Reynolds stress, respectively, and are the same as those for the single-phase flows, i.e.,

$$D_{ij} = -\frac{\partial}{\partial x_m}\left[\rho\overline{v_i v_j v_m} + \overline{p' v_j}\delta_{im} + \overline{p' v_i}\delta_{jm} - \mu\left(\frac{\partial}{\partial x_m}\overline{v_i v_j}\right)\right]$$

$$P_{ij} = -\rho\left(\overline{v_i v_m}\frac{\partial V_j}{\partial x_m} + \overline{v_j v_m}\frac{\partial V_i}{\partial x_m}\right),$$

$$G_{ij} = \beta\rho(g_i\overline{v_j T'} + g_j\overline{v_i T'}) \quad \varepsilon_{ij} = -2\mu\left(\overline{\frac{\partial v_i'}{\partial x_m}\frac{\partial v_j'}{\partial x_m}}\right) \quad \Pi_{ij} = p'\left(\overline{\frac{\partial v_i'}{\partial x_j} + \frac{\partial v_j'}{\partial x_i}}\right)$$

The last two terms on the right-hand side of Eq. (6.2)

$$G_{p,ij} = \sum_k \frac{\rho_k}{\tau_{rk}}(\overline{v_{ki} v_j} + \overline{v_{kj} v_i} - 2\overline{v_i v_j}), \quad G_{R,ij} = \overline{v_i v_j}S$$

are source terms of the gas Reynolds stress due to the drag force and particle reaction (evaporation, devolatilization, and combustion), respectively. The four terms on the right-hand side of Eq. (6.3) are diffusion, shear production, buoyancy production, and drag force terms of particle Reynolds stress, respectively, i.e.,

$$D_{k,ij} = -\frac{\partial}{\partial x_m}\left[N_k \overline{v_{km} v_{ki} v_{kj}}\right]$$

$$P_{k,ij} = -(V_{km}\overline{n_k v_{kj}} + N_k\overline{v_{km} v_{kj}})\frac{\partial V_{ki}}{\partial x_m} - (V_{km}\overline{n_k v_{ki}} + N_k\overline{v_{km} v_{ki}})\frac{\partial V_{kj}}{\partial x_m}$$

$$G_{k,ij} = \overline{n_k v_{kj}}g_i + \overline{n_k v_{ki}}g_j$$

$$\varepsilon_{k,ij} = \left(\frac{1}{\tau_{rk}} + \frac{\dot{m}_k}{m_k}\right)\left[N_k(\overline{v_{ki} v_j} + \overline{v_{kj} v_i} - 2\overline{v_{ki} v_{kj}}) + (V_i - V_{ki})\overline{n_k v_{kj}} + (V_j - V_{kj})\overline{n_k v_{ki}}\right]$$

The transport equations of gas and particle turbulent kinetic energy are

$$\frac{\partial}{\partial t}(\rho k) + \frac{\partial}{\partial x_m}(\rho V_m k) = -\frac{\partial}{\partial x_m}(\overline{\rho v_m v_i^2}/2 + \overline{p' v_m} - \mu \frac{\partial k}{\partial x_m}) - \rho \overline{v_m v_i} \frac{\partial V_i}{\partial x_m} + \beta \rho g_i \overline{v_i T'}$$

$$- \mu \left(\frac{\partial v_i}{\partial x_m}\right)^2 + \sum_k \frac{N_k m_k}{\tau_{rk}}(\overline{v_{ki} v_i} - 2k) + kS$$

$$\frac{\partial}{\partial t}(N_k k_k) + \frac{\partial}{\partial x_m}(N_k V_{km} k_k) = -\frac{\partial}{\partial x_m}(N_k \overline{v_{km} v_{ki}^2}/2) - \left(N_k \overline{v_{km} v_{ki}} \frac{\partial V_{ki}}{\partial x_m}\right)$$

$$+ \left(\frac{1}{\tau_{rk}} + \frac{\dot{m}_k}{m_k}\right)[N_k(\overline{v_{ki} v_i} - 2k_k) + (V_i - V_{ki})\overline{n_k v_{ki}}]$$

$$(6.4)$$

The closed gas Reynolds stress equation for turbulent gas-particle flows is

$$\frac{\partial}{\partial t}(\rho \overline{v_i v_j}) + \frac{\partial}{\partial x_k}(\rho V_k \overline{v_i v_j}) = D_{ij} + P_{ij} + G_{p,ij} + \Pi_{ij} - \varepsilon_{ij} \qquad (6.5)$$

where D_{ij}, P_{ij}, Π_{ij}, ε_{ij} are

$$D_{ij} = \frac{\partial}{\partial x_k}\left(c_s \frac{k}{\varepsilon} \overline{v_k v_l} \frac{\partial \overline{v_i v_j}}{\partial x_k}\right), \quad P_{ij} = -\rho\left(\overline{v_i v_k} \frac{\partial V_j}{\partial x_k} + \overline{v_j v_k} \frac{\partial V_i}{\partial x_k}\right)$$

$$\Pi_{ij} = \Pi_{ij,1} + \Pi_{ij,2}, \quad \Pi_{ij,1} = -c_1 \frac{\varepsilon}{k} \rho\left(\overline{v_i v_j} - \frac{2}{3}\delta_{ij} k\right)$$

$$\Pi_{ij,2} = -c_2\left(P_{ij} - \frac{2}{3}\delta_{ij} G\right), \quad \varepsilon_{ij} = \frac{2}{3}\delta_{ij}\varepsilon,$$

$$G = -\rho \overline{v_i v_k} \frac{\partial V_i}{\partial x_k}$$

The source term expressing the effect of particles on gas Reynolds stress is

$$G_{p,ij} = \sum_p \frac{\rho_p}{\tau_{rp}}(\overline{v_{pi} v_j} + \overline{v_{pj} v_i} - 2\overline{v_i v_j})$$

The closed transport equation of the dissipation rate of the gas turbulent kinetic energy is

$$\frac{\partial}{\partial t}(\rho \varepsilon) + \frac{\partial}{\partial x_k}(\rho V_k \varepsilon) = \frac{\partial}{\partial x}\left(c_\varepsilon \frac{k}{\varepsilon} \overline{v_k v_l} \frac{\partial \varepsilon}{\partial x_l}\right) + \frac{\varepsilon}{k}[c_{\varepsilon 1}(G + G_p) - c_{\varepsilon 2} \rho \varepsilon] \quad (6.6)$$

where $G_p = \sum_p \frac{\rho_p}{\tau_{rp}}(\overline{v_{pi} v_i} - \overline{v_i v_i})$

The closed particle Reynolds stress equation is

$$\frac{\partial}{\partial t}(N_p \overline{v_{pi} v_{pj}}) + \frac{\partial}{\partial x_k}(N_p V_{pk} \overline{v_{pi} v_{pj}}) = D_{p,ji} + P_{p,ij} + \varepsilon_{p,ij} \tag{6.7}$$

where

$$D_{p,ij} = \frac{\partial}{\partial x_k}\left[N_p c_p^s \frac{k_p}{\varepsilon_p} \overline{v_{pk} v_{pl}} \frac{\partial}{\partial x_l}(\overline{v_{pi} v_{pj}}) \right]$$

$$P_{p,ij} = -(V_{pk}\overline{n_p v_{pj}} + N_p \overline{v_{pk} v_{pj}})\frac{\partial V_{pi}}{\partial x_k} - (V_{pk}\overline{n_p v_{pi}} + N_p \overline{v_{pk} v_{pi}})\frac{\partial V_{pj}}{\partial x_k} + \overline{n_p v_{pj}} g_i + \overline{n_p v_{pi}} g_j$$

$$\varepsilon_{p,ij} = \frac{1}{\tau_{rp}}\left[N_p(\overline{v_{pi} v_j} + \overline{v_{pj} v_i} - 2\overline{v_{pi} v_{pj}}) + (V_i - V_{pi})\overline{n_p v_{pj}} + (V_j - V_{pj})\overline{n_p v_{pi}} \right]$$

The closed transport equations for $\overline{n_p v_{pi}}$, $\overline{n_p v_{pj}}$, $\overline{n_p n_p}$, $\overline{v_{pi} v_j}$, $\overline{v_{pj} v_i}$ are

$$\frac{\partial}{\partial t}(n_p \overline{n_p v_{pi}}) + \frac{\partial}{\partial x_k}(n_p V_{pk} \overline{n_p v_{pi}}) = D_{nv_i} + P_{nv_i} + \varepsilon_{nv_i} \tag{6.8}$$

where

$$D_{nv_i} = \frac{\partial}{\partial x_k}\left[n_p c_n^v \frac{k_p}{\varepsilon_p} \overline{v_{pk} v_{pl}} \frac{\partial}{\partial x_l}(\overline{n_p v_{pi}}) \right]$$

$$P_{nv_i} = -(n_p \overline{v_{pk} v_{pi}} + V_{pk}\overline{n_p v_{pi}} + 2V_{pi}\overline{n_p v_{pk}})\frac{\partial n_p}{\partial x_k} - (n_p \overline{n_p v_{pk}} + V_{pk}\overline{n_p n_p})\frac{\partial V_{pi}}{\partial x_k}$$

$$- 2(n_p \overline{n_p v_{pi}} + V_{pi}\overline{n_p n_p})\frac{\partial V_{pk}}{\partial x_k}$$

$$\varepsilon_{nvi} = \frac{1}{\tau_{rp}}\left[(V_i - V_{pi})\overline{n_p n_p} + n_p \overline{n_p v_{pi}} \right]$$

$$\frac{\partial}{\partial t}(n_p \overline{n_p n_p}) + \frac{\partial}{\partial x_k}(n_p V_{pk} \overline{n_p n_p}) = D_{nn} - 2n_p \overline{n_p n_p}\frac{\partial V_{pk}}{\partial x_k} - 2n_p \overline{n_p v_{pk}}\frac{\partial n_p}{\partial x_k}$$

$$\tag{6.9}$$

where

$$D_{nn} = \frac{\partial}{\partial x_k}\left[n_p c_n^n \frac{k_p}{\varepsilon_p} \overline{v_{pk} v_{pl}} \frac{\partial}{\partial x_l}(\overline{n_p n_p}) \right]$$

$$\frac{\partial}{\partial t}(\overline{v_{pi} v_j}) + (V_{pk} + V_k)\frac{\partial}{\partial x_k}(\overline{v_{pi} v_j}) = \frac{\partial}{\partial x_k}\left[(\nu_e + \nu_p)\frac{\partial}{\partial x_k}(\overline{v_{pi} v_j}) \right]$$

$$+ \frac{1}{\rho \tau_{rp}}\left[\rho_p \overline{v_{pi} v_{pj}} + \rho \overline{v_i v_j} - (\rho_p + \rho)\overline{v_{pi} v_j} \right] - \left(\overline{v_{pk} v_j}\frac{\partial V_{pi}}{\partial x_k} + \overline{v_{pi} v_k}\frac{\partial V_j}{\partial x_k} \right) - \frac{\varepsilon}{k}\overline{v_{pi} v_j}\delta_{ij}$$

$$\tag{6.10}$$

$$\frac{\partial}{\partial t}(n_p k_p) + \frac{\partial}{\partial x_k}(n_p V_{pk} k_p) = \frac{\partial}{\partial x_k}\left(n_p c_p^s \frac{k_p}{\varepsilon_p}\overline{v_{pk}v_{pl}}\frac{\partial k_p}{\partial x_l}\right) + P_p - n_p \varepsilon_p \qquad (6.11)$$

where

$$P_p = -(n_p \overline{v_{pk}v_{pi}} + V_{pk}\overline{n_p v_{pi}})\frac{\partial V_{pi}}{\partial x_k}; \quad \varepsilon_p = -\frac{1}{\tau_{rp}}\left[\overline{v_{pi}v_i} - \overline{v_{pi}v_{pi}} + \frac{1}{n_p}(V_i - V_{pi})\overline{n_p v_{pi}}\right]$$

In some cases the transport equation for the two-phase velocity correlation can be simplified into an algebraic expression as

$$\overline{v_{pi}v_j} = -\frac{\rho \tau_{rp}}{\rho + \rho_p}\left(\overline{v_{pi}v_k}\frac{\partial V_j}{\partial x_k} + \overline{v_j v_{pk}}\frac{\partial V_{pi}}{\partial x_k}\right) + \frac{\rho}{\rho + \rho_p}\overline{v_i v_j} + \frac{\rho_p}{\rho + \rho_p}\overline{v_{pi}v_{pj}} - \frac{\rho \tau_{rp}}{\rho + \rho_p}\frac{1}{T_e}\overline{v_{pi}v_i}\delta_{ij}$$

$$(6.12)$$

Similarly, the transport equations of two-phase Reynolds stresses can be simplified into the following algebraic expressions

$$\overline{v_i v_j} = (1 - \lambda)\frac{2}{3}k\delta_{ij} + \lambda\frac{k}{\varepsilon}\left(\overline{v_i v_k}\frac{\partial V_j}{\partial x_k} + \overline{v_j v_k}\frac{\partial V_i}{\partial x_k}\right) + \frac{k}{c_1 \rho \varepsilon}(\overline{v_{pi}v_j} + \overline{v_i v_{pj}} - 2\overline{v_i v_j})$$

$$(6.13)$$

$$\overline{v_{pi}v_{pj}} = -\frac{\tau_{rp}}{2}\left(\overline{v_{pi}v_{pk}}\frac{\partial V_{pj}}{\partial x_k} + \overline{v_{pj}v_{pk}}\frac{\partial V_{pi}}{\partial x_k}\right) + \frac{1}{2}(\overline{v_i v_{pj}} + \overline{v_{pi}v_j}) \qquad (6.14)$$

Eqs. (6.12), (6.13), and (6.14), together with the k, k_p, ε, and k_{pg} equations (by setting $i = j$ in Eq. (6.10), the k_{pg} equation can be obtained) constitute the algebraic model of two-phase Reynolds stress.

6.4   THE $K - \varepsilon - K_p$ AND $K - \varepsilon - A_p$ TWO-PHASE TURBULENCE MODEL

The USM model can simulate anisotropic turbulent two-phase flows. For nearly isotropic flows or weakly swirling flows it is not necessary to use the complex USM model. In this case, by introducing the Boussinesq expression and scalar turbulent viscosity, the USM model is reduced to a $k - \varepsilon - k_p$ model or a two-phase turbulent kinetic energy equation model. The closed equations and expressions are [18]

$$\overline{v_i v_j} = \frac{2}{3}k\delta_{ij} - \nu_T\left(\frac{\partial V_i}{\partial x_j} + \frac{\partial V_j}{\partial x_i}\right) \qquad (6.15)$$

$$\overline{v_{pi}v_{pj}} = \frac{2}{3}k_p\delta_{ij} - \nu_p\left(\frac{\partial V_{pi}}{\partial x_j} + \frac{\partial V_{pj}}{\partial x_i}\right) \qquad (6.16)$$

$$\overline{n_p v_{pi}} = -\frac{\nu_p}{\sigma_p}\frac{\partial n_p}{\partial x_i} \tag{6.17}$$

$$\overline{n_p v_{pj}} = -\frac{\nu_p}{\sigma_p}\frac{\partial n_p}{\partial x_j} \tag{6.18}$$

$$\frac{\partial}{\partial t}(\rho k) + \frac{\partial}{\partial x_j}(\rho V_j k) = \frac{\partial}{\partial x_j}\left(\frac{\mu_e}{\sigma_k}\frac{\partial k}{\partial x_j}\right) + G + G_p - \rho \varepsilon \tag{6.19}$$

$$\frac{\partial}{\partial t}(\rho \varepsilon) + \frac{\partial}{\partial x_j}(\rho V_j \varepsilon) = \frac{\partial}{\partial x_j}\left(\frac{\mu_e}{\sigma_\varepsilon}\frac{\partial \varepsilon}{\partial x_j}\right) + \frac{\varepsilon}{k}\left[c_{\varepsilon 1}(G + G_p) - c_{\varepsilon 2}\rho \varepsilon\right] \tag{6.20}$$

$$\frac{\partial}{\partial t}(n_p k_p) + \frac{\partial}{\partial x_j}(n_p V_{pj} k_p) = \frac{\partial}{\partial x_j}\left(\frac{n_p \nu_p}{\sigma_p}\frac{\partial k_p}{\partial x_j}\right) + P_p + P_g - n_p \varepsilon_p \tag{6.21}$$

where

$$G = \mu_t\left(\frac{\partial V_i}{\partial x_j} + \frac{\partial V_j}{\partial x_i}\right)\frac{\partial V_i}{\partial x_j} \qquad G_p = \sum_p \frac{2m_p n_p}{\tau_{rp}}(c_p^k \sqrt{k k_p} - k)$$

$$P_p = n_p \nu_p\left(\frac{\partial V_{pi}}{\partial x_j} + \frac{\partial V_{pj}}{\partial x_i}\right)\frac{\partial V_{pi}}{\partial x_j} \qquad P_g = \frac{2m_p n_p}{\tau_{rp}}(c_p^k \sqrt{k k_p} - k_p)$$

$$\varepsilon_p = -\frac{1}{\tau_{rp}}\left[2(c_p^k \sqrt{k_p k} - k_p) + \frac{1}{n_p}\overline{n_p v_{pi}}(V_i - V_{pi})\right]$$

$$\mu_e = \mu + \mu_T \qquad \nu_T = c_\mu \frac{k^2}{\varepsilon} \qquad \mu_T = \rho \nu_T \qquad \nu_p = c_{\mu p}\frac{k_p^2}{|\varepsilon_p|}$$

Eqs. (6.15)–(6.21) are called a $k - \varepsilon - k_p$ two-phase turbulence model. Both USM and $k - \varepsilon - k_p$ models can simulate the convection, diffusion, production, and dissipation of two-phase turbulences and their interaction well. The Hinze–Tchen algebraic particle turbulence model together with the $k - \varepsilon$ gas turbulence model, that is Eqs. (6.1), (6.15)–(6.20), is called a $k - \varepsilon - A_p$ model. The USM model, $k - \varepsilon - k_p$ model and $k - \varepsilon - A_p$ model are three levels of two-phase turbulence models.

6.5   THE APPLICATION AND VALIDATION OF USM, $K - \varepsilon - K_p - K_{PG}$ AND $K - \varepsilon - A_p$ MODELS

The USM, $k - \varepsilon - k_p - k_{pg}$ and $k - \varepsilon - A_p$ models were applied to predict gas-particle jets, sudden-expansion, and swirling gas-particle flows. Fig. 6.1 shows the predicted particle mass flux for an enclosed gas-particle jet [15]. It can be seen that the $k - \varepsilon - k_p$ model properly predicts the diffusion of larger particles faster than that of smaller particles, but the $k - \varepsilon - A_p$ model

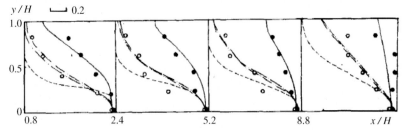

FIGURE 6.1   Particle mass flux (Exp.: •165 μm, o 26 μm; Predic.: $k - \varepsilon - k_p$ 165 μm, ······$k - \varepsilon - A_p$ 165 μm, · $k - \varepsilon - k_p$ 26 μm, ·· $k - \varepsilon - A_p$ 26 μm).

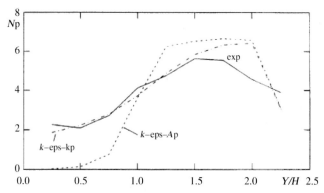

FIGURE 6.2   Particle number density.

gives the opposite results, and hence is unreasonable. Fig. 6.2 gives the predicted particle number density in dust-air flows using both $k - \varepsilon - k_p$ and $k - \varepsilon - A_\mathrm{p}$ models [19] in comparison with the measurement results. Obviously, the $k - \varepsilon - k_p$ model gives much better results than the $k - \varepsilon - A_p$ model.

Figs. 6.3−6.5 give the predicted particle axial velocity, particle tangential velocity, and particle mass flux of swirling gas-particle flows, respectively [20], in comparison with the phase Doppler particle anemometer (PDPA) measurement results made in Reference [21]. The predictions using the $k - \varepsilon - k_p$ model are much better than those using other models. Figs. 6.6−6.9 show the predicted vertical and horizontal components of liquid and bubble normal Reynolds stresses in bubble-liquid flows using the USM (SOM) two-phase turbulence model in comparison with the PIV measurement results [22]. The agreement is pretty good. It can be seen that the bubble turbulence is stronger than the liquid turbulence, i.e., bubbles induce the liquid turbulence and on average the vertical components of two-phase normal Reynolds stresses are larger than their horizontal components, indicating that the two-phase turbulence is anisotropic.

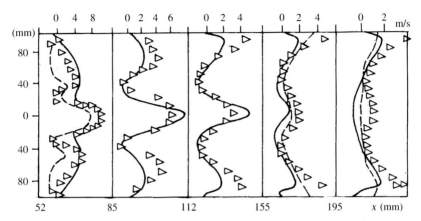

FIGURE 6.3   Particle axial velocity (Δ- Exp.; — k_p Pred.; --- Wennberg's Pred.).

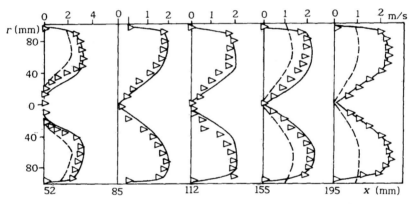

FIGURE 6.4   Particle tangential velocity (Δ- Exp.; —k_p Pred.; --- Simonin's Pred.).

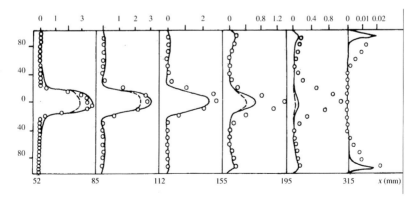

FIGURE 6.5   Particle mass flux (o- Exp.; — k_p Pred.; --- Simonin's Pred.).

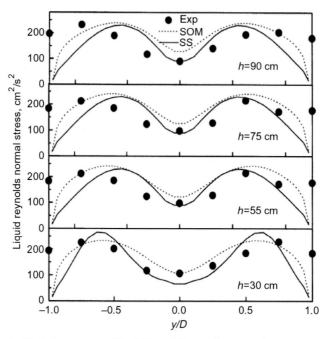

FIGURE 6.6   Vertical component of liquid Reynolds normal stress.

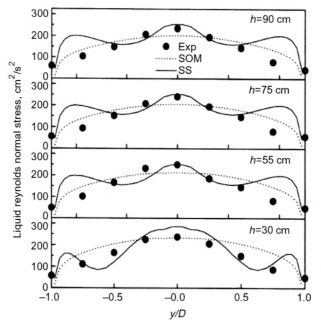

FIGURE 6.7   Horizontal component of liquid Reynolds normal stress.

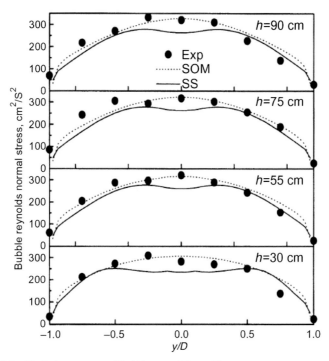

FIGURE 6.8 Vertical component of bubble normal Reynolds stress.

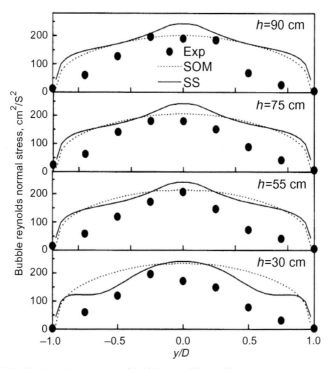

FIGURE 6.9 Horizontal component of bubble normal Reynolds stress.

6.6 AN IMPROVED SECOND-ORDER MOMENT TWO-PHASE TURBULENCE MODEL

It has been found that for swirling gas-particle flows the $k - \varepsilon - k_p$ and USM models can well simulate the time-averaged values, but the two-phase fluctuation velocities are underpredicted. There are two possible reasons for this. First, in processing the LDV/PDPA data, the particle sizes of each size group cannot be monodispersed, it has a certain size range, the measured "particle fluctuation velocity" includes not only the effect of turbulence, but also the effect of nonuniform sizes; hence the measured value is greater than the real value of the particle fluctuation velocity, particularly for swirling gas-particle flows, where the particle concentration distribution is more nonuniform. Second, the closure of USM and $k - \varepsilon - k_p$ models remains to be improved. In analyzing the two-phase Reynolds stress equations, it can be seen that the closure of the term of two-phase velocity correlation is very important. It has three levels of closure: transport equation, algebraic expression, and dimensional analysis. The transport equation model is more reasonable, but computationally is expensive. The algebraic model is more convenient. A closure model based on Lagrangian analysis has been proposed [23]. Zaichik [24] started from a Lagrangian PDF equation, assuming a Gaussian PDF distribution for isotropic homogeneous turbulence, and obtained a simple expression of two-phase velocity correlation

$$\overline{v_i v_{pj}} = \alpha \overline{v_i v_j} - \beta \overline{v_i v_j} \frac{\partial V_j}{\partial x_j} \tag{6.22}$$

where $\alpha = T_L/(\tau_p + T_L); \beta = T_L$, τ_p is the particle relaxation time, and T_L is the Lagrangian time scale of single-phase turbulence. An improvement was made by using the gas turbulence scale seen by particles instead of the single-phase turbulence scale, based on the Lagrangian analysis of Wang and Stock [25] and Huang and Stock [26], accounting for the crossing trajectory effect, inertial effect, continuity effect, and anisotropy. The result gives

$$\alpha_{ii} = T_L/(\tau_p + T_L) \quad \beta_{ii} = \alpha_{ii} T_L, \tag{6.23}$$

where $\quad T_L = \min\left(\tau_{eii}, \frac{l_{eii}}{|V_{rel}|}\right) \quad T_{eii} = 2T_{mEii}\left(1 - \frac{1 - T_{Lii}/T_{mEii}}{(1+St_{ii})^{0.4(1+0.01St_{ii})}}\right)$

$T_L/T_{mE} = 0.356$

$St = \dfrac{\tau_p}{T_{mE}}, \quad T_{Lii} = 0.235\dfrac{\overline{v_i v_i}}{\varepsilon}, \quad l_{eii} = \dfrac{L_{fii}}{2}\left(1 + \cos^2\theta\right) \quad L_{fii} = 2.5 T_{Lii}\sqrt{\overline{v^2}},$

$\overline{v^2} = \dfrac{1}{3}\left(\overline{v_1^2} + \overline{v_2^2} + \overline{v_3^2}\right)$

Figs. 6.10 and 6.11 are the predicted axial components of particle fluctuation velocity and gas-particle two-phase velocity correlation, respectively,

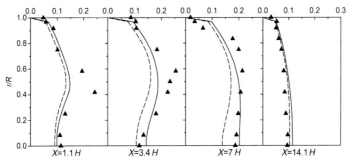

FIGURE 6.10   Axial fluctuation velocity of 50-μm particles (m/s, —Present Model; ----Zaichik Model, ■ Exp.).

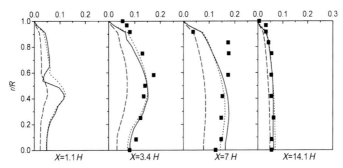

FIGURE 6.11   Axial component of gas-particle (50 μm) velocity correlation (m/s, — Present Model; ----Zaichik model, ▲ Exp.).

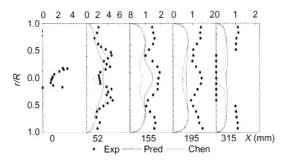

FIGURE 6.12   Particle axial fluctuation velocity ($s = 0.47$, 45 μm)(m/s).

for sudden-expansion gas-particle flows using the present model, Zaichik model and their comparison with the PDPA measurement results. Although there is still some discrepancy between predictions and measurements, the present model is better than the Zaichik model.

Figs. 6.12 and 6.13 give the predicted particle axial and tangential fluctuation velocities for swirling gas-particle flows using the original USM

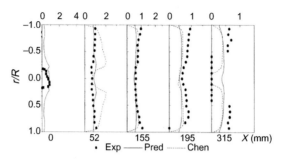

FIGURE 6.13   Particle tangential fluctuation velocity ($s = 0.47$, 45 μm, m/s).

(notation: Chen) and improved USM (notation: Pred) models in comparison with the PDPA measurement results. The improved USM model is somewhat better than the original USM model; however, the particle fluctuation velocities are still underpredicted.

6.7   THE MASS-WEIGHTED AVERAGED USM TWO-PHASE TURBULENCE MODEL

The above discussed is a time-averaged USM model, where the transport equations related to the particle number density fluctuation are introduced, hence there is a large number of equations, making the model very complex. On the other hand, the fluid and particle velocities measured by LDV/PDPA are close to the number-weighted averaged. If a mass-weighted averaging, similar to the Favre averaging used in compressible flows, is used instead of simple time averaging, it should be more reasonable. In that case no equations related to number density fluctuation are required; the number of equations will be reduced. Therefore a mass-weighted averaged USM (MUSM) model has been proposed [27]. Let us define $\rho_p = \alpha_p \rho_{pm}$; $\rho_g = \alpha_p \rho_{gm}$; $\alpha_p + \alpha_g = 1$, where ρ_p, ρ_g are the particle and gas apparent density, respectively, ρ_{pm}, ρ_{gm} are their material density, respectively, neglecting the buoyancy force, added-mass force, Basset force, Magnus force, Saffman force, and accounting for only the drag force and the gravitational force, introducing the apparent-density weighted averaging by defining $\phi_k = \tilde{\phi}_k + \phi''_k$;

$$\tilde{\phi}_k = \overline{\alpha_k \rho_{km} \phi_k} / \overline{\alpha_k \rho_{km}}; \quad k = p, g. \text{ We have}$$

$$\overline{\phi''_k} = \overline{\alpha_k \rho_{km} \phi''_k} / \overline{\alpha_k \rho_{km}} = 0; \quad \rho_k = \tilde{\rho}_k + \rho''_k; \quad \tilde{\rho}_k = \overline{\alpha_k \rho_{km}} / \overline{\alpha_k}$$

For the volume fraction, pressure and shear stress, the time averaging is still used as: $\alpha_k = \overline{\alpha}_k + \alpha'_k$; $p = \overline{p} + p'$; $\tau_{ji} = \overline{\tau}_{ji} + \tau'_{ji}$. For isothermal nonreacting gas-particle flows, the gas and particle material densities are constants; the apparent-density weighted averaging is equivalent to the volume-fraction-weighted averaging. The obtained generalized form of mass-weighted averaged two-phase Reynolds stress equations is

$$\frac{\partial \overline{\alpha}_k \rho_{mk} \overline{u''_{ki} u''_{kj}}}{\partial t} + \frac{\partial \overline{\alpha}_k \rho_{km} \tilde{u}_{kk} \overline{u''_{ki} u''_{kj}}}{\partial x_k} = D_{k,ij} + P_{k,ij} + \Pi_{k,ij} - \varepsilon_{k,ij} + S_{k,ij}$$

(6.24)

where the subscript k denotes the gas phase g or the particle phase p, also a sign of summation, the superscript bar denotes the time-averaging and the wave and double-prime denote the apparent-density-weighted averaging. $D_{k,ij}$, $P_{k,ij}$, $\Pi_{k,ij}$, $\varepsilon_{k,ij}$, and $S_{k,ij}$ are diffusion, shear production, pressure$-$strain, dissipation, and phase interaction terms, respectively. Their closed forms are

$$P_{k,ij} = -\overline{\alpha}_k \rho_{km} \left(\overline{u''_{kk} u''_{ki}} \frac{\partial \tilde{u}_{kj}}{\partial x_k} + \overline{u''_{kk} u''_{kj}} \frac{\partial \tilde{u}_{ki}}{\partial x_k} \right);$$

$$D_{k,ij} = \frac{\partial}{\partial x_k} \left(C_k \overline{\alpha}_k \rho_{km} \frac{k_k}{\varepsilon_k} \overline{u''_{kk} u''_{kl}} \frac{\partial \overline{u''_{ki} u''_{kj}}}{\partial x_l} \right);$$

$$\Pi_{g,ij} = \Pi_{g,ij,1} + \Pi_{g,ij,2}; \quad \Pi_{p,ij} = 0; \quad \varepsilon_{p,ij} = 0; \quad \varepsilon_{g,ij} = \frac{2}{3} \delta_{ij} \overline{\alpha}_g \rho_{gm} \varepsilon_g$$

$$\Pi_{g,ij,1} = -C_{g1} \left(\varepsilon_g / k_g \right) \alpha_g \rho_{gm} \left(\overline{u''_{gi} u''_{gj}} - \frac{2}{3} \delta_{ij} k_g \right); \quad \Pi_{g,ij,2} = -C_{g2} \left(P_{g,ij} - \frac{2}{3} \delta_{ij} G_g \right)$$

$$S_{k,ij} = \frac{\overline{\alpha}_p \rho_{pm}}{\tau_{rp}} \left(\overline{u''_{gi} u''_{kj}} + \overline{u''_{gj} u''_{ki}} - \overline{u''_{pi} u''_{kj}} - \overline{u''_{pj} u''_{ki}} \right)$$

For the two-phase velocity correlation equation, the main problem is to close the dissipation term. It is assumed that the dissipation is isotropic and is proportional to the summation of its normal components divided by a time scale. This time scale is taken as the minimum value from the particle relaxation time and the gas fluctuation time. The closed two-phase velocity correlation equation is

$$\frac{\partial}{\partial t} \left(\overline{\alpha}_p \tilde{\rho}_p \overline{u''_{gi} u''_{pj}} \right) + \frac{\partial}{\partial x_k} \left(\overline{\alpha}_p \tilde{\rho}_p \tilde{u}_{pk} \overline{u''_{gi} u''_{pj}} \right) = D_{gp,ij} + P_{gp,ij} + G^p_{gp,ij} + G^R_{gp,ij}$$
$$+ \Pi_{gp,ij} - \varepsilon_{gp,ij}$$

(6.25)

where

$$D_{gp} = \frac{\partial}{\partial x_k} \left(C_{gp3} \overline{\alpha}_p \tilde{\rho}_p \frac{k}{\varepsilon} \overline{u''_{pk} u''_{pm}} \frac{\partial \overline{u''_{gi} u''_{pj}}}{\partial x_m} \right)$$

$$P_{gp,ij} = -\overline{\alpha}_p \rho_{pm} \left(\overline{u''_{pk} u''_{gi}} \frac{\partial \tilde{u}_{pj}}{\partial x_k} + \overline{u''_{gk} u''_{pj}} \frac{\partial \tilde{u}_{gi}}{\partial x_k} \right)$$

$$G^p_{gp,ij} = -\frac{\overline{\alpha}_p \rho_{pm}}{\tau_{rp}} \left[\left(\overline{u''_{gi} u''_{pj}} - \overline{u''_{gi} u''_{gj}} \right) - \frac{\overline{\alpha}_p \rho_{pm}}{\overline{\alpha}_g \rho_{gm}} \left(\overline{u''_{gi} u''_{pj}} - \overline{u''_{pi} u''_{pj}} \right) \right]$$

$$\Pi_{gp,ij} = \Pi_{gp,ij,1} + \Pi_{gp,ij,2} \quad \Pi_{gp,ij,1} = -\frac{C_{gp1}}{\tau_{rp}} \alpha_p \rho_{pm} \left(\overline{u''_{gi} u''_{pj}} - \frac{1}{3} \delta_{ij} \overline{u''_{gk} u''_{pk}} \right)$$

$$\Pi_{gp,ij,2} = -C_{gp2} \left(P_{gp,ij} - \frac{2}{3} \delta_{ij} \sqrt{G_{gk} G_{pk}} \right)$$

$$\varepsilon_{gp,ij} = \frac{1}{3} \frac{\overline{\alpha}_p \rho_{pm}}{\tau_e} \overline{u''_{gk} u''_{pk}} \delta_{ij}; \quad \tau_e = \min[\tau_T, \tau_{rp}]; \quad \tau_T = k/\varepsilon$$

The above equations constitute the MUSM model. This model was used to simulate swirling gas-particle flows with a swirl number of 0.47 with different time scales in the algebraic expressions of two-phase velocity correlation, in comparison with the time-averaged USM modeling results and the PDPA measurement results.

Figs. 6.14–6.21 give the predicted particle-averaged and fluctuation velocities using different averaging methods and different closure models for the dissipation term in the two-phase velocity correlation equation (Eq. 6.25). The legends in these figures are: USM-0: time averaging with

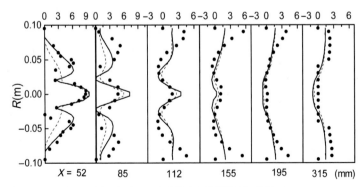

FIGURE 6.14   Particle axial velocity (— MUSM-1, □ USM-0, ● Exp).

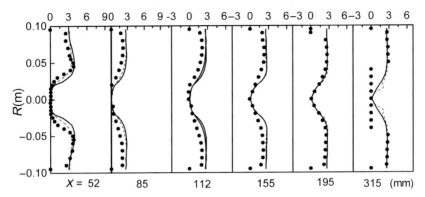

FIGURE 6.15   Particle tangential velocity (— MUSM-1, □USM-0, ● Exp).

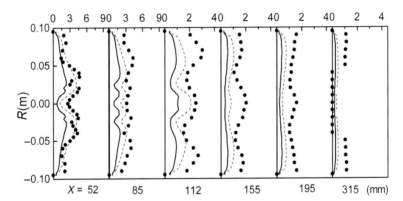

FIGURE 6.16   Particle axial fluctuation velocity (— MUSM-1, □ USM-0, ● Exp).

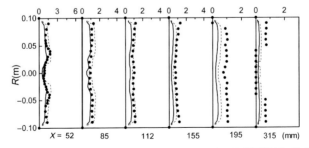

FIGURE 6.17   Particle tangential fluctuation velocity (—MUSM-1, □USM-0, ● Exp).

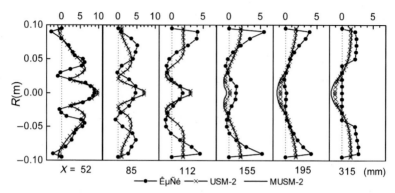

FIGURE 6.18   Particle Axial Velocity (—MUSM-1, □ USM-0, ● Exp).

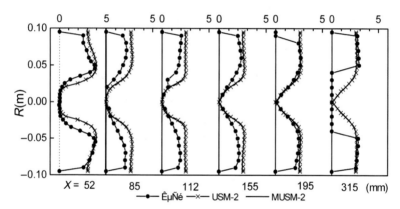

FIGURE 6.19   Particle Tangential Velocity (— MUSM-1, □ USM-0, ● Exp).

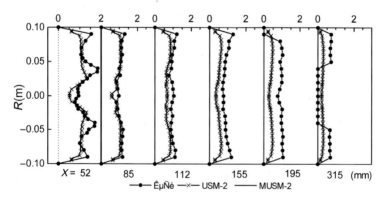

FIGURE 6.20   Particle Axial Fluctuation Velocity (— MUSM-1, □ USM-0, ● Exp).

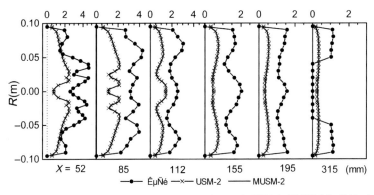

FIGURE 6.21 Particle Tangential Fluctuation Velocity (— MUSM-1, □ USM-0, ● Exp).

zero dissipation in Eq. (6.25); USM-2: time averaging with a time scale $\tau_T = k/\varepsilon$ in the dissipation term of Eq. (6.25); MUSM-1: mass-weighted averaging with a time scale τ_{rp} in the dissipation term of Eq. (6.25); MUSM-2: mass-weighted averaging with a time scale $\tau_T = k/\varepsilon$ in the dissipation term of Eq. (6.25). It can be seen that for the particle-averaged axial and tangential velocities (Figs. 6.14 and 6.15), different averaging methods and closure models give almost the same results, in general agreement with the experimental results, but different averaging methods and closure models underpredict the particle axial and tangential fluctuation velocities (Figs. 6.16 and 6.17). When comparing the prediction results using the same closure model but different averaging methods (Figs. 6.18−6.21), no improvement is made. In general, both MUSM and USM models give slightly different results, but the MUSM model saves a great deal of computation time.

6.8   THE DSM-PDF AND $K-\varepsilon$-PDF TWO-PHASE TURBULENCE MODELS

In Chapter 4, the transport equations of probability density function (PDF) for turbulent gas-particle flows are derived and from the PDF equations the two-phase Reynolds stress equations can be obtained. In the framework of two-fluid models, a group of investigators developed the particle Reynolds stress equation, based only on the derivation and closure of PDF transport equations and then the integration over PDF. Zaichik [28], Reeks [29], Simonin [30], and Zhou [31] independently derived and closed PDF transport equations for turbulent gas-particle flows using different closure methods for the phase interaction term. The first three investigators used PDF transport equations to construct the particle Reynolds stress equation in two-fluid modeling. Zhou's approach in contrast directly solves the particle PDF equation without using the particle Reynolds stress equation. Two methods are used to solve the particle PDF equation. The first uses the finite-difference method, and the other uses the Monte-Carlo method. For the first method, the gas-phase k-ε or DSM model is still used, and the particle Reynolds stresses are obtained by integration over PDF, so the two-phase turbulence models are called k-ε-PDF and DSM-PDF models, whereas for the second method the PDF values and the integration over PDF are not required, the particle Reynolds stresses are directly obtained from the Monte-Carlo simulation, which is a Lagrangian method of particle motion in the phase space.

The transport equation for the joint PDF of the fluctuation velocity p_{fi} in the two-phase velocity coordinates proposed by Li and Zhou [31] is obtained as

$$
\begin{aligned}
\frac{\partial p_{fi}}{\partial t} &+ (<U_j> + <U_{pj}>)\frac{\partial p_{fi}}{\partial x_j} + (u_j + u_{pj})\frac{\partial p_{fi}}{\partial x_j} = \frac{\partial}{\partial v_i}<\frac{p'_{fi}}{\rho}\frac{\partial P'}{\partial x_i}> \\
&- <\frac{p'_{fi}}{\rho}\frac{\partial \tau'_{ij}}{\partial x_j}\delta_{ij}> + \frac{\partial p_{fi}}{\partial v_i}\left(u_j\frac{\partial <U_i>}{\partial x_j}\right) + \frac{\partial p_{fi}}{\partial v_{pi}}\left(u_{pj}\frac{\partial <U_{pi}>}{\partial x_j}\right) \\
&- <p'_{fi}\left(\frac{\rho'}{\rho}g_i\right)> - (<U_{pj}> + u_{pj})\frac{\partial u_i}{\partial x_j}\frac{\partial p_{fi}}{\partial v_i} - (<U_j> + u_j)\frac{\partial u_{pi}}{\partial x_j}\frac{\partial p_{fi}}{\partial v_{pi}} \\
&- \frac{\partial}{\partial v_i}<p'_{fi}\left(\sum f_{rpi}\right)> - \frac{\partial}{\partial v_{pi}}<p'_{fi}\left(\sum f_{ri}\right)>
\end{aligned}
$$

$$(6.26)$$

where $f_{rpi} = \dfrac{\rho_p}{\rho \tau_{rp}}(u_{pi} - u_i)$, $f_{ri} = \left(\dfrac{1}{\tau_{rp}} + \dfrac{\dot{m}_p}{m_p}\right)(u_i - u_{pi})$ express the fluctuat-
ing drag forces acting on the gas and particle phases, respectively, and U_j, U_{pj} are gas and particle instantaneous velocities, respectively, and $U_j = <U_j> + u_j$, $U_{pj} = <U_{pj}> + u_{pj}$. u_j and u_{pj} are gas and particle fluctuation velocities, respectively. Since the modeling of the particle phase is more important in the simulation of turbulent gas-particle flows, and solving PDF equations is more expensive, the PDF transport equation model can be applied only to the particle phase, and the k-ε model or Reynolds stress equation (DSM) model can still be used for the gas phase. Hence, k-ε-PDF and DSM-PDF two-phase turbulence models have been proposed by Li and Zhou [31,32]. Integrating Eq. (6.26) into the gas fluctuating velocity space, and using the gradient modeling for the term of fluctuating drag force, the closed form of particle-phase PDF transport equation of p_{fp} can be obtained as:

$$
\frac{\partial p_{fp}}{\partial t} + (<U_{pj}> + u_{pj})\frac{\partial p_{fp}}{\partial x_j} = \left(u_{pj}\frac{\partial <U_{pi}>}{\partial x_j}\right)\frac{\partial p_{fp}}{\partial v_{pi}} - \left(\frac{1}{\tau_{rp}} + \frac{\dot{m}_p}{m_p}\right)\frac{2c_p^k(k - k_p)}{1 + \rho/\rho_p}\frac{\partial^2 p_{fp}}{\partial v_{pi}^2}
$$

$$(6.27)$$

If p_{fp} is obtained by solving Eq. (6.27), then the particle Reynolds stress and turbulent kinetic energy can be obtained directly by the integration over PDF without solving their transport equations, that is:

$$<u_{pi}u_{pj}> = \left(\int p_{fpi}u_{pi}\mathrm{d}v_{pi} \right)\left(\int p_{fpj}u_{pj}\mathrm{d}v_{pj} \right) \tag{6.28}$$

$$k_p = \frac{1}{2}\sum_i <u_{pi}^2> = \frac{1}{2}\sum_i \left(\int p_{fpi}u_{pi}\mathrm{d}v_{pi} \right)^2 \tag{6.29}$$

Eqs. (6.27)–(6.29) constitute the PDF model of the particle phase. The DSM-PDF two-phase turbulence model, that is a combination of the second-order-moment model for gas-phase turbulence in two-phase flows with the PDF model for particle phase, proposed by Li and Zhou [32], is used to simulate swirling gas-particle flows. Figs. 6.22 and 6.23 show the predicted particle tangential time-averaged and RMS fluctuation velocities, respectively, by the DSM-PDF model and their comparison with experimental results and the k-ε-k_p modeling results. It can be seen that in case of weakly swirling flows, the DSM-PDF model cannot make many improvements for the results obtained using the k-ε-k_p model.

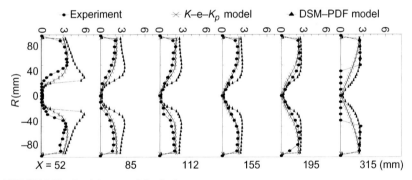

FIGURE 6.22   Particle tangential velocity.

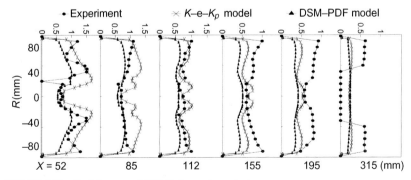

FIGURE 6.23   Particle tangential RMS fluctuation velocity.

6.9 AN SOM-MC MODEL OF SWIRLING GAS-PARTICLE FLOWS

An SOM-MC model proposed by Liu and Zhou et al. [33] is a combination of a gas-phase second-order moment (Reynolds stress equation) model with a Monte-Carlo method to solve the particle Lagrangian PDF equation. The stochastic equations of particle motion and the gas-eddy motion seen by particles in Lagrangian coordinate can be written as

$$d\tilde{x}_{pi}/dt = \tilde{u}_{pi} \tag{6.30}$$

$$d\tilde{u}_{pi}/dt = A_{pi}(\tilde{x}_{pi}, t) = g_i + (\tilde{u}_{gi,p} - \tilde{u}_{pi})/\tau_{rp} \tag{6.31}$$

$$d\tilde{u}_{gi}/dt = A_{gi}(\tilde{x}_{pi}, t) \tag{6.32}$$

The Langevin equation of gas-eddy motion, seen by particles is

$$d\tilde{u}_{gi}/dt = g_i - \frac{1}{\rho}\frac{\partial p}{\partial x_i} + \frac{\partial}{\partial x_j}\left[\nu \frac{\partial U_{gi}}{\partial x_j}\right] + (\tilde{u}_{pj} - \tilde{u}_{gj})\frac{\partial U_{gi}}{\partial x_j} + G_{gp,ij}(\tilde{u}_{gj} - U_{gj}) + B_{gp}^{1/2}\omega_i \tag{6.33}$$

where the first three terms on the right-hand side of Eq. (6.33) reflect the effect of the time-averaged gas velocity field on the stochastic motion of gas eddies, the fourth term is the additional force caused by the difference between the trajectories of particles and gas eddies, and the fifth and final terms express the effect of viscosity, fluctuating pressure, and particle motion, while the coefficient $G_{gp,ij}$ needs further consideration. $G_{gp,ij}$ can be modeled as

$$G_{gp,ij} = -\delta_{ij}/\tau_{Lp} \tag{6.34}$$

where τ_{Lp} is the Lagrangian integral time scale of gas seen by particles, which is different from τ_L, the Lagrangian integral time scale of gas itself. Considering the crossing-trajectory effect and continuity effect, $G_{gp,ij}$ can be written as follows

$$G_{gp,ij} = -\frac{1}{\tau_{Lp,\perp}}\delta_{ij} - \left(\frac{1}{\tau_{Lp,//}} - \frac{1}{\tau_{Lp,\perp}}\right)p_i p_j; \quad p_i = \frac{V_{r,i}}{|V_r|} \tag{6.35}$$

where $\tau_{Lp,//}$ and $\tau_{Lp,\perp}$ are the Lagrangian integral time scales of gas seen by particles parallel and perpendicular to the trajectories of particles, respectively. According to the experiments of particle dispersion in homogeneous turbulence, we have

$$\tau_{Lp,//} = \tau_L\left(1 + C_\beta \xi_r^2\right)^{-1/2}; \quad \tau_{Lp,//} = \tau_L\left(1 + 4C_\beta \xi_r^2\right)^{-1/2}; \quad \xi_r^2 = \frac{3}{2}\frac{|V_r|^2}{k_f} \tag{6.36}$$

where $C_\beta = 0.45$, and τ_L is the Lagrangian integral time scale of gas itself and is defined as

$$\tau_L = k/(\beta_1 \varepsilon) \quad (\beta_1 = 2.075)$$

For the gas-phase second-order moment (SOM) model in the SOM-MC two-phase turbulence model, the gas-phase Reynolds stress equation in two-phase flows in the USM model is used. The SOM-MC model is used to simulate swirling gas-particle flows. Figs. 6.24 and 6.25 show the predicted particle tangential time-averaged and RMS fluctuation velocities, respectively, by the SOM-MC model and their comparison with the experimental results and the USM modeling results. Obviously, there is only a slight difference between these two modeling results. Hence the PDF equation model validates the USM model.

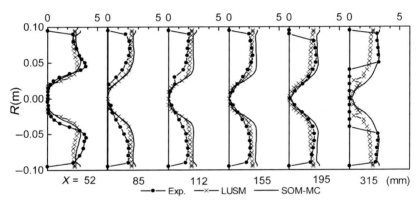

FIGURE 6.24 Particle tangential averaged velocity.

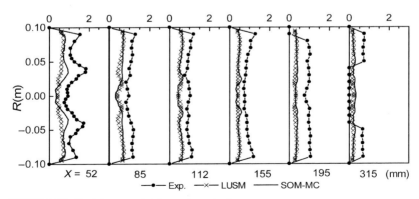

FIGURE 6.25 Particle tangential fluctuation velocity.

6.10 THE NONLINEAR $K-\varepsilon-K_p$ TWO-PHASE TURBULENCE MODEL

It has been found that the conventional or linear $k-\varepsilon-k_p$ model is rather simple and can simulate nonswirling and weakly swirling gas-particle flows well. However, for strongly swirling flows the USM model should be better, however it is rather complex and is not convenient for engineering applications. The best compromise between reasonableness and simplicity is either an implicit algebraic two-phase Reynolds stress model, or a nonlinear $k-\varepsilon-k_p$ two-phase turbulence model, i.e., an explicit algebraic two-phase Reynolds stress model. Since the algebraic Reynolds stress models frequently cause some divergence problem due to lack of diffusion terms in the momentum equation, particularly in 3-D flows, a nonlinear $k-\varepsilon-k_p$ two-phase turbulence model has been developed by Zhou and Gu [34]. This is because the momentum equations and k, ε, k_p equations have the same form for both linear and nonlinear $k-\varepsilon-k_p$ models, making it easier to obtain the convergent results. When using the full second-order moment model for three-dimensional flows, we should solve 26 differential equations, including six gas Reynolds stress equations, six particle Reynolds stress equations, nine two-phase velocity correlation equations, one dissipation-rate equation for gas turbulent kinetic energy, three particle diffusion mass-flux equations, and one equation for the mean square value of particle number density fluctuation.

In order to reduce the computation time and simultaneously to retain the anisotropic features of the turbulence model, as is done in single-phase turbulence models, an algebraic two-phase stress model is obtained by simplifying the stress transport equations. Neglecting the convection and diffusion terns in the two-phase Reynolds stress and two-phase velocity correlation equations of the USM model, the algebraic expressions of two-phase Reynolds stresses and the two-phase velocity correlation can be obtained as:

$$\overline{v_i v_j} = (1-\lambda)\frac{2}{3}k\delta_{ij} + \lambda\frac{k}{\varepsilon}\left(\overline{v_i v_k}\frac{\partial V_j}{\partial x_k} + \overline{v_j v_k}\frac{\partial V_i}{\partial x_k}\right) + \frac{k}{c_1\rho\varepsilon}\left(\overline{v_{pi}v_j} + \overline{v_i v_{pj}} - 2\overline{v_i v_j}\right)$$

(6.37)

$$\overline{v_{pi}v_{pj}} = -\frac{\tau_{rp}}{2}\left(\overline{v_{pi}v_{pk}}\frac{\partial V_{pj}}{\partial x_k} + \overline{v_{pj}v_{pk}}\frac{\partial V_{pi}}{\partial x_k}\right) + \frac{1}{2}\left(\overline{v_i v_{pj}} + \overline{v_{pi}v_j}\right)$$

(6.38)

$$\overline{v_{pi}v_j} = -\frac{\rho\tau_{rp}}{\rho+\rho_p}\left(\overline{v_{pi}v_k}\frac{\partial V_j}{\partial x_k} + \overline{v_j v_{pk}}\frac{\partial V_{pi}}{\partial x_k}\right) + \frac{\rho}{\rho+\rho_p}\overline{v_i v_j} + \frac{\rho_p}{\rho+\rho_p}\overline{v_{pi}v_{pj}} - \frac{\rho\tau_{rp}}{\rho+\rho_p}\frac{1}{\tau_e}\overline{v_{pi}v_i}\delta_{ij}$$

(6.39)

To construct a nonlinear $k-\varepsilon-k_p$ two-phase turbulence model, we can transform Eqs. (6.37), (6.38), and (6.39) into the following explicit form, on the right-hand side of which there are no terms containing $\overline{v_{pi}v_j}$, $\overline{v_{pi}v_{pj}}$, or

$\overline{v_i v_j}$. The obtained nonlinear stress−strain-rate relationships written to quadratic-power terms of the strain rates are:

$$
\overline{v_i v_j} = G_1 \delta_{ij} + G_2 \left(\frac{\partial V_i}{\partial x_j} + \frac{\partial V_j}{\partial x_i} \right) + G_3 \left(\frac{\partial V_{pi}}{\partial x_j} + \frac{\partial V_{pj}}{\partial x_i} \right)
$$

$$
+ G_4 \left(\frac{\partial V_i}{\partial x_k} \left(\frac{\partial V_j}{\partial x_k} + \frac{\partial V_k}{\partial x_j} \right) + \frac{\partial V_j}{\partial x_k} \left(\frac{\partial V_i}{\partial x_k} + \frac{\partial V_k}{\partial x_i} \right) \right)
$$

$$
+ G_5 \left(\frac{\partial V_{pi}}{\partial x_k} \left(\frac{\partial V_{pj}}{\partial x_k} + \frac{\partial V_{pk}}{\partial x_j} \right) + \frac{\partial V_{pj}}{\partial x_k} \left(\frac{\partial V_{pi}}{\partial x_k} + \frac{\partial V_{pk}}{\partial x_i} \right) \right)
$$

$$
+ G_6 \left(\frac{\partial V_i}{\partial x_k} \left(\frac{\partial V_{pj}}{\partial x_k} + \frac{\partial V_{pk}}{\partial x_j} \right) + \frac{\partial V_j}{\partial x_k} \left(\frac{\partial V_{pi}}{\partial x_k} + \frac{\partial V_{pk}}{\partial x_i} \right) \right)
$$

$$
+ G_7 \left(\frac{\partial V_{pi}}{\partial x_k} \left(\frac{\partial V_j}{\partial x_k} + \frac{\partial V_k}{\partial x_j} \right) + \frac{\partial V_{pj}}{\partial x_k} \left(\frac{\partial V_i}{\partial x_k} + \frac{\partial V_k}{\partial x_i} \right) \right)
$$

$$
+ G_8 \left(\left(\frac{\partial V_i}{\partial x_k} - \frac{\partial V_{pi}}{\partial x_k} \right) \left(\frac{\partial V_j}{\partial x_k} - \frac{\partial V_{pj}}{\partial x_k} \right) \right)
$$

(6.40)

$$
\overline{v_{pi} v_{pj}} = P_1 \delta_{ij} + P_2 \left(\frac{\partial V_i}{\partial x_j} + \frac{\partial V_j}{\partial x_i} \right) + P_3 \left(\frac{\partial V_{pi}}{\partial x_j} + \frac{\partial V_{pj}}{\partial x_i} \right)
$$

$$
+ P_4 \left(\frac{\partial V_i}{\partial x_k} \left(\frac{\partial V_j}{\partial x_k} + \frac{\partial V_k}{\partial x_j} \right) + \frac{\partial V_j}{\partial x_k} \left(\frac{\partial V_i}{\partial x_k} + \frac{\partial V_k}{\partial x_i} \right) \right)
$$

$$
+ P_5 \left(\frac{\partial V_{pi}}{\partial x_k} \left(\frac{\partial V_{pj}}{\partial x_k} + \frac{\partial V_{pk}}{\partial x_j} \right) + \frac{\partial V_{pj}}{\partial x_k} \left(\frac{\partial V_{pi}}{\partial x_k} + \frac{\partial V_{pk}}{\partial x_i} \right) \right)
$$

$$
+ P_6 \left(\frac{\partial V_i}{\partial x_k} \left(\frac{\partial V_{pj}}{\partial x_k} + \frac{\partial V_{pk}}{\partial x_j} \right) + \frac{\partial V_j}{\partial x_k} \left(\frac{\partial V_{pi}}{\partial x_k} + \frac{\partial V_{pk}}{\partial x_i} \right) \right)
$$

$$
+ P_7 \left(\frac{\partial V_{pi}}{\partial x_k} \left(\frac{\partial V_j}{\partial x_k} + \frac{\partial V_k}{\partial x_j} \right) + \frac{\partial V_{pj}}{\partial x_k} \left(\frac{\partial V_i}{\partial x_k} + \frac{\partial V_k}{\partial x_i} \right) \right)
$$

$$
+ P_8 \left(\left(\frac{\partial V_i}{\partial x_k} - \frac{\partial V_{pi}}{\partial x_k} \right) \left(\frac{\partial V_j}{\partial x_k} - \frac{\partial V_{pj}}{\partial x_k} \right) \right)
$$

(6.41)

$$\overline{v_{pi}v_j} = T_1\delta_{ij} + T_2\left(\frac{\partial V_i}{\partial x_j} + \frac{\partial V_j}{\partial x_i}\right) + T_3\left(\frac{\partial V_{pi}}{\partial x_j} + \frac{\partial V_{pj}}{\partial x_i}\right) + T_4\left(\frac{\partial V_j}{\partial x_i} + \frac{\partial V_{pi}}{\partial x_j}\right)$$

$$+ T_5\frac{\partial V_i}{\partial x_k}\left(\frac{\partial V_j}{\partial x_k} + \frac{\partial V_k}{\partial x_j}\right) + T_6\frac{\partial V_j}{\partial x_k}\left(\frac{\partial V_i}{\partial x_k} + \frac{\partial V_k}{\partial x_i}\right)$$

$$+ T_7\frac{\partial V_{pi}}{\partial x_k}\left(\frac{\partial V_{pj}}{\partial x_k} + \frac{\partial V_{pk}}{\partial x_j}\right) + T_8\frac{\partial V_{pj}}{\partial x_k}\left(\frac{\partial V_{pi}}{\partial x_k} + \frac{\partial V_{pk}}{\partial x_i}\right)$$

$$+ T_9\frac{\partial V_{pi}}{\partial x_k}\left(\frac{\partial V_j}{\partial x_k} + \frac{\partial V_k}{\partial x_j}\right) + T_{10}\frac{\partial V_{pj}}{\partial x_k}\left(\frac{\partial V_i}{\partial x_k} + \frac{\partial V_k}{\partial x_i}\right)$$

$$+ T_{11}\frac{\partial V_i}{\partial x_k}\left(\frac{\partial V_{pj}}{\partial x_k} + \frac{\partial V_{pk}}{\partial x_j}\right) + T_{12}\frac{\partial V_j}{\partial x_k}\left(\frac{\partial V_{pi}}{\partial x_k} + \frac{\partial V_{pk}}{\partial x_i}\right)$$

$$+ T_{13}\left(\frac{\partial V_i}{\partial x_k}\left(\frac{\partial V_{pj}}{\partial x_k} - \frac{\partial V_j}{\partial x_k}\right) + \frac{\partial V_{pj}}{\partial x_k}\left(\frac{\partial V_i}{\partial x_k} - \frac{\partial V_{pi}}{\partial x_k}\right)\right)$$

$$+ T_{14}\left(\frac{\partial V_j}{\partial x_k}\left(\frac{\partial V_{pi}}{\partial x_k} - \frac{\partial V_i}{\partial x_k}\right) + \frac{\partial V_{pi}}{\partial x_k}\left(\frac{\partial V_j}{\partial x_k} - \frac{\partial V_{pj}}{\partial x_k}\right)\right)$$

$$(6.42)$$

where all of the coefficients $G1-G8$, $P1-P8$, and $T1-T14$ are functions of k, ε, k_p, k_{pg}, ρ_p, ρ, and τ_{rp}.

The variables k, ε, k_p, and k_{pg} are determined by the following governing equations

$$\frac{\partial}{\partial t}(\rho k) + \frac{\partial}{\partial x_j}(\rho V_j k) = \frac{\partial}{\partial x_j}\left(\frac{\mu_e}{\sigma_k}\frac{\partial k}{\partial x_j}\right) - \rho\overline{v_i v_k}\frac{\partial v_i}{\partial x_k} - \rho\varepsilon + \frac{\rho_p}{\tau_{rp}}(2k_{pg} - 2k)$$

$$(6.43)$$

$$\frac{\partial}{\partial t}(\rho\varepsilon) + \frac{\partial}{\partial x_k}(\rho V_k\varepsilon) = \frac{\partial}{\partial x_k}\left(\rho c_\varepsilon\frac{k}{\varepsilon}\overline{v_k v_l}\frac{\partial \varepsilon}{\partial x_l}\right)$$

$$+ \frac{\varepsilon}{k}\left(c_{\varepsilon 1}\left(-\rho\overline{v_i v_k}\frac{\partial v_i}{\partial x_k} + \frac{\rho_p}{\tau_{rp}}(2k_{pg} - 2k)\right) - c_{\varepsilon 2}\rho\varepsilon\right)$$

$$(6.44)$$

$$\frac{\partial}{\partial t}\left(\rho_p k_p\right) + \frac{\partial}{\partial x_k}\left(\rho_p V_{pk}k_p\right) = \frac{\partial}{\partial x_k}\left(\rho_p c_{kp}\frac{k_p}{\varepsilon_p}\overline{v_{pk}v_{pl}}\frac{\partial k_p}{\partial x_l}\right) - \rho_p\overline{v_{pi}v_{pk}}\frac{\partial v_{pi}}{\partial x_k}$$

$$+ \frac{\rho_p}{\tau_{rp}}(2k_{pg} - 2k_p)$$

$$(6.45)$$

$$\frac{\partial}{\partial t}\left(k_{pg}\right) + \left(V_k + V_{pk}\right)\frac{\partial}{\partial x_k}\left(k_{pg}\right) = \frac{\partial}{\partial x_k}\left(\left(c_s\frac{k}{\varepsilon}\overline{v_k v_l} + c_{kp}\frac{k_p}{\varepsilon_p}\overline{v_{pk}v_{pl}}\right)\frac{\partial k_{pg}}{\partial x_l}\right)$$

$$+ \frac{1}{\rho\tau_{rp}}\left(\rho_p k_p + \rho k - (\rho + \rho_p)k_{pg}\right)$$

$$-\frac{1}{2}\left(\overline{v_i v_{pk}}\frac{\partial v_{pi}}{\partial x_k} + \overline{v_{pi}v_k}\frac{\partial v_i}{\partial x_k}\right) - \frac{1}{\tau_e}k_{pg}$$

$$(6.46)$$

The nonlinear $k-\varepsilon-k_p$ (NKP) model was used to simulate swirling gas-particle flows with a swirl number of 0.47 and was compared with the USM model. Figs. 6.26 and 6.27 show the predicted particle tangential time-averaged and RMS fluctuation velocities by the NKP and USM models, respectively, and their comparison with the experimental results. It can be seen that in most regions of the flow field, the difference between two model predictions for the time-averaged velocity is small and both are in agreement with experiments. In general, the NKP model can predict what the USM model can predict, but the former can save almost 50% computational time for a 2-D flow with small geometrical sizes. It is expected that for 3-D flows

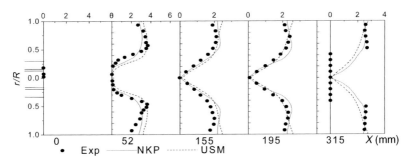

FIGURE 6.26 Tangential velocity of 45-μm particles (m/s, $s = 0.47$).

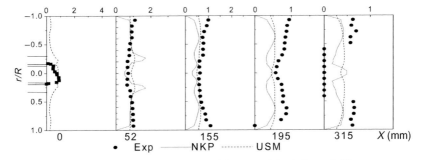

FIGURE 6.27 Tangential fluctuation velocity of 45-μm particles (m/s, $s = 0.47$).

with large geometrical sizes, the NKP model can save much more computational time. Keeping in mind that in engineering applications the accuracy of predicting the two-phase averaged velocities is more important, the NKP model can be considered for use instead of the USM model.

6.11 THE KINETIC THEORY MODELING OF DENSE PARTICLE (GRANULAR) FLOWS

Dense gas-particle flows are encountered in dense pneumatic conveying, dense cyclone separator, risers and downers of circulating fluidized beds, pulverized-coal-injected blast furnaces. Particle dispersion in dense gas-particle flows is dominated by both large-scale fluctuation due to turbulence and small-scale fluctuation due to interparticle collisions. To simulate dense gas-particle flows, both Eulerian–Eulerian (two-fluid) and Eulerian–Lagrangian approaches are adopted. The former needs models of particle-collisional pressure and shear stress and particle turbulence, and the latter needs particle collision models, known as hard-sphere and soft-sphere models. In the E–E approach, the particles are simulated using a pseudo-fluid model. The advantage of the E–E approach is its ability to simulate large-scale engineering processes with acceptable computational requirements, but the main problem is its complex closure models, which need further improvement and experimental validation. In the E–E approach, the kinetic theory of dense gas-particle flows proposed by Gidaspow [35] was widely adopted in the 1990s to simulate particle–particle collision without accounting for the effect of gas-particle turbulence. For high-velocity dense gas-particle flows in risers, downers, and channels, besides gas turbulence, account is necessarily taken for both small-scale particle fluctuations due to particle–particle collision and large-scale particle fluctuations due to particle turbulence. As mentioned earlier, Zhou proposed a $k-\varepsilon-k_p$ model and a unified second-order moment (USM) model to simulate particle turbulence for dilute gas–particle flows. In recent years, different investigators have developed two-phase turbulence models for simulating dense gas-particle flows. Zheng et al. [36] proposed a $k-\varepsilon-k_p-\varepsilon_p-\Theta$ (Θ—particle pseudo-temperature) model.

Gidaspow used a kinetic theory model of granular flows for two-fluid modeling of fluidized beds. In the frame of two-fluid modeling, the particle momentum equation is

$$\frac{\partial}{\partial t}(\alpha_p \rho_p \overline{v_{pi}}) + \frac{\partial}{\partial x_j}(\alpha_p \rho_p \overline{v_{pi} v_{pj}}) = -\frac{\partial p_p}{\partial x_j} - \frac{\partial \tau_{p,ij}}{\partial x_j} + \frac{\alpha_p \rho_p}{\tau_r}(\overline{v_{gi}} - \overline{v_{pi}}) \quad (6.47)$$

When neglecting the particle large-scale fluctuation (particle turbulence) and using the kinetic theory, the particle shear stress due to particle collision is determined by

$$\tau_p = \left\{ -\alpha_p \rho_p \theta \left[1 + 2(1+e)g_0 \alpha_p \right] + \alpha_p \xi_p \nabla \cdot \overline{v_p} \right\} \delta_{ij} - 2\alpha_p \mu_p \overline{S}_p \quad (6.48)$$

In Eq. (6.48), the so-called "particle pseudo-temperature" θ is given by a transport equation as

$$\frac{3}{2}\left[\frac{\partial}{\partial t}\left(\alpha_p \rho_p \theta\right) + \nabla\cdot\left(\alpha_p \rho_p \theta \overline{v}_p\right)\right] = \nabla\cdot\left(\kappa_p \nabla\theta\right) - \gamma_p - 3\beta\theta \tag{6.49}$$

The particle collision pressure is

$$p_p = \varepsilon_p \rho_p \theta\left[1 + 2(1+e)g_0\alpha_p\right] \tag{6.50}$$

The particle dynamic viscosity μ_p is

$$\mu_p = \frac{4}{5}\alpha_p^2\rho_p d_p g_o(1+e)\sqrt{\frac{\theta}{\pi}} + \frac{10\rho_p d_p\sqrt{\pi\theta}}{96(1+e)\alpha_p g_o}\left[1 + \frac{4}{5}g_o\alpha_p(1+e)\right]^2 \tag{6.51}$$

The particle apparent viscosity ξ_p is

$$\xi_p = \frac{4}{5}\alpha_p\rho_p d_p g_0(1+e)\left(\frac{\theta}{\pi}\right)^{\frac{1}{2}} \tag{6.52}$$

The radial distribution function g_0 is

$$g_0 = \left[1 - \left(\frac{\alpha_p}{\alpha_{pmax}}\right)^{\frac{1}{3}}\right]^{-1} \tag{6.53}$$

The transport coefficient k_p is

$$k_p = 2\alpha_p^2\rho_p d_p g_0(1+e)\left(\frac{\theta}{\pi}\right)^{\frac{1}{2}} \tag{6.54}$$

The collision dissipation rate γ_p is

$$\gamma_p = 3(1-e^2)\alpha_p^2\rho_p g_0\theta\left[\frac{4}{d_p}\left(\frac{\theta}{\pi}\right)^{\frac{1}{2}} - \nabla\cdot\overline{v}_p\right] \tag{6.55}$$

Goldschmidt et al. [37] used the kinetic theory to simulate a bubbling fluidized bed. The simulation results give the shapes and sizes of gas bubbles, showing that the coefficient of restitution significantly affects the predicted bubble shapes and sizes. The predicted particle volume fraction and velocity have no experimental validation. Mineto et al. [38] also used the kinetic theory to simulate a bubbling fluidized bed. The simulated gas bubbles for different boundary conditions and different θ equations are obtained. There is no experimental validation of the predicted time-averaged gas and particle velocities. Cheng et al. [39] used a $k-\varepsilon-k_p-\Theta$ model by combining the $k-\varepsilon-k_p$ model for dilute gas-particle flows with the Θ model of kinetic theory to simulate downer flows in a circulating fluidized bed. Fig. 6.28 gives

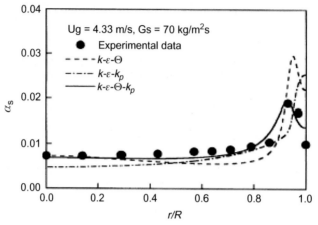

FIGURE 6.28 Particle volume fraction.

the predicted particle volume fraction. It can be seen that the modeling results using the $k-\varepsilon-k_p-\Theta$ model are in agreement with the experimental results, whereas the prediction results using the $k-\varepsilon-\Theta$ model (gas $k-\varepsilon$ turbulence model combined with the particle Θ model) and only the $k-\varepsilon-k_p$ model are not in agreement with the experimental results. It points out that for dense gas-particle flows in circulating fluidized beds beside the gas turbulence and interparticle collision the particle turbulence (large-scale fluctuations) should be taken into account.

The kinetic theory model of granular flows was used to simulate risers of a circulating fluidized bed by Neri and Gidaspow [40], not accounting for gas and particle turbulence. The predicted particle volume fraction and particle velocity were not in agreement with the experimental results. It is interesting to note that, recently, the kinetic theory model of granular flows was used together with a Smagorinsky subgrid scale model of gas turbulence to simulate both dense gas-particle flows in a riser by Lu and Gidaspow et al. [41] and dilute gas-particle flows in a backward-facing step by Liu and Zhou et al. [42]. Although these simulations are called "large-eddy simulations," due to the simulations being two-dimensional and the grid sizes being too coarse, the simulations are in fact unsteady RANS (URANS) modeling. Fig. 6.29 shows the predicted particle RMS fluctuation velocities in a backward-facing step flow. Both these URANS modeling results and the results using the steady USM two-phase turbulence model are in good agreement with the experimental results. These results indicate that unsteady simulation is important, even where the gas turbulence model is very simple (like a mixing-length model) and the particle turbulence is not taken into account.

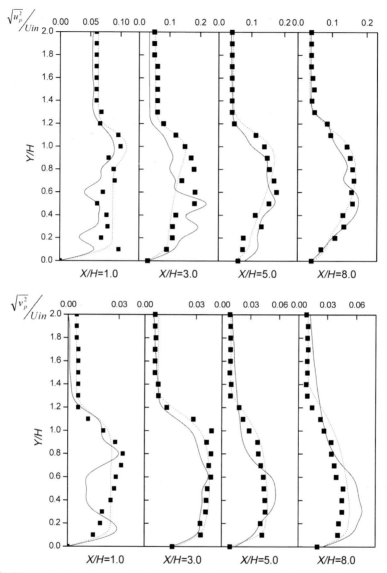

FIGURE 6.29 Particle RMS fluctuation velocities in a backward-facing step (left—longitudinal; right—transverse) (m/s; ■ Exp.; — URANS; ... USM).

6.12 TWO-PHASE TURBULENCE MODELS FOR DENSE GAS-PARTICLE FLOWS

In dense gas-particle flows with higher velocities, there are both large-scale particle fluctuations due to particle turbulence and small-scale particle fluctuations due to interparticle collisions. A USM-Θ two-phase turbulence model for dense gas-particle flows was proposed by Yu and Zhou et al. [43].

In this model the gas turbulence and particle large-scale fluctuation are predicted using the USM two-phase turbulence model, and the particle small-scale fluctuation due to interparticle collisions is predicted using the particle pseudo-temperature equation—Θ equation, given by Gidaspow's kinetic theory. It is not a simple superposition, since there are interaction terms in the particle Reynolds stress equations and the Θ equation. Some of the closed USM-Θ model equations include the following.

The gas Reynolds stress equation

$$\frac{\partial\left(\overline{\alpha_g}\rho_{gm}\overline{u_{gi}u_{gj}}\right)}{\partial t} + \frac{\partial\left(\overline{\alpha_g}\rho_{gm}U_{gk}\overline{u_{gi}u_{gj}}\right)}{\partial x_k} = D_{g,ij} + P_{g,ij} + \Pi_{g,ij} - \varepsilon_{g,ij} + G_{g,gp,ij}$$

(6.56)

where $G_{g,gp,ij} = \beta\left(\overline{u_{pi}u_{gj}} + \overline{u_{gi}u_{pj}} - 2\overline{u_{gi}u_{gj}}\right)$

The particle Reynolds stress equation

$$\frac{\partial\left(\overline{\alpha_p}\rho_{pm}\overline{u_{pi}u_{pj}}\right)}{\partial t} + \frac{\partial\left(\overline{\alpha_p}\rho_{pm}U_{pk}\overline{u_{pi}u_{pj}}\right)}{\partial x_k} = D_{p,ij} + P_{p,ij} + \Pi_{p,ij} - \varepsilon_{p,ij} + G_{p,gp,ij}$$

(6.57)

where $G_{p,gp,ij} = \beta\left(\overline{u_{pi}u_{gj}} + \overline{u_{pj}u_{gi}} - 2\overline{u_{pi}u_{pj}}\right)$

The equations of dissipation rate of turbulent kinetic energy for gas and particle phases:

$$\frac{\partial\left(\overline{\alpha_g}\rho_{gm}\varepsilon_g\right)}{\partial t} + \frac{\partial\left(\overline{\alpha_g}\rho_{gm}U_{gk}\varepsilon_g\right)}{\partial x_k} = \frac{\partial}{\partial x_k}\left(C_g\overline{\alpha_g}\rho_{gm}\frac{k_g}{\varepsilon_g}\overline{u_{gk}u_{gl}}\frac{\partial\varepsilon_g}{\partial x_l}\right)$$
$$+ \frac{\varepsilon_g}{k_g}\left[c_{\varepsilon1}\left(P_g + G_{g,gp}\right) - c_{\varepsilon2}\overline{\alpha_g}\rho_{gm}\varepsilon_g\right]$$

(6.58)

where $G_{g,gp} = 2\beta\left(k_{gp} - k_g\right)$, $c_{\varepsilon3} = 1.8$

$$\frac{\partial\left(\overline{\alpha_p}\rho_{pm}\varepsilon_p\right)}{\partial t} + \frac{\partial\left(\overline{\alpha_p}\rho_{pm}U_{pk}\varepsilon_p\right)}{\partial x_k} = \frac{\partial}{\partial x_k}\left(\overline{\alpha_p}\rho_{pm}C_p^d\frac{k_p}{\varepsilon_p}\overline{u_{pk}u_{pl}}\frac{\partial\varepsilon_p}{\partial x_l}\right)$$
$$+ \frac{\varepsilon_p}{k_p}\left[C_{\varepsilon p,1}\left(P_p + G_{p,gp}\right) - C_{\varepsilon p,2}\overline{\alpha_p}\rho_{pm}\varepsilon_p\right]$$

(6.59)

where β is the inverse relaxation time, $G_{p,gp} = 2\beta\left(k_{pg} - k_p\right)$

The two-phase velocity correlation equation:

$$\frac{\partial\overline{u_{pi}u_{gj}}}{\partial t} + \left(U_{gk} + U_{pk}\right)\frac{\partial\overline{u_{pi}u_{gj}}}{\partial x_k} = D_{g,p,ij} + P_{g,p,ij} + \Pi_{g,p,ij} - \varepsilon_{g,p,ij} + T_{g,p,ij}$$

(6.60)

The particle pseudo-temperature transport equation:

$$
\frac{3}{2}\left[\frac{\partial\left(\overline{\alpha_p}\,\rho_{pm}\Theta\right)}{\partial t} + \frac{\partial\left(\overline{\alpha_p}\,\rho_{pm}U_{pk}\tilde{u}_{pk}\Theta\right)}{\partial x_k}\right] = -\frac{\partial}{\partial x_k}\left(\frac{3}{2}\overline{\alpha_p\rho_{pm}u_{pk}\theta} + \Gamma_\Theta\frac{\partial\Theta}{\partial x_k}\right)
$$

$$
+ \mu_p\left(\frac{\partial U_{pk}}{\partial x_i} + \frac{\partial U_{pi}}{\partial x_k}\right)\frac{\partial U_{pi}}{\partial x_k} + \mu_p\varepsilon_p
$$

$$
- P_p\frac{\partial U_{pl}}{\partial x_l} + \left(\xi_p - \frac{2}{3}\mu_p\right)\left(\frac{\partial U_{pl}}{\partial x_l}\right)^2 - \overline{\gamma}
$$

$$(6.61)$$

where the notations in Eq. (6.61) are the same as that given by Gidaspow. The interaction between the large-scale and small-scale particle fluctuations is the third term on the right-hand-side of Eq. (6.61), expressing the effect of the dissipation rate of particle turbulent kinetic energy on the particle pseudo-temperature. Simulation results for dense gas-particle flows in a downer indicate that for predicting the particle volume fraction (Fig. 6.30) and particle velocity (Fig. 6.31) the USM-Θ model is much better than the DSM-Θ model, not accounting for particle turbulence, the USM model, not accounting for interparticle collision and the $k-\varepsilon-k_p-\Theta$ model, not accounting for the anisotropy of turbulence. Fig. 6.32 shows the predicted particle horizontal RMS fluctuation velocity for horizontal gas-particle pipe flows measured in experiments. It can be seen that the USM-Θ model is better than other models.

6.13   THE EULERIAN−LAGRANGIAN SIMULATION OF GAS-PARTICLE FLOWS

In the Eulerian−Lagrangian simulation the particles are treated as a dispersed system, and are divided into several size groups. Each particle size

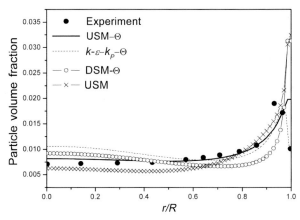

FIGURE 6.30   Particle volume fraction.

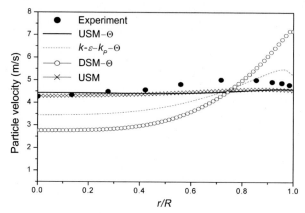

FIGURE 6.31 Particle time-averaged velocity.

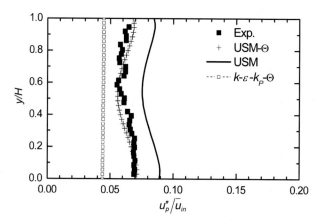

FIGURE 6.32 Particle horizontal RMS fluctuation velocity.

group moves, starting from their initial position, and has the same size, velocity, and temperature at any point in time. The particle velocity, temperature, and mass change are tracked along their trajectories. The effects of particles on the gas mass, momentum, and energy are expressed as source terms in the gas computation cell. The algorithm in the numerical procedure is called the "particle-source-in cell" (PSIC) method, which is discussed in Chapter 8.

6.13.1 Governing Equations for the Deterministic Trajectory Model

In the early stages of developing the trajectory model, i.e., at beginning of the 1980s, Crowe et al. proposed the deterministic trajectory model [44], not accounting for the effect of gas turbulence on the particle motion. In the mid-1980s, Gosman et al. proposed the stochastic trajectory model [45], accounting for the particle diffusion/dispersion caused by gas fluctuation. For the deterministic trajectory model, the particle conservation equations correspond to the time-averaged equations with zero correlation terms, as described here.

Particle continuity equation

$$\frac{\partial \rho_k}{\partial t} + \frac{\partial}{\partial x_j}(\rho_k v_{kj}) = S_k \tag{6.62}$$

Particle momentum equation

$$\frac{\partial}{\partial t}(\rho_k v_{ki}) + \frac{\partial}{\partial x_j}(\rho_k v_{kj} v_{ki}) = \rho_k g_i + \frac{\rho_k}{\tau_{rk}}(v_i - v_{ki}) + v_i S_k \tag{6.63}$$

Particle energy equation

$$\frac{\partial}{\partial t}(\rho_k c_k T_k) + \frac{\partial}{\partial x_j}(\rho_k v_{kj} c_k T_k) = n_k(Q_h - Q_k - Q_{rk}) + c_p T S_k \tag{6.64}$$

where S_k is source term due to particle evaporation/devolatilization/reaction

$$S = -\sum_k S_k = -\sum_k n_k \dot{m}_k \quad \dot{m}_k = \frac{dm_k}{dt}$$

These equations can be written in the form of a Lagrangian coordinate as

$$\int_A n_k v_{kn} dA = N_k = \text{const}$$

$$\frac{dv_{ki}}{dt_k} = \left(\frac{1}{\tau_{rk}} + \frac{\dot{m}_k}{m_k}\right)(v_i - v_{ki}) + g_i \tag{6.65}$$

$$\frac{dT_k}{dt_k} = \left[Q_h - Q_k - Q_{rk} + \dot{m}_k(c_p T - c_k T_k)\right] / (m_k c_k)$$

6.13.2 Modification for Particle Turbulent Diffusion

In the original particle deterministic trajectory model the particle turbulent diffusion is not taken into account. In fact, the particle turbulent diffusion cannot be neglected. The simplest method is to introduce the concepts of "particle drift velocity" by Smoot et al. [46] or "particle drift force" by Lockwood et al. [47]. For example, it is assumed that the particle velocity is composed of two parts

$$v_{kj} = v_{kc,j} + v_{kd,j} \tag{6.66}$$

where $v_{kc,j}$ is the particle convective velocity, determined by the particle momentum equation $v_{kd,j}$ is the particle drift velocity due to turbulent diffusion, determined by the Fick Law

$$-\rho_k v_{kd,j} = -n_k m_k v_{kd,j} = D_k m_k \frac{\partial n_k}{\partial x_j} \tag{6.67}$$

or

$$v_{kd,j} = -\frac{D_k}{n_k} \frac{\partial n_k}{\partial x_j} \tag{6.68}$$

It is seen that this is actually a concept of two-fluid modeling. The particle continuity equation in two-fluid modeling is

$$\frac{\partial n_k}{\partial t} + \frac{\partial}{\partial x_j}(n_k v_{kj}) = -\frac{\partial}{\partial x_j}(\overline{n_k' v_{kj}'})$$

If we take a gradient modeling, then we have

$$-\overline{n_k' v_{kj}'} = -n_k v_{kd,j} = D_k \frac{\partial n_k}{\partial x_j}$$

Hence Eq. (6.68) is obtained. The particle continuity equation becomes

$$\frac{\partial n_k}{\partial t} + \frac{\partial}{\partial x_j}(n_k v_{kj}) = \frac{\partial}{\partial x_j}\left(D_k \frac{\partial n_k}{\partial x_j}\right) \tag{6.69}$$

If the total particle velocity is taken as

$$v_{kj,0} = v_{kj} + v_{kd,j} \tag{6.70}$$

then we have

$$\frac{\partial n_k}{\partial t} + \frac{\partial}{\partial x_j}(n_k v_{kj,0}) = 0 \tag{6.71}$$

These equations imply that the concept of particle drift velocity is based purely on the particle diffusion source term in two-fluid modeling. However, it is not introduced in the particle momentum equation. In order to find the particle drift velocity, the particle diffusivity D_k and the particle number density gradient $\partial n_k/\partial x_j$ need to be known. These two quantities cannot be given by the particle trajectory model. Therefore, it has to be found using two-fluid modeling. The particle diffusivity is given by the Hinze−Tchen model

$$\nu_p/\nu_T = D_p/D_T = (k_p/k)^2 = (1 + \tau_{r1}/\tau_T)^{-1}$$

The particle number density is obtained by solving a particle continuity equation of the no-slip or single-fluid model, i.e., assuming the particle velocity is equal to the gas velocity

$$\frac{\partial n_k}{\partial t} + \frac{\partial}{\partial x_j}(n_k v_j) = \frac{\partial}{\partial x_j}\left(\frac{\nu_T}{\sigma_{kp}} \frac{\partial n_k}{\partial x_j}\right) \tag{6.72}$$

This "particle drift velocity modification" was introduced in the computer codes PCGC-2 and PCGC-3 for simulation of coal combustion, developed in the Advanced Combustion Engineering Research Center at Brigham Young University.

6.13.3 The Stochastic Trajectory Model

The above-stated modification for the trajectory model can correct only the position of the trajectories by particle turbulent diffusion, but cannot give the continuous distribution of the particle velocity and concentration. Furthermore, this modification uses the concept of two-fluid modeling, and is not a pure Lagrangian approach. In the 1980s, Gosman et al. [45] proposed a stochastic trajectory model to account for the particle diffusion due to the effect of gas turbulence. This model actually is a semidirect numerical simulation, i.e., for the gas phase the Eulerian macroscopic turbulence modeling (Reynolds averaged modeling) is still used; for the particle phase it is a direct numerical simulation. The stochastic trajectory model is based on the instantaneous particle momentum equation

$$\frac{du_p}{dt} = (\bar{u} + u' - u_p)/\tau_r \tag{6.73}$$

$$\frac{dv_p}{dt} = (\bar{v} + v' - v_p)/\tau_r + w_p^2/r_p + g \tag{6.74}$$

$$\frac{dw_p}{dt} = (\bar{w} + w' - w_p)/\tau_r - v_p w_p/r_p \tag{6.75}$$

where u_p, v_p, w_p are the instantaneous particle axial, radial, and tangential velocities, respectively, and $\bar{u}, \bar{v}, \bar{w}$ and $u', v', w'$are the gas time-averaged and fluctuation velocities in three directions, respectively. Assuming the gas turbulence is isotropic and locally homogeneous, the gas stochastic velocity obeys the Gaussian PDF distribution, and the gas fluctuation velocity can be randomly sampled as

$$u' = \varsigma\left(\overline{u'^2}\right)^{1/2}, v' = \varsigma\left(\overline{v'^2}\right)^{1/2}, w' = \varsigma\left(\overline{w'^2}\right)^{1/2}, (\varsigma = 0, 1, 2, 3, \ldots), \overline{u'^2} = \overline{v'^2} = \overline{w'^2} = \frac{2}{3}k$$

where ς is a random number. Substituting the random gas fluctuation velocities into Eqs. (6.73)−(6.75), the particle instantaneous velocities can be found and the trajectories can be obtained as

$$x_p = \int u_p dt, \quad r_p = \int v_p dt, \quad \theta_p = \int (w_p/r_p) dt$$

The particle-gas eddy interaction time is taken as $\tau_e = \min[\tau_r, \tau_T]$. In computations, the particle stochastic trajectories are solved using the

Monte-Carlo method. Frequently, more than tens of thousands of trajectories are needed and the computation time can be very large. The particle trajectory model does not need the particle turbulence model, but needs the gas instantaneous velocity model. The above-stated Gaussian distribution is the simplest model. Cen and Fan [48] proposed a Fourier random series model for the fluctuation frequency. Many investigators, including Sommerfeld [49], have adopted the Langevin equation model, originally proposed by Pope for solving the PDF equation. To save on computation time, Baxter and Smith [50] and Pereira and Chen [51] have proposed a method of maximum probable trajectory plus presumed empirical PDF, which can give Eulerian particle velocity and concentration distributions and decrease remarkably the predicted number of trajectories. For any Eulerian particle property, for example, the particle number density is expressed as

$$< n_p(x_i, t) > \, = \dot{q} \int_{t_0}^{t} p(x_i, \tau) \mathrm{d}\tau \qquad (6.76)$$

where $\dot{q}$ is the particle number flux and p is the PDF, given as a Gaussian distribution

$$p(x_i, t) = \frac{(\sigma)^{-3}}{(2\pi)^{3/2}} \exp\left\{ -\frac{1}{2\sigma^2} \left[(x - <Ut>)^2 + (y - <Vt>)^2 + (z - <Wt>)^2 \right] \right\}$$

$$(6.78)$$

where $\sigma^2 = <v_{pi}^2>$ is the mean square of particle fluctuation, using the Hinze−Tchen model. The particle trajectory model is widely used in engineering applications and is incorporated in a great deal of commercial software, including CFX, ANSYS-FLUENT, STAR-CD, etc., and also as the computer code PCGC-3 for simulating coal combustion developed at the Brigham Young University. It was used in the simulation of spray combustion in gas-turbine combustors [52] and pulverized coal combustion [46,53,54]. The merits of the particle trajectory model include not requiring the particle turbulence model, easiness in obtaining the complex detailed history of reacting particles, and being free of particle numerical diffusion. Its shortcoming is that large computation storage and time are needed, if sufficient information of particle flow field is required for experimental validation. In addition, it was found by Zhang [55] that the predicted particle fluctuation velocity by the trajectory model is smaller than that measured by PDPA. The particle concentration distribution predicted by the trajectory model is much more nonuniform than that measured by PDPA [56]. Fig. 6.33 shows the predicted particle concentration distribution in a cold model of a tangentially fired furnace using both a stochastic trajectory model and a two-fluid $k-\varepsilon-k_p$ two-phase turbulence model. Obviously, the former remarkably underpredicts the particle turbulent diffusion due to the underprediction of the particle turbulent fluctuation. These results imply that the particle trajectory model cannot fully predict or underpredicts the particle turbulent fluctuation.

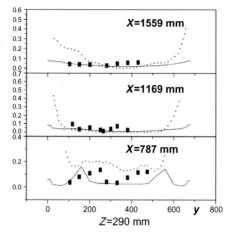

FIGURE 6.33 Particle concentration distribution (■ PDPA measurements; — two-fluid model; - - - stochastic trajectory model).

Fig. 6.34 gives the oil droplet trajectories in an oil−water hydrocyclone predicted by the stochastic trajectory model together with a Reynolds stress gas turbulence model [57]. It can be seen that larger droplets are more impinged on the wall and separated. However, no experimental validation of the predicted droplet velocity and concentration was made.

6.13.4   The DEM Simulation of Dense Gas-Particle Flows

For Eulerian−Lagrangian simulation of dense-gas particle flows, the most frequently used is called "discrete element modeling" (DEM). It was originally proposed by Tsuji et al. [58] and further developed by Chu and Yu [59]. The governing equations for DEM are:

$$m_i \frac{dv_i}{dt} = f_{f,i} + \sum_{j=1}^{k_i} (f_{c,ij} + f_{d,ij} + f_{ij}^v) + , m_i g \qquad (6.79)$$

$$I_i \frac{d\omega_i}{dt} = \sum_{j=1}^{k_i} \mathbf{T}_{ij} \qquad (6.80)$$

where the terms on the right-hand side of Eqs. (6.79) and (6.80) are the drag force, interparticle collisional contact force, friction force, and torque. For the interparticle collision, a hard-sphere or a soft-sphere model can be used. Figs. 6.35 and 6.36 give the prediction results using the DEM for gas−solid flows in a circulating fluidized bed (CFB) [59]. The particle clusters and core−annulus flow structure can be seen in dense gas-particle flows.

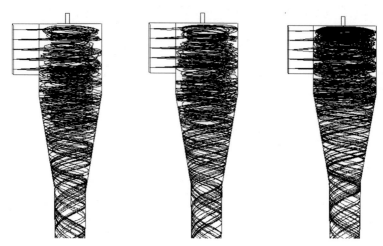

FIGURE 6.34 Oil droplet trajectories in an oil–water hydrocyclone. (Droplet sizes from left to right: 50, 100, 200 μm).

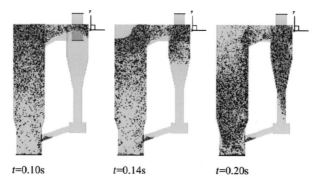

$t=0.10$s $t=0.14$s $t=0.20$s

FIGURE 6.35 Particle position in CFB (different colors indicate different particle sizes) [59].

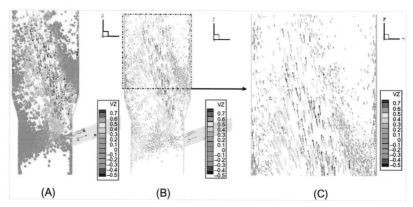

(A)   (B)   (C)

FIGURE 6.36 Flow structure in CFB (A, particle position; B and C, particle velocity vectors) [59].

6.14 THE LARGE-EDDY SIMULATION OF TURBULENT GAS-PARTICLE FLOWS

Large-eddy simulation (LES) is under rapid development and is recognized as a second generation of CFD methods used in engineering. Originally it was developed for single-phase flows, and subsequently was extended to two-phase flows. As mentioned Chapter 5, Modeling of Single-Phase Turbulence in LES the governing equations are filtered and the filtered variables of large scales are directly solved. For variables of small scales (unresolved scales) it is necessary to use subgrid scale (SGS) models. Most LES of gas-particle flows are Eulerian−Lagrangian LES. The filtered gas continuity and momentum equations for the gas phase in isothermal two-phase flows can be obtained as:

$$\frac{\partial \rho}{\partial t} + \frac{\partial}{\partial x_i}(\rho \bar{u}_i) = 0 \tag{6.81}$$

$$\frac{\partial}{\partial t}(\rho \bar{u}_i) + \frac{\partial}{\partial x_j}\left(\rho \bar{u}_i \bar{u}_j\right) = \frac{\partial}{\partial x_j}\left(\mu\left(\frac{\partial \bar{u}_i}{\partial x_j} + \frac{\partial \bar{u}_j}{\partial x_i}\right) - \frac{2}{3}\left(\mu \frac{\partial \bar{u}_j}{\partial x_j}\right)\delta_{ij}\right) - \frac{\partial \bar{p}}{\partial x_i} - \frac{\partial \tau_{ij}}{\partial x_j}$$
$$+ \sum_k \overline{\frac{\rho_k}{\tau_{rk}}(u_{ki} - u_i)}$$

$$\tag{6.82}$$

For the gas-phase in two-phase flows the most frequently used SGS stress models are those for single-phase flows, not accounting for the effect of particles, i.e., the Smagorinsky eddy viscosity model, Germano dynamic eddy viscosity model, and Kim SGS energy equation model. The SGS stress in the momentum equation is defined by:

$$\tau_{ij} \equiv \rho \overline{u_i u_j} - \rho \bar{u}_i \bar{u}_j \tag{6.83}$$

The Smagorinsky eddy viscosity model is given by

$$\tau_{ij} - \frac{1}{3}\tau_{kk}\delta_{ij} = -2\mu_t \bar{S}_{ij} \quad \mu_t = \rho(C_s \Delta)^2 |\bar{S}| \tag{6.84}$$

$$\bar{S}_{ij} \equiv \frac{1}{2}\left(\frac{\partial \bar{u}_i}{\partial x_j} + \frac{\partial \bar{u}_j}{\partial x_i}\right) \quad |\bar{S}| \equiv \sqrt{2\bar{S}_{ij}\bar{S}_{ij}} \tag{6.85}$$

where $C_s = 0.16$ is the Smagorinsky constant, and Δ is the filtered scale. The shortcoming of the Smagorinsky model is its too large dissipation rate for the kinetic energy. Germano proposed a dynamic Smagorinsky eddy viscosity model, in which the coefficient C_s was not a constant, but was determined by the following filtration process:

$$C_s = L_{ij}M_{ij}/(M_{ij}M_{ij})$$

$$M_{ij} = 2\rho C_s \overline{\Delta}^2 \left|\hat{\bar{S}}\right|\hat{\bar{S}}_{ij} - 2\rho C_s \hat{\overline{\Delta}}^2 \left|\hat{\bar{S}}\right|\hat{\bar{S}}_{ij} \quad L_{ij} = 2\rho C_s \hat{\overline{\Delta}}^2 \left|\hat{\bar{S}}\right|\hat{\bar{S}}_{ij} - 2\rho C_s \overline{\Delta}^2 \left|\hat{\bar{S}}\right|\hat{\bar{S}}_{ij}$$

$$\tag{6.86}$$

It was indicated by many investigators that in many cases the dynamic viscosity model gives better results than that given by the standard Smagorinsky model. Kim proposed a SGS kinetic energy model as follows:

$$\tau_{ij} = -2\rho v_t (\bar{S}_{ij} - \frac{1}{3}\bar{S}_{kk}\delta_{ij}) + \frac{2}{3}\rho k^{sgs}\delta_{ij} \qquad (6.87)$$

$$\frac{\partial \bar{\rho}k^{sgs}}{\partial t} + \frac{\partial}{\partial x_i}(\bar{\rho}\bar{u}_i k^{sgs}) = P^{sgs} - D^{sgs} + \frac{\partial}{\partial x_i}\left(\frac{\bar{\rho}v_t}{\text{Pr}_t}\frac{\partial k^{sgs}}{\partial x_i}\right) + \dot{W}_s \qquad (6.88)$$

where $k^{sgs} = \frac{1}{2}(\overline{u_i u_j} - \bar{u}_i\bar{u}_j)$ $v_t = C_v(k^{sgs})^{1/2}\overline{\Delta}$ $D^{sgs} = C_\varepsilon(K^{sgs})^{3/2}/\overline{\Delta}$

The merit of the SGS kinetic energy model is that it avoids the negative value of eddy viscosity; hence it is better than the eddy viscosity models. In application it was reported that the results of the SGS kinetic energy model are close to those obtained using the dynamic eddy viscosity model. For the particle phase in Eulerian–Lagrangian LES, the instantaneous continuity and momentum equations are used as those in RANS modeling. In the framework of Eulerian–Lagrangian LES, Yuu et al. considered the effect of particles on the gas SGS stress [60] by adding a particle source term in the SGS energy equation and then taking the local equilibrium, i.e., the production equals the dissipation. The algebraic gas SGS tress model is obtained as

$$v_t = C_{vt}C_\varepsilon^{1/3}k_{sgs}^{1/2}\Delta \qquad (6.89)$$

$$\bar{k}_{sgs}^{1/2} = \frac{-A_2 + \sqrt{A_2^3 + 4A_1(A_3 + A_4)}}{2A_1} \qquad (6.90)$$

$$A_1 = \frac{C_\varepsilon}{\Delta} \quad A_2 = \frac{6\pi DD_p^2}{\text{Re}}(1 + 0.15\overline{\text{Re}}_p^{0.687})\frac{\bar{n}}{aT_L + 1} \quad A_3 = 2C_{vt}C_\varepsilon^{1/3}\Delta\overline{D}_D^2$$

$$A_4 = \frac{3\pi C_{vt}C_\varepsilon^{1/3}\Delta D_p D^2(1 + 0.15\overline{\text{Re}}_p^{0.687})}{\sigma\text{Re}}\frac{\partial\bar{n}}{\partial x_i}(\bar{u}_i - \overline{u}_{pi})$$

This model is used in E-L LES of an air-particle jet, and the predicted particle turbulence intensity is in agreement with the LDV measurement results. It was reported that small particles reduce the gas subgrid scale turbulence. However, no comparison was made between the SGS models with and without the effect of particles. H. S. Zhou et al. [61] abandoned the concept of local equilibrium, proposing an SGS stress energy equation with the particle source term. The gas SGS stress model is expressed as

$$\tau_{ij}^{sgs} = \rho_f \overline{u_i' u_j'} = \rho_f v_t \left(\tilde{S}_{i,j} - \frac{2}{3}\tilde{S}_{kk}\delta_{ij}\right) - \frac{2}{3}\rho_f \bar{k}^{sgs}\delta_{ij} \qquad (6.91)$$

$$v_t = C_k \Delta \bar{k}_{sgs}^{1/2}$$

$$\frac{\partial(\varepsilon\rho_f\overline{k}_{sgs})}{\partial t} + \frac{\partial(\varepsilon_f\rho_f\overline{u}_{f,j}\overline{k}_{sgs})}{\partial x_j} = \frac{\partial}{\partial x_j}\left[\varepsilon\rho_f\left(\nu_f + \frac{\nu_t}{\text{Pr}}\right)\frac{\partial\overline{k}_{sgs}}{\partial x_j}\right] + \frac{1}{2}\varepsilon\rho_f\nu_t\left(\tilde{S}_{f,j}\tilde{S}_{f,j}\right)$$

$$- \varepsilon\rho_f C_\varepsilon\frac{\overline{k}_{sgs}^{3/2}}{\Delta} - 3\pi d_p\mu_f\left(1+0.15\overline{\text{Re}}_p^{0.687}\right) \times \left[\overline{n}\frac{2\overline{k}_{sgs}}{aT_L+1} - \frac{\nu_t}{\sigma}\frac{\partial\overline{n}}{\partial x_i}\left(\overline{u}_{f,i}-\overline{v}_{p,i}\right)\right]$$

$$(6.92)$$

This model is used to simulate gas-particle flows in a fluidized bed. However, no theoretical or experimental validation was made, and there is also no comparison between the SGS stress models with and without the effect of particles

6.14.1 Eulerian−Lagrangian LES of Swirling Gas-Particle Flows

Apte et al. [62] carried out the LES of swirling gas-particle flows, measured by Sommerfeld and Qiu, using Germano's dynamic Smagorinsky SGS stress model for gas flows and Lagrangian particle tracking approach. Fig. 6.37 gives the instantaneous particle distribution in the swirl chamber. The particles are concentrated in the shear region. Fig. 6.38 shows the LES statistically averaged particle RMS fluctuation velocities in comparison with the measurement results, where (A), (B), and (C) correspond to (D), (E), and (F) of the original figures, respectively. The agreement is sufficiently good, and is much better than that obtained using the RANS modeling.

The Eulerian−Lagrangian LES of gas-particle flows was also used to validate the two-fluid models by Simonin et al. [63]. The gas-particle velocity correlation predicted by LES using a Smagorinsky SGS stress model is in fair agreement with that obtained using Simonin's two-fluid model. In addition, it was shown that the interparticle collision makes the particle velocity distribution more uniform and particles of small St numbers reduce the gas turbulence.

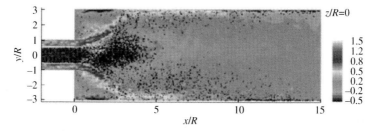

FIGURE 6.37 Instantaneous particle distribution in a swirl chamber [62].

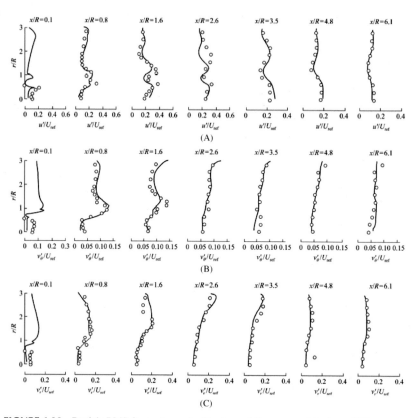

FIGURE 6.38 Particle RMS fluctuation velocities for swirling gas-particle flows [62].

6.14.2 Eulerian–Lagrangian LES of Bubble-Liquid Flows

The study of bubble-liquid confined two-phase jets containing a large number of bubbles was made using Eulerian–Lagrangian LES by the present author and his colleagues [64]. For LES, the liquid large-eddy continuity and momentum equations are obtained using two-way coupling, and the effect of drag force and buoyancy force are taken into account. The Smagorinsky–Lilly's sub-grid scale model is used, in which the effect of bubbles on the subgrid scale stress is neglected. In the bubble motion equation the drag force, gravitational and buoyancy forces, and the added-mass force are taken into account. The filtered and volume-averaged large-eddy continuity and momentum equations for the liquid phase are:

$$\frac{\partial}{\partial t}(\alpha_l \rho) + \frac{\partial}{\partial x_i}(\alpha_l \rho \bar{u}_i) = 0 \tag{6.93}$$

$$\frac{\partial}{\partial t}(\alpha_l \rho \overline{u_i}) + \frac{\partial}{\partial x_j}\left(\alpha_l \rho \overline{u_i u_j}\right) = \frac{\partial}{\partial x_j}\left(\alpha_l \mu \frac{\partial \overline{u_i}}{\partial x_j}\right) - \alpha_l \frac{\partial \overline{p}}{\partial x_i} - \alpha_l \frac{\partial \tau_{ij}}{\partial x_j} + \frac{\alpha_b \rho_g}{\tau_r}(u_{bi} - \overline{u_i})$$
$$+ \alpha_l \rho g_i + \alpha_b \rho g_i$$

$$(6.94)$$

The bubble motion equation in the Lagrangian reference frame, accounting for the drag force, the gravitational force with buoyancy effect, and the added-mass force, is written as:

$$\alpha_b \rho_g \frac{\mathrm{d}u_{bi}}{\mathrm{d}t} = \frac{\alpha_b \rho_g}{\tau_r}(\overline{u_i} - u_{bi}) + \alpha_b(\rho_g - \rho)g_i + \frac{\alpha_b}{2}\rho\frac{\mathrm{d}}{\mathrm{d}t}(\overline{u_i} - u_{bi}) \qquad (6.95)$$

The obtained instantaneous velocity vectors for the single-phase liquid and liquid in bubble-liquid flows are given in Fig. 6.39A and 6.39B. This shows the shear-produced large-eddy vortex structure and its development. It can be seen that under the effect of both shear generation and bubble generation, the liquid eddies become stronger in size and intensity than in the single liquid phase, showing the enhancement of liquid turbulence by bubbles. Fig. 6.39C gives the instantaneous bubble trajectories, indicating the large-eddy structure of the bubble motion. Figs. 6.40 and 6.41 show the liquid RMS fluctuation velocities with and without bubbles, respectively. It can be seen that the bubbles enhance the liquid turbulence. The liquid RMS fluctuation velocity for two-phase flows is increased 50% compared with that of single-phase liquid flows.

6.14.3   Two-Fluid LES of Swirling Gas-Particle Flows

Most LES adopt a Eulerian–Lagrangian approach, which requires a long computation time. A few investigators have reported two-fluid LES, adopting

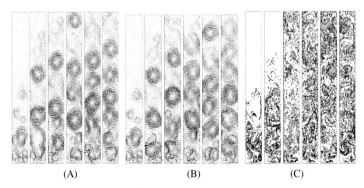

(A)   (B)   (C)

FIGURE 6.39   The instantaneous velocity vectors and bubble trajectories (A, single-phase liquid; B, liquid in bubble-liquid flows; C, bubble trajectories).

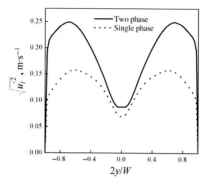

FIGURE 6.40   Liquid vertical fluctuation velocity.

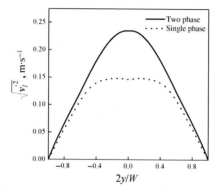

FIGURE 6.41   Liquid horizontal fluctuation velocity.

the Smagorinsky SGS eddy viscosity model for both gas and particle phases without theoretical justification, and the interaction between two phases and the anisotropy of the two-phase SGS stresses has not taken into account. Extending the idea of the two-phase turbulence models in RANS modeling, a unified second-order moment (USM) two-phase SGS model is proposed by the author and his colleagues for two-fluid LES of gas-particle flows [65], accounting for the interaction between the gas and particle SGS stresses and their anisotropy. The LES-USM is used to simulate swirling gas-particle flows. For a two-fluid LES, the filtered continuity and momentum equations for gas and particle phases can be obtained as:

$$\frac{\partial}{\partial t}\left(\alpha_k \rho_k\right) + \frac{\partial}{\partial x_j}\left(\alpha_k \rho_k \overline{u_{ki}}\right) = 0 \quad (k = g, p) \tag{6.96}$$

$$\frac{\partial}{\partial t}\left(\alpha_g \rho_g \overline{u_{gi}}\right) + \frac{\partial}{\partial x_j}\left(\alpha_g \rho_g \overline{u_{gi} u_{gj}}\right) = -\frac{\partial \overline{p_g}}{\partial x_j} + \frac{\partial \tau_{g,ij}}{\partial x_j} + \frac{\partial \tau_{gs,ij}}{\partial x_j} + \frac{\alpha_g \rho_g}{\tau_r}\left(\overline{u_{pi}} - \overline{u_{gi}}\right) \tag{6.97}$$

$$\frac{\partial}{\partial t}\left(\alpha_p \rho_p \overline{u_{pi}}\right) + \frac{\partial}{\partial x_j}\left(\alpha_p \rho_p \overline{u_{pi} u_{pj}}\right) = \frac{\partial \tau_{p,ij}}{\partial x_j} + \frac{\partial \tau_{ps,ij}}{\partial x_j} + \frac{\alpha_g \rho_g}{\tau_r}\left(\overline{u_{gi}} - \overline{u_{pi}}\right)$$

(6.98)

where the filtered gas and particle viscous forces are:

$$\tau_{g,ij} = \mu_{gl}\left(\frac{\partial u_{gi}}{\partial x_j} + \frac{\partial u_{gj}}{\partial x_i}\right) - \frac{2}{3}\mu_{gl}\frac{\partial u_{gj}}{\partial x_j}\delta_{ij}; \quad \tau_{p,ij} = \mu_p\left(\frac{\partial u_{pi}}{\partial x_j} + \frac{\partial u_{pj}}{\partial x_i}\right) - \frac{2}{3}\mu_p\frac{\partial u_{pj}}{\partial x_j}\delta_{ij}$$

The gas and particle subgrid scale (SGS) stresses are defined as:

$$\tau_{gs,ij} = -\rho_g R_{gs,ij} = -\rho_g\left(\overline{u_{gi} u_{gj}} - \overline{u_{gi}}\,\overline{u_{gj}}\right); \quad \tau_{ps,ij} = -\rho_p R_{ps.ij} = -\rho_p\left(\overline{u_{pi} u_{pj}} - \overline{u_{pi}}\,\overline{u_{pj}}\right)$$

The SGS stresses of gas and particle phases can be given by the following transport equations:

$$\frac{\partial}{\partial t}\left(\alpha_g \rho_g R_{gs,ij}\right) + \frac{\partial}{\partial x_k}\left(\alpha_g \rho_g \overline{u}_{gk} R_{gs,ij}\right) = D_g^{sgs} + P_g^{sgs} + G_{pg}^{sgs} + \Pi_g^{sgs} - \varepsilon_g^{sgs} \quad (6.99)$$

$$\frac{\partial}{\partial t}\left(\alpha_p \rho_p R_{ps,ij}\right) + \frac{\partial}{\partial x_k}\left(\alpha_p \rho_p \overline{u}_{pk} R_{ps,ij}\right) = D_p^{sgs} + P_p^{sgs} + \varepsilon_p^{sgs} \quad (6.100)$$

The meanings of the source terms on the right-hand side of Eqs. (6.99) and (6.100) and other equations, like the SGS two-phase velocity correlation and SGS enery dissipation equations, can be found in Reference [65]. Figs. 6.42 and 6.43 give the predicted two-phase tangential time-averaged and RMS fluctuation velocities, respectively. It can be seen that the LES-USM gives better results than those obtained by the RANS-USM, although there is still some discrepancy between predictions and experiments. The results are not as good as those obtained by Apte et al., since a 2-D two-fluid LES and a second-order numerical scheme are used by Liu, whereas Apte et al. used 3-D Eulerian−Lagrangian LES and a higher-order numerical scheme.

The instantaneous gas and particle streamlines are shown in Figs. 6.44 and 6.45. There are more complicated multiple recirculation zones of the gas flows (Fig. 6.44), including a corner recirculation zone and more recirculation zones in the near axis and intermediate regions, than those of the time-averaged gas flows, where only a corner recirculation zone and a central recirculation zone are observed. The particle flow field (Fig. 6.45) is different from the gas flow field. Particles at first concentrate in the near-axis zone and enter the corner recirculation zone, then gradually move to the wall under the effect of centrifugal force and turbulent diffusion, and finally concentrate in a thin layer adjacent to the wall. There are almost no recirculating flows of particles in the near-axis and downstream regions due to different inertia of the two phases.

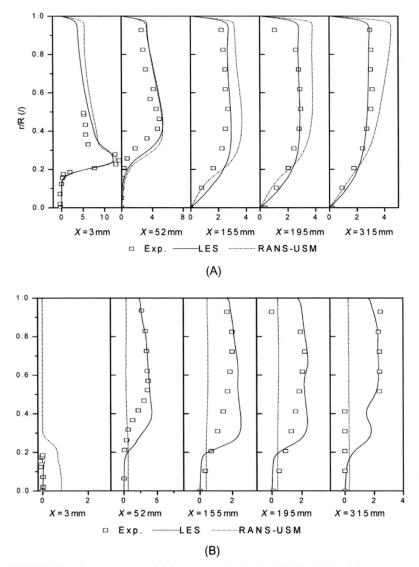

FIGURE 6.42 Two-phase tangential time-averaged velocities. (A) Gas. (B) Particle.

6.14.4 Application of LES in Engineering Gas-Particle Flows

Derksen et al. carried out LES of gas-particle flows in gas−solid cyclones [66]. Instantaneous particle deposition and particle positions were obtained. However, only the predicted gas flow field was validated by experiments. No predicted particle flow field and its validation were reported.

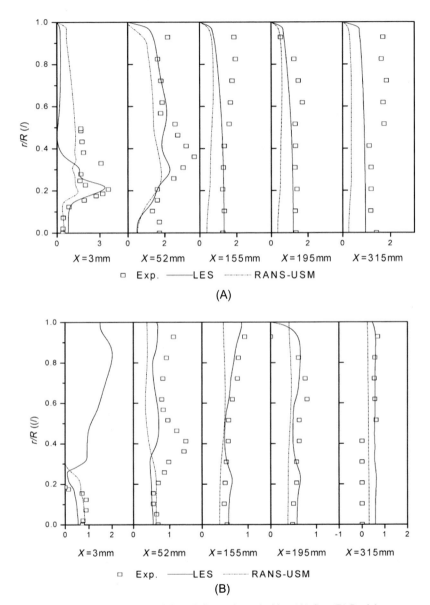

FIGURE 6.43 Two-phase tangential RMS fluctuation velocities. (A) Gas. (B) Particle.

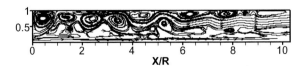

FIGURE 6.44 Instantaneous gas streamlines for swirling gas-particle flows.

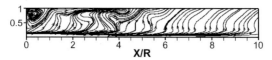

FIGURE 6.45 Instantaneous particle streamlines for swirling gas-particle flows.

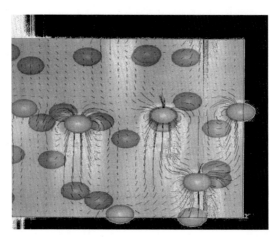

FIGURE 6.46 Instantaneous bubble positions.

6.15   THE DIRECT NUMERICAL SIMULATION OF DISPERSED MULTIPHASE FLOWS

In recent DNS studies of dispersed multiphase flows, special attention is paid to fully resolved DNS (FDNS). This is done using the lattice Boltzmann or finite-difference methods to solve 3-D instantaneous N-S equation with finite-size particles and using VOF, front-tracking, level-set or immerse-boundary methods to treat the moving interface. Bagchi and Balachandar [67] simulated steady and unsteady turbulent flows passing a stagnant and a moving particle with $Re_P = 10-300$ using FDNS. The FDNS database is used to obtain the drag, lift, and other forces. The FDNS of flows passing a single particle with a size of double Kolmogorov scales was reported in Reference [68]. The gas vorticity isolines for cases with and without the particle were obtained. The result is that the turbulence is enhanced near the particle, but generally is reduced. The velocity vectors of flows around two interacting cylinders using FDNS are given in Reference [69].The instantaneous bubble position and fluid velocity isolines for fluid flows around several bubbles made by FDNS [70] are shown in Figs. 6.46 and 6.47, respectively. From these results the bubble rising velocity, bubble drag, and fluid subgrid scale stress are obtained. The vorticity

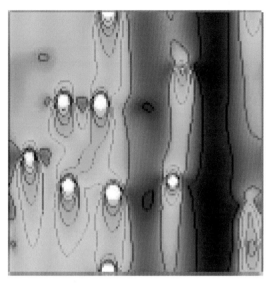

FIGURE 6.47   Instantaneous fluid velocity isolines.

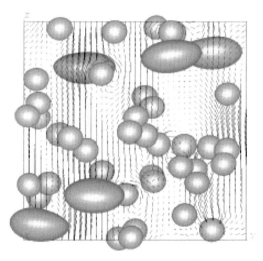

FIGURE 6.48   Instantaneous velocity vectors around 45 bubbles.

tubes behind 78 particles in isotropic homogeneous turbulent flows made by FDNS are given in Reference [71]. The results show an enhancement of fluid turbulence by particles.

The fluid flows around 45 bubbles of different sizes given by FDNS are reported in Reference [72]. The instantaneous velocity vectors (Fig. 6.48) and bubble fluctuation velocities (Fig. 6.49) are obtained. It can be seen that the fluctuation velocities (in terms of particle Reynolds number) of larger particles are stronger than those of smaller particles. The separation, falling, and deposition of 147 colliding particles given by FDNS are reported in Reference [73]. Fig. 6.50 gives the species concentration distribution and gas velocity field around the particles with gas−solid mass transfer [74] using a ghost-cell-based immersed boundary method. From these results the effect of Re number and the particle−particle distance on the mass transfer rate were studied.

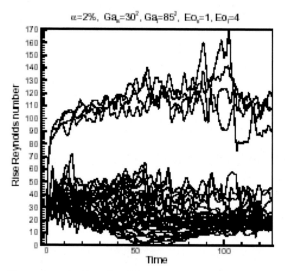

FIGURE 6.49 Fluctuation velocities of particles.

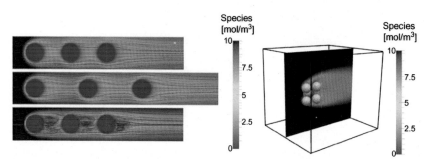

FIGURE 6.50 Species concentration and gas velocities around particles.

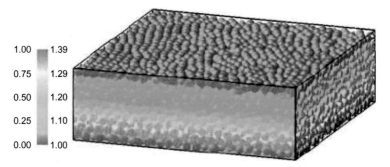

FIGURE 6.51 Instantaneous flow field and thermal fields around 5487 particles. (Particles colored by the surface temperature and temperature contours and velocity in the front cross-section.)

Fig. 6.51 shows the instantaneous flow field and thermal fields around 5487 particles with liquid−solid heat transfer in dense liquid-particle flows using the immerse boundary method [75]. It can be seen that the surface temperature of each particle is uniform rather than a gradual variation, suggesting that the solid temperature distribution for each particle is almost uniform.

For point-particle DNS (PDNS), the instantaneous gas velocity vectors and particle position in isotropic homogeneous turbulent flows are shown in Figs. 6.52 [72], indicating that the particles are located in the periphery of vortex structures.

The PDNS of a gas-particle jet [72] gives the instantaneous particle position for particles of (a) $St = 0.01$, (b) $St = 1$, and (c) $St = 50$, as shown in Fig. 6.53. These results show the dispersion of particles of different sizes under the effect of gas turbulence, similar to the results obtained by other investigators using large-eddy simulation and discrete vortex simulation. However, the modeling results for particles have not been validated by experiments and the results show only the reduction of gas turbulence, even for large particles with $St = 50$, indicating the limitation of the PDNS.

In conclusion, although different PDNS results have been used to validate the LES and RANS models, as mentioned above, unlike the DNS of single-phase flows, which can give an exact database for validating the LES and RANS models, the results of PDNS for gas-particle flows are questionable, since it cannot fully reflect the real situation due to the relationships between the particle size and the Kolmogorov scale. As for FDNS, many studies give only qualitative results and more detailed results, such as the turbulence modulation by the particle wake effect, still remain to be studied.

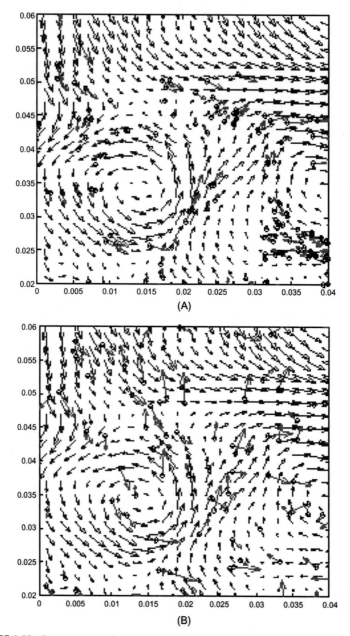

FIGURE 6.52   Instantaneous velocity vectors and particle position.

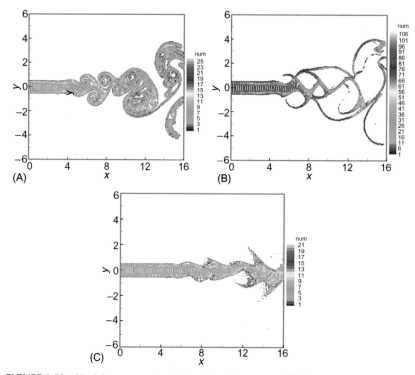

FIGURE 6.53 Vorticity maps and particle position for a gas-particle jet.

REFERENCES

[1] C.T. Crowe. The state-of-the-art in the numerical models for dispersed phase flows. in: G. Matsui, A. Serizawa, Y. Tsuji, (Eds.), Proceedings of the First International Conference on Multiphase Flows, Tsukuba, Japan, Vol. 3, pp. 49–60, 1991.

[2] S.L. Soo, Fluid Dynamics of Multiphase Systems., Blaisdell (Ginn), New York, 1967.

[3] L.X. Zhou, Multiphase fluid dynamics of gas-particle systems with phase change., Adv. Mech. (in Chin.) 12 (1982) 141–150.

[4] S.E. Elghobashi, T.W. Abou-Arab, A two-equation model for two-phase flows., Phys. Fluids 26 (1983) 931–938.

[5] S.E. Elghobashi, T.W. Abou-Arab, M. Risk, A. Mostafa, Prediction of the particle-laden jet with a two-equation turbulence model., Int. J. Multiphase Flow 10 (1984) 697–710.

[6] M.A. Rizk, S.E. Elghobashi, A two-equation turbulence model for dispersed dilute confined two-phase flows., Int. J. Multiphase Flow 15 (1989) 119–133.

[7] W.K. Melville, K.N.C. Bray, A model of the two-phase turbulent jet., Int. J. Heat Mass Transfer 22 (1979) 647–656.

[8] C.P. Chen, P.E. Wood, A turbulence closure model for dilute gas-particle flows., Can. J. Chem. Eng. 63 (1985) 349–360.

[9] A.A. Mostafa, H.C. Mongia, On the interaction of particles and turbulent flow., Int. J. Heat Mass Transfer 31 (1988) 2063–2075.

[10] C.M. Tchen. Mean value and correlation problems connected with the motion of small particles. in: a turbulent field. Ph.D. Dissertation, Delft University, Martinus Nijhoff, Hague, 1947.

[11] J.O. Hinze, Turbulence., McGraw Hill, New York, 1975.

[12] L.X. Zhou, H.Q. Zhao, X.Q. Huang. Numerical and experimental studies of an enclosed gas-particle jet. in: Sato, H. Ed., Proceedings of 3rd Asian Congress of Fluid Mechanics, Tokyo, pp.471-474, 1986.

[13] Th. Borner, F. Durst. LDA measurements of gas-particle confined jet. Flow and Digital Data Processing, LSTM Report, LSTM 153/E/86, University of Erlangen-Nurnberg, 1986.

[14] L.X. Zhou, X.Q. Huang. Particle turbulent kinetic energy transport equation in suspension two-phase flows. in: F.G. Zhuang (Ed.), Proceedings of 1st International Conference on Fluid Mechanics, Beijing, pp. 791−793, 1987.

[15] L.X. Zhou, X.Q. Huang, Prediction of confined gas-particle jets by an energy equation model of particle turbulence., Sci. China Engl. Ed. 33 (1990) 53−59.

[16] L.X. Zhou, C.M. Liao, T. Chen, A unified second-order moment two-phase turbulence model for simulating gas-particle flows. Numerical Methods in Multiphase Flows, ASME-FED 185 (1994) 307−313.

[17] J.Y. Tu, C.A.J. Fletcher, Numerical computation of turbulent gas-particle flow in a 90-degree bend., AIChE J. 41 (1995) 2187−2197.

[18] L.X. Zhou, T. Chen, Simulation of strongly swirling gas-particle flows using USM and k-ε-kp two-phase turbulence models., Powder Technol. 114 (2001) 1−11.

[19] L.X. Zhou, Advances in studies on two-phase turbulence in dispersed multiphase flows, Int. J. Multiphase Flow 36 (2010) 100−108.

[20] L.X. Zhou, W.Y. Lin, K.M. Sun, Simulation of swirling gas-particle flows using a k-ε-kp model., J. Eng. Thermophys. (in Chin.) 16 (1995) 481−485.

[21] M. Sommerfeld, H.H. Qiu, Detailed measurement of a swirling particulate two-phase flow by a phase-Doppler anemometer., Int. J. Heat Fluid Flow 12 (1991) 20−28.

[22] L.X. Zhou, M. Yang, C.Y. Lian, L.S. Fan, D.J. Lee, On the second-order moment turbulence model for simulating a bubble column., Chem. Eng. Sci. 57 (2002) 3269−3281.

[23] L.X. Zhou, Y. Xu, Simulation of swirling gas-particle flows using an improved second-order moment two-phase turbulence model., Powder Technol. 116 (2001) 178−189.

[24] I.V. Derevich, L.I. Zaichik, The equation for the probability density of the particle velocity and temperature in a turbulent flow simulated by the Gauss stochastic field., Prikl. Mat. Mekh, (in Russian) 54 (1990) 767..

[25] L.P. Wang, D.E. Stock, Dispersion of heavy particles by turbulent motion., J. Atmos. Sci. 50 (1993) 1897−1913.

[26] X.Y. Huang, D.E. Stock, Using the Monte-Carlo process to simulate two-dimensional heavy particle dispersion in a uniformly sheared turbulent flow. Numerical Methods in Multiphase Flows, ASME FED 185 (1994) 243−249.

[27] Y. Yu, L.X. Zhou, C.G. Zheng, Z.H. Liu, Simulation of swirling gas-particle flows using different time scales for the closure of two-phase velocity correlation in the second-order moment two-phase turbulence model, J. Fluid Eng. Trans. ASME 125 (2003) 247−250.

[28] L.I. Zaichik, V.M. Aplichenkov, A statistical model for transport and deposition of high-inertia colliding particles in turbulent flow., Int. J. Heat Fluid Flow 22 (2001) 365−371.

[29] M.W. Reeks, On the kinetic equation for the transport of particles in turbulent flows., Phys. Fluids A3 (1991) 446−456.

[30] O. Simonin. Continuum modeling of dispersed two-phase flows. in: Combustion and Turbulence in Two-Phase Flows, Lecture Series 1996-02, von Karman Institute for Fluid Dynamics, 1996.

[31] Y. Li, L.X. Zhou, A k-ε-PDF two-phase turbulence model for simulating sudden-expansion particle laden flow., ASME-FED 236 (1996) 311−315.

[32] L.X. Zhou, Y. Li, Simulation of strongly swirling gas-particle flows using a DSM-PDF two-phase turbulence model., Powder Technol. 113 (2000) 70−79.

[33] Z.H. Liu, C.G. Zheng, L.X. Zhou, A second-order-moment-Monte-Carlo (SOM-MC) model for simulating swirling gas-particle flows., Powder Technol. 120 (2001) 216−222.

[34] L.X. Zhou, H.X. Gu, Simulation of swirling gas-particle flows using a nonlinear k-ε-k_p two-phase turbulence model., Powder Technol. 128 (2002) 47−55.

[35] D. Gidaspow, Multiphase Flow and Fluidization: Continuum and Kinetic Theory Descriptions., Academic Press, New York, 1994.

[36] L.X. Zhou, Y. Yu, et al., Two-phase turbulence models for simulating dense gas-particle flows., Particuology 16 (2014) 100−107.

[37] M.J.V. Goldschmidt, J.A.M. Kuipers, W.P.M. van Swaaij, Hydrodynamic modeling of dense gas-fluidized beds using the kinetic theory of granular flow: effect of coefficient of restitution on bed dynamics., Chem. Eng. Sci. 56 (2001) 571−578.

[38] A.T. Mineto, M.P.S. Braun, et al., Influence of the granular temperature in the numerical simulation of gas-solid flow in a bubbling fluidized bed., Chem. Eng. Commun. 201 (2014) 1003−1020.

[39] Y. Cheng, F. Wei, Y.C. Guo, Y. Jin, W.Y. Lin, Modeling the hydrodynamics of downer reactors based on kinetic theory., Chem. Eng. Sci. 54 (1999) 2019−2027.

[40] A. Neri, D. Gidaspow, Riser hydrodynamics: simulation using kinetic theory., AIChE J. 46 (2000) 52−67.

[41] H.L. Lu, D. Gidaspow, J. Bouillard, W.T. Liu, Hydrodynamic simulation of gas-solid flow in a riser using kinetic theory of granular flow., Chem. Eng. J 95 (2003) 1−13.

[42] Y. Liu, L.X. Zhou, C.X. Xu, L.Y. Hu, Two fluid large-eddy simulation of backward-facing step gas-particle flows and validation of second-order-moment two-phase turbulence model., J. Chem. Ind. Eng. (in Chin.) 59 (2008) 2485−2489.

[43] Y. Yu, L.X. Zhou, B.G. Wang, et al., A USM-Theta two-phase turbulence model for simulating dense gas-particle flows., Acta Mech. Sin. 21 (2005) 228−234.

[44] C.T. Crowe, M.P. Sharma, D.E. Stock, The particle-source-in-cell (PSIC) method for gas--droplet flows., J. Fluid Eng. 99 (1977) 325−332.

[45] A.D. Gosman, E. Ioannides, AIAA 19th Aerospace Science Meeting, Paper 81-0323, 1981.

[46] L.D. Smoot, P.J. Smith, Coal Combustion and Gasification., Plenum Press, New York, 1985.

[47] F.C. Lockwood, C. Papadopoulos, A new method for the prediction of particulate dispersion in turbulent two-phase flows., Combustion Flame 75 (1989) 403−413.

[48] K.F. Cen, J.R. Fan, Theory and Computation of Engineering Gas-Solid Flows, (in Chinese), Zhejiang University Press, 1990.

[49] M. Sommerfeld, G. Kohnen, M. Ruger, Some open questions and inconsistencies of Lagrangian particle dispersion models. in: Proceedings of Ninth Symposium on Turbulent Shear Flows, Kyoto, Japan, Paper 5.1, 1993.

[50] L.L. Baxter, P.J. Smith. Turbulent dispersion of particles. in: Proceedings of Western Sec. of Comb. Inst., Salt Lake City, 1988.

[51] X.Q. Chen, J.C.F. Pereira, Efficient computation of particle dispersion in turbulent flows with a stochastic-probabilistic method., Int. J. Heat Mass Transfer 40 (1997) 1727–1741.

[52] F. Boysan, W.H. Ayers, J. Swithenbank, Z.G. Pan, Modeling of Spray Combustion in Gas Turbine Combustors, Report BIC 354., University of Sheffield, 1980.

[53] F. Boysan, R. Weber, J. Swithenbank, C. Lawn, Modeling coal-fired cyclone combustor., Combustion Flame 63 (1986) 73–85.

[54] K. Gorner, Prediction of turbulent flow, heat release and heat transfer in utility boiler furnaces, coal combustion., in: Junkai Feng (Ed.), Proceedings of First International Symposium on Coal Combustion, Hemisphere, 1988, pp. 273–282.

[55] K.C. Chang, J.C. Yang, Transient effects of drag coefficient in the Eulerian-Lagrangian calculation of two-phase flows., ASME-FED 236 (1996) 5–10.

[56] L.X. Zhou, L. Li, R.X. Li, Simulation of 3-D gas-particle flows and coal combustion in a tangentially fired furnace using a two-fluid-trajectory model., Powder Technol. 125 (2002) 226–233.

[57] Y.J. Lu, 3-D Numerical Investigation on the Mechanisms of Strongly Swirling Two-phase Turbulent Flow and Separation in Liquid-Liquid Hydro-cyclones. Postdoctoral Research Report, Xi'an Jiaotong University, 2000.

[58] Y. Tsuji, T. Kawaguchi, T. Tanaka, Discrete particle simulation of a two-dimensional fluidized bed., Powder Technol. 77 (1993) 79–87.

[59] K.W. Chu, A.B. Yu, Numerical simulation of complex particle–fluid flows., Powder Technol. 179 (2008) 104–114.

[60] S. Yuu, T. Ueno, T. Umekage, Numerical simulation of the high Reynolds number slit nozzle gas-particle jet using sub-grid-scale coupling large eddy simulation, Chem. Eng. Sci. 56 (2001) 4293–4307.

[61] H.S. Zhou, G. Flamant, D. Gauthier, DEM-LES of coal combustion in a bubbling fluidized bed, Part I: gas-particle turbulent flow structure., Chem. Eng. Sci. 59 (2004) 4193–4203.

[62] L.X. Zhou, K. Li, F. Wang, Advances in large-eddy simulation of two-phase combustion (I): LES of spray combustion., Chin. J. Chem. Eng. 20 (2012) 205–211.

[63] O. Simonin, E. Deutsch, M. Boivin, Comparison of large-eddy simulation and second-moment closure model of particle fluctuating motion in two-phase turbulent shear flows., in: Durst, et al. (Eds.), Turbulent Shear Flow 9, Springer-Verlag, 1995, pp. 85–115.

[64] M. Yang, L.X. Zhou, L.S. Fan, Large-eddy simulation of bubble-liquid confined jets., Chin. J. Chem. Eng. 10 (2002) 381–384.

[65] Y. Liu, L.X. Zhou, C.X. Xu, Large-eddy simulation of swirling gas-particle flows using a USM two-phase SGS stress model, Powder Technol. 198 (2010) 183–188.

[66] J.J. Derksen, S. Sundaresan, van den Hea Akker, Simulation of mass loading effect in gas-solid cyclone separators., Powder Technol. 163 (2006) 55–68.

[67] P. Bagchi, S. Balachandar. Unsteady motion and forces on a spherical particle. in: nonuniform flows. ASME-FED, 251, Paper FEDSM2000-11128, 2000.

[68] T.M. Burton, J.K. Eaton. Fully resolved simulations of stationary particles in turbulent flow. in: Proceedings of ASME FED, Paper FEDSM2003-45721, 2003.

[69] A. Prosperetti, Z. Zhang. Simulation of two-dimensional particle flows. in: Proceedings of ASME-FED, Paper FEDSM2002-31216, 2002.

[70] L.X. Zhou, Dynamics of Multiphase Turbulent Reacting Fluid Flows (in Chinese)., Defense Industry Press, Beijing, 2002.

[71] T. Tsuji, T. Yokomine, A. Shimizu. Direct numerical simulation of interactions between particles relative larger than Kolmogorov micro scale and isotropic homogeneous

turbulence. in: Y. Matsumoto (Ed.), in: Proceedings of 5th International Conference on Multiphase Flow, ICMF'04, Yokohama, 2004-05-31-06-04, Paper No. 326.

[72] L.X. Zhou, Advances in studies on turbulent dispersed multiphase flows., Chin. J. Chem. Eng. 18 (2010) 1−10.

[73] T.N. Randrianarivelo, S. Vincent, G. Pianet, O. Simonin, J.P. Caltagirone, Towards direct numerical simulation of fluidized beds. in: Y. Matsumoto (Ed.), Proceedings of 5th International Conference on Multiphase Flow, ICMF'04, Yokohama, 2004-05-31-06-04, Paper No. 207.

[74] J.T. Lu, E.A.J.F. Peters, J.A.M. Kuipers, Direct numerical simulation of fluid-solid mass transfer using a ghost-cell based immersed boundary method. in: 9th International Conference on Multiphase Flow, May 22−27, Firenze, Italy, 2016.

[75] J. Gu, K. Kondo, S. Takeuchi, T. Kajishima, Direct numerical simulation of heat Transfer in dense particle-liquid two-phase media. in: 9th International Conference on Multiphase Flow, May 22−27, Firenze, Italy, 2016.

Chapter 7

Modeling of Turbulent Combustion

7.1 INTRODUCTION

Turbulent combustion and reaction are widely encountered in air and water pollution, chemical reactors, iron- and steel-making furnaces, industrial and utility furnaces, gas-turbines, rocket engines, ramjet engines, and internal engine combustors, plasma chemical reactors, and reentry of spacecraft. There are strong turbulence–chemistry interactions. The chemical reaction may affect turbulence by the density variation due to heat release. In the 1940s, Landau, based on the theory of nonviscous gas dynamics, analyzed the instability caused by a combustion wave [1]. Hence, many investigators thought that the flame induces turbulence. However, subsequent laser Doppler measurements and numerical simulation show that, in contrast, combustion or the gas expansion due to compressibility frequently reduce but do not increase turbulence. The turbulence, whether increased or reduced, is determined by the temperature gradient. For the effect of turbulence on combustion, in laminar flows, the chemical reaction rate is determined by the intermolecular collision, or mixing on the molecular level. In turbulent flows, the chemical reaction rate is determined not only by molecular-level mixing, but also by mixing due to the collision between the randomly moved turbulent eddies. The turbulence not only enhances heat and mass transfer, but also increases the time-averaged reaction rate by intensifying the mixing among different reactants and combustion products. Turbulence causes species and temperature fluctuations, increasing the time-averaged reaction rate. To simulate turbulent combustion, the key problem is that the reaction rate in terms of averaged variables is not equal to the time-averaged value of the reaction rate.

7.2 THE TIME-AVERAGED REACTION RATE

The instantaneous global/elementary reaction rate can be expressed by the Arrhenius expression

$$w_s = B\rho^m \exp(-E/RT) \prod_s Y_s^{m_s} \tag{7.1a}$$

Theory and Modeling of Dispersed Multiphase Turbulent Reacting Flows.
DOI: https://doi.org/10.1016/B978-0-12-813465-8.00007-7

In the simplest case of second-order reactions, the reaction rate can be expressed by

$$w_s = B\rho^2 Y_1 Y_2 \exp(-E/RT) = k\rho^2 Y_1 Y_2 \tag{7.1b}$$

Using the Reynolds expansion, taking

$$k = \bar{k} + k', \quad Y_1 = \bar{Y}_1 + Y'_1, \quad Y_2 = \bar{Y}_2 + Y'_2$$

When neglecting the density fluctuation and the third-order correlation, after taking the time averaging, we have

$$\overline{w}_s = \overline{k\rho^2 Y_1 Y_2} = \rho^2(\overline{k}\,\overline{Y}_1\overline{Y}_2 + k\overline{Y'_1 Y'_2} + \overline{Y}_1\overline{k'Y'_2} + \overline{Y}_2\overline{k'Y'_1})$$

$$\text{or } \overline{w}_s = \rho^2 \overline{k}\,\overline{Y}_1\overline{Y}_2(1 + F) \tag{7.2}$$

where $\bar{k} = \overline{B\exp(-E/RT)}, \quad F = \dfrac{\overline{Y'_1 Y'_2}}{\overline{Y}_1\overline{Y}_2} + \dfrac{\overline{k'Y'_1}}{\overline{k}\,\overline{Y}_1} + \dfrac{\overline{k'Y'_2}}{\overline{k}\,\overline{Y}_2} > 0$

obviously, in the case of turbulent flows, the time-averaged reaction rate, namely the time-averaged value of the reaction rate, is not equal to the reaction rate in terms of the time-averaged values, namely

$$\overline{w}_s \neq \rho^2 \overline{k}\,\overline{Y}_1\overline{Y}_2$$

The function F is composed of correlations due to temperature and concentration fluctuations. The physical meaning is enhancing the time-averaged reaction rate by correlations of $Y-Y$ fluctuations and $k-Y$ fluctuations. If we take the reaction rate in terms of the time-averaged values, corresponding to taking $F = 0$, obviously underpredict the time-averaged reaction rate in turbulent flows. The simplest idea is the so-called "moment method," i.e., making a Reynolds expansion and then taking an average, like that in turbulence modeling. However, the difficulty in Reynolds expansion is the highly nonlinear exponential function of temperature k (T). In the following text we will go back to this treatment.

7.3 THE EDDY-BREAK-UP (EBU) MODEL/EDDY DISSIPATION MODEL (EDM)

Due to the difficulty in the moment method, turbulent combustion modeling was developed based on different physical mechanistic understandings. The simplest turbulent combustion model is the so-called "Eddy-Break-Up" (EBU) model, proposed by Spalding in the 1970s [2], and based on an intuitive concept, similar to the mixing-length model for turbulent flows. Assume that the turbulent reaction rate is determined by two mechanisms: the Arrhenius reaction mechanism and the turbulent fluctuation mechanism.

There are two time scales: the reaction time and the diffusion time. In the case of turbulent flows the latter is the fluctuation time. The reaction time is

$$\tau_c = Y_s / \overline{w}_{sA} \tag{7.3}$$

where $\overline{w}_{sA}$ is the reaction rate due to the laminar reaction mechanism, namely, the reaction rate in terms of the time-averaged values. For a one-step second-order reaction, the Arrhenius expression is

$$\overline{w}_{sA} = \rho^2 \overline{k} \overline{Y}_1 \overline{Y}_2 = B \rho^2 \overline{Y}_1 \overline{Y}_2 \exp(-E/R\overline{T}) \tag{7.4}$$

The fluctuation time is

$$\tau_T = c_T k / \varepsilon \tag{7.5}$$

Assume $\overline{w}_s = f_1(\tau_c, \tau_T)$, or $\overline{w}_s = f_2(\overline{w}_{sA}, \overline{w}_{sT})$, where $\overline{w}_{sA}$ is given by Eq. (7.4). The problem is how to determine $\overline{w}_{sT}$. Spalding supposed that it is proportional to the mean square value of concentration $g = \overline{Y'^2}$ and the fluctuation frequency ε/k, this is the concept of "Eddy Break Up" (EBU), hence it gives $\overline{w}_{sT} \sim g\varepsilon/k$.

The dimensional analysis gives

$$\overline{w}_{sT} = c_E \rho g \varepsilon / k \tag{7.6}$$

where the empirical constant c_E is taken as $0.35-0.4$. The value of g is determined by solving a transport equation or by an algebraic expression. Alternatively, we may take $g \sim Y_1$ or $g \sim Y_2$. A modification given by Magnussen and Hjertager [3] is

$$\overline{w}_{sT} = -A_{\mathrm{EBU}} \overline{\rho} \frac{\varepsilon}{k} \min \left\{ Y_{\mathrm{Fu}}, \frac{Y_{O_2}}{\beta}, B_{\mathrm{EBU}} \frac{Y_P}{1+\beta} \right\} \tag{7.7}$$

This modified EBU model is called "Eddy Dissipation model" (EDM) in the commercial software FLUENT [4]. Finally the time-averaged reaction rate can be determined by

$$\overline{w}_s = \min[\overline{w}_{sA}, \overline{w}_{sT}] \tag{7.8}$$

Eq. (7.8) is called the EBU-Arrhenius (E-A) model, or a "finite-rate-chemistry/eddy dissipation model" (FRCM/EDM) in the FLUENT code. The E-A model is widely adopted in much commercial software, like FLUENT, CFX, STAR-CD, etc. Its advantage is that it is simple, intuitive, and robust. However, in practical turbulent combustion, in most regions of the flow field, the temperature is sufficiently high, $\overline{w}_{sT}$ is frequently much smaller than $\overline{w}_{sA}$, so only $\overline{w}_{sT}$ plays an important role, and in fact the chemical kinetics are not taken into account, or the turbulence−chemistry interaction is not taken into account. This problem will be analyzed in more detail in the following text. In the FLUENT code, there is a model called the "Eddy dissipation concept (EDC) model" [4], it is claimed that the detailed chemical mechanism can be incorporated in this model. However, it is assumed that the reactions take place on small scales, so the time-averaged reaction rate is affected only by small-scale fluctuations. In fact, the time-averaged reaction rate is affected mainly by large-scale structures.

7.4 THE PRESUMED PDF MODELS

Other kinds of combustion models, different from the idea of the EBU model, are related to the statistic description—the "probability density distribution function (PDF)." Originally, the presumed or simplified PDF methods were proposed to simulate the fast-chemistry diffusion-controlled turbulent flames. Subsequently, for simulating a multiple-species turbulent reaction in coal combustion, a presumed PDF-local instantaneous equilibrium model was proposed. However, it was found that this model could only simulate hydrogen–air non-premixed combustion well with very fast reactions. For hydrocarbon–air, CO–air combustion, and NO_x formation in turbulent flows, the finite-rate chemical kinetics must be taken into account. Therefore, a model of finite-rate chemistry with presumed PDF was later proposed to simulate NO_x formation in turbulent flows. It is difficult to simulate detailed chemistry in the presumed PDF models. Subsequently, in the 1980s, a PDF transport equation model was proposed. In this section, only the presumed PDF models will at first be discussed.

7.4.1 The Probability Density Distribution Function

Assume a random function f changing in the range of 0 to 1 with time (Fig. 7.1). The probability of this function in the interval of "f" to "$f + df$" is $p(f)df$. Then $p(f)$ is called the probability density distribution function (PDF). Obviously, there should be

$$\int_0^1 p(f)df = 1 \tag{7.9}$$

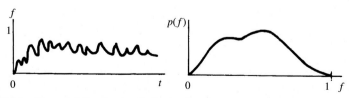

FIGURE 7.1   The probability density distribution function.

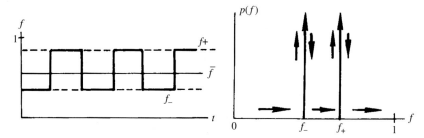

FIGURE 7.2 The PDF of a top-hat fluctuation.

For a given PDF the statistically averaged value and mean square value of the fluctuation of f are determined by

$$\bar{f} = \int_0^1 fp(f)\mathrm{d}f$$

$$\overline{f'^2} = \overline{f^2} - (\bar{f})^2 = \int_0^1 f^2 p(f)\mathrm{d}f - \left(\int_0^1 fp(f)\mathrm{d}f\right)^2 \tag{7.10}$$

The statistically averaged and mean square fluctuation values of any function $\phi(f)$ are

$$\overline{\phi(f)} = \int_0^1 \phi(f)p(f)\mathrm{d}f \quad \overline{\phi'^2} = \int_0^1 \phi^2(f)p(f)\mathrm{d}f - \left(\int_0^1 \phi(f)p(f)\mathrm{d}f\right)^2 \tag{7.11}$$

The simplest method is to assume a top-hat fluctuation, as shown in Fig. 7.2. In this case there are two peaks of $p(f)$ at $f = f_-$ and $f = f_+$. When $f \neq f_-$ and $f \neq f_+$, $p(f) = 0$. The expression of this PDF is

$$p(f) = \alpha\delta(f_-) + (1 - \alpha)\delta(f_+) \tag{7.12}$$

When taking $\alpha = 0.5$, we have

$$\bar{f} = (f_- + f_+)/2 \tag{7.13}$$

$$g = \overline{f'^2} = (\bar{f} - f_-)^2 = (f_+ - \bar{f})^2 \tag{7.14}$$

$$f_- = \bar{f} - g^{1/2}, \quad f_+ = \bar{f} + g^{1/2} \tag{7.15}$$

7.4.2 The Simplified PDF-Local Instantaneous Nonpremixed Fast-Chemistry Model

This model was first proposed by Spalding [5]. For a one-step reaction of fuel with oxygen, the stoichiometric relationship is

$$\begin{array}{cccc} \text{Fuel} & + & \text{Oxygen} & \rightarrow & \text{Combustion Products} \\ w_F & & w_{ox}/\beta & & -w_{pr}/(1 + \beta) \end{array}$$

Using the Zeldovich transformation, a combined mass fraction is obtained as

$$X = Y_F - Y_{ox}/\beta$$

X is a conservative scalar, since there are no source terms in the conservation equation of X. For the nonpremixed or diffusion-controlled combustion, the so-called "mixture fraction" can be defined as

$$f \equiv (X - X_2)/(X_1 - X_2)$$

where X_1 and X_2 denote the values of X at the fuel side and the oxygen side of an instantaneous laminar diffusion flame (Burk–Shuman flame), respectively. There should be

$$X_1 = Y_{F1} = 1, \quad X_2 = -Y_{ox2}/\beta = -1/\beta,$$
$$f_1 = 1, \quad f_2 = 0$$

In general, the mixture fraction expresses the degree of mixing of two or more species in space or time. Its conservation equation is:

$$\frac{\partial}{\partial t}(\rho f) + \frac{\partial}{\partial x_j}(\rho v_j f) = \frac{\partial}{\partial x_j}\left(D\rho\frac{\partial f}{\partial x_j}\right) \tag{7.16}$$

Assume a local instantaneous nonpremixed flame with fast chemistry, i.e., the fuel and oxygen do not coexist at the same time, but their time-averaged value may coexist at the same location. Therefore, the relationship among the mass fraction spatial distributions of fuel, oxygen, and combustion products in a laminar diffusion flame is taken as the possibly realizable instantaneous relationships between Y_F and Y_{pr}, or Y_{ox} and Y_{pr} in the turbulent diffusion flame (Fig. 7.3). When $1 > f > f_F$ there is no oxygen, $Y_F = (f - f_F)/(1 - f_F)$. When $0 < f < f_F$ there is no fuel, $Y_{ox} = 1 - f/f_F$. It should be noticed that these relationships cannot be taken as time-averaged values, namely for all the time, $\overline{Y}_F \neq (\overline{f} - f_F)/(1 - f_F)$, $\overline{Y}_{ox} \neq 1 - \overline{f}/f_F$, because these instantaneous relationships do not hold all the time, and the occurrence of these instantaneous values has different probabilities. Therefore,

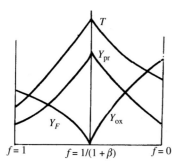

FIGURE 7.3 The relationships among different instantaneous species mass fractions.

the assumption of instantaneous nonpremixing gives the instantaneous species mass fractions and temperature in terms of the same conservative scalar—the mixture fraction f.

The time-averaged values are determined by solving the following equations:

$$\frac{\partial \rho}{\partial t} + \frac{\partial}{\partial x_j}(\rho v_j) = 0 \tag{7.17}$$

$$\frac{\partial}{\partial t}(\rho v_i) + \frac{\partial}{\partial x_j}(\rho v_i v_j) = \frac{\partial}{\partial x_j}\left(\mu_e \frac{\partial v_i}{\partial x_j}\right) + S_{vi} \tag{7.18}$$

$$\frac{\partial}{\partial t}(\rho k) + \frac{\partial}{\partial x_j}(\rho v_j k) = \frac{\partial}{\partial x_j}\left(\frac{\mu_e}{\sigma_k}\frac{\partial k}{\partial x_j}\right) + G_k - \rho \varepsilon \tag{7.19}$$

$$\frac{\partial}{\partial t}(\rho \varepsilon) + \frac{\partial}{\partial x_j}(\rho v_j \varepsilon) = \frac{\partial}{\partial x_j}\left(\frac{\mu_e}{\sigma_\varepsilon}\frac{\partial \varepsilon}{\partial x_j}\right) + \frac{\varepsilon}{k}(c_1 G_k - c_2 \rho \varepsilon) \tag{7.20}$$

$$\frac{\partial}{\partial t}(\rho \bar{f}) + \frac{\partial}{\partial x_j}(\rho v_j \bar{f}) = \frac{\partial}{\partial x_j}\left(\frac{\mu_e}{\sigma_f}\frac{\partial \bar{f}}{\partial x_j}\right) \tag{7.21}$$

$$\frac{\partial}{\partial t}(\rho g) + \frac{\partial}{\partial x_j}(\rho v_j g) = \frac{\partial}{\partial x_j}\left(\frac{\mu_e}{\sigma_g}\frac{\partial g}{\partial x_j}\right) + c_{g1}\mu_T\left(\frac{\partial \bar{f}}{\partial x_j}\right)^2 - c_{g2}\rho g \varepsilon / k \tag{7.22}$$

where $g = \overline{f'^2}$. By solving these equations, the time-averaged velocity, turbulent kinetic energy its dissipation rate, and the mixture fraction and its mean square fluctuation value can be obtained. Hence, this is called a $k - \varepsilon - f - g$ model. Using Eqs. (7.14) and (7.15), from the time-averaged values of f and g, the values of f_+ and f_- at each location of the flow field can be found. Finally, using Eqs. (7.11) and (7.12) the time-averaged species mass fraction and temperature can be found as

$$\bar{\phi} = \phi(f_-)/2 + \phi(f_+)/2 \tag{7.23}$$

$$\overline{\phi'^2} = [\phi(f_-) - \bar{\phi}]^2 - [\phi(f_+) - \bar{\phi}]^2 \tag{7.24}$$

Other kinds of PDF, such as the clipped Gaussian PDF, can also be used, but its computation time is much greater than that for the top-hat PDF. The predicted axial velocity profiles and the velocity change along the axial direction for a coaxial sudden-expansion diffusion combustion using the k-ε-f-g model and the EBU model [6] are shown in Figs. 7.4 and 7.5, respectively. It can be seen that the results obtained by both the $k - \varepsilon - f - g$ model and the EBU model are close to the experimental results, and there is only a slight difference between the top-hat PDF and the clipped Gaussian PDF predictions.

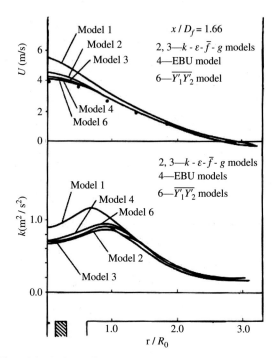

FIGURE 7.4   The axial velocity profiles.

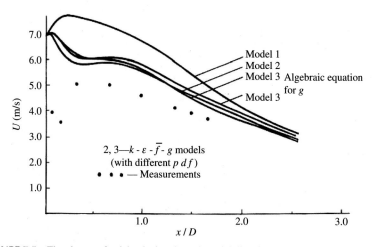

FIGURE 7.5   The change of axial velocity along the axial direction.

7.4.3   The Simplified PDF-Local Instantaneous Equilibrium Model

The above-discussed model is appropriate for only two-component reactions, such as gaseous fuel−oxygen diffusion combustion. For more complicated diffusion combustion, such as coal-volatile combustion, a model of simplified PDF with local instantaneous equilibrium was proposed by Smoot and Smith [7]. For a sudden-expansion combustor with a coaxial jet inlet, the central primary jet is gas fuel plus air, and the annular secondary jet is pure air, the mixture fraction is defined as

$$f = \frac{m_p}{m_p + m_s} \tag{7.25}$$

where m_p is the atom mass coming from the primary gas stream, and m_s is the atom mass coming from the secondary gas stream. In this case the mixture fraction f expresses the degree of mixing at any instant in time and any location, namely the local instantaneous stoichiometric ratio. Any other conservative scalar (not including the species mass fraction Y, since Y is not a conservative scalar) can be expressed as

$$\phi = f\phi_p + (1 - f)\phi_s \tag{7.26}$$

where ϕ_p and ϕ_s are the values of ϕ in the primary and secondary gas streams, respectively. Since in ordinary reactions any element cannot be produced or destroyed, the local instantaneous mass fraction of the element "b_k" is also a conservative scalar. For equal diffusivities of elements, we have

$$b_k = fb_{kp} + (1 - f)b_{ks} \tag{7.27}$$

For an adiabatic system with no radiative and heat conductive losses, the gas enthalpy is also a conservative scalar, thus, we have

$$h = fh_p + (1 - f)h_s \tag{7.28}$$

The time-averaged mixture fraction $\bar{f}$ and the mean square fluctuation g still can be obtained by solving Eqs. (7.21) and (7.22). From the obtained $\bar{f}$ and g and the assumed form of PDF, the PDF at each location can be found. For any scalar function $\phi(f)$, its time-averaged value is

$$\bar{\phi} = \int_0^1 \phi p(f) \mathrm{d}f$$

$$\text{or } \bar{\phi} = \alpha_p \phi_p + \alpha_s \phi_s + \int_{0_+}^{1_-} \phi p(f) \mathrm{d}f \tag{7.29}$$

where α_p and α_s are the values of $p(f)$ at $f = 1$ and $f = 0$, respectively. ϕ_p and ϕ_s are the values of ϕ in the primary and secondary gas streams, respectively. Now, the relationships among the instantaneous temperature, species concentration, and density are obtained by using the concept of local instantaneous equilibrium, based on the given enthalpy and element kinds (for negligible pressure change)

$$T = T(b_k, h), \quad \rho = \rho(b_k, h), \quad Y_s = Y_s(b_k, h)$$

The above-stated functions can be found by the calculation of chemical equilibrium, using the Newton–Rapson iteration, or the method of equilibrium constants, or the Gibbs minimum energy method can be used. For an adiabatic system, the enthalpy and element mass fraction are only the function of the mixture fraction, i.e.,

$$b_k = b_k(f); \quad h = h(f)$$

Hence, $T = T(f)$ $\rho = \rho\ (f)$ $Y_s = Y_s(f)$ and the time-averaged temperature, density, and species concentration can be found directly by using the $p(f)$

$$\overline{T}(f); \quad \overline{\rho}(f); \quad \overline{Y}_s(f)$$

For the nonadiabatic system, it is necessary to solve the energy equation to find the enthalpy. In this case we have

$$T = T(f, h), \quad r = r(f, h), \quad Y_s = Y_s(f, h)$$

The time-averaged temperature, density, and species mass fraction are obtained by the integration of their instantaneous values over PDF. For a given clipped Gaussian PDF plus two δ functions at $f = 0$ and $f = 1$, the time-averaged value is

$$\overline{\phi} = \alpha_p \phi_p + \alpha_s \phi_s + \int_{0_+}^{1_-} \phi p(f, h) df dh \tag{7.30}$$

However, the joint PDF $p(f, h)$ is difficult to be presumed in advance. An approximate approach is to assume $p(f, h) = p(f)p(h)$. Another approximate approach is to assume

$$h = h_f + h_r; \quad h_f = h_f(h); \quad h_r = \overline{h}_r$$

Hence, $p(f, h) = p(h)$. The prediction results for hydrogen–air diffusion combustion in a coaxial jet sudden-expansion combustor [6] using the simplified PDF-local instantaneous equilibrium model are shown in Fig. 7.6. Except for the RMS fluctuation velocity, the predicted time-averaged velocity, temperature, H_2 O_2, and H_2O mass fractions are in good agreement with the experimental results. These results imply that the mechanism of hydrogen–air combustion is approaching the local instantaneous equilibrium model. However, the predicted species mass fractions for methane–air diffusion combustion (Fig. 7.7) [6] are far from the experimental results; there is qualitative discrepancy. These results indicate that the hydrocarbon–air combustion is not in the local instantaneous equilibrium

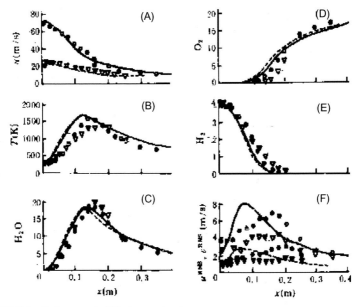

FIGURE 7.6 Hydrogen−air diffusion combustion.

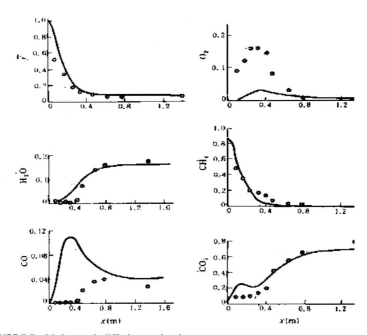

FIGURE 7.7 Methane−air diffusion combustion.

state. As regards the nitric oxide formation, due to high activation energy, the chemical reaction rate is low. The finite-rate chemical kinetics must be taken into account.

7.4.4   The Simplified-PDF Finite-Rate Model

For many cases the reactions are not infinitively fast or not in local instantaneous equilibrium. A typical case is nitric oxide formation in turbulent flows. When using the simplified PDF, the finite reaction rate has to be incorporated. A review for modeling the NO_x turbulent reaction rate has been given by the present author [8]. In the FLUENT code, a simplified PDF with finite-rate reaction model is taken to simulate NO_x formation [4]. The reaction mechanism for thermal NO formation is determined by the Zeldovich formula

$$
\begin{aligned}
w_{NO}^{+} &= 1.35 \times 10^{16} T^{-1} \rho Y_{N_2} Y_{O_2}^{0.5} \exp(-69160/T) \\
w_{NO}^{-} &= 22.6 T^{-1} \rho Y_{N_2}^2 Y_{O_2}^{-0.5} \exp(-47355/T)
\end{aligned}
\tag{7.31}
$$

To simulate the NO turbulent reaction rate, different forms of PDF can be used. One approach is using the product of the PDF for temperature fluctuation with the PDF for the mixture fraction instead of the joint PDF $p(f, T)$. The time-averaged NO reaction rate is

$$
\overline{w}_{NO} = \int_{T_0}^{T_m} \int_{f_1}^{f_2} w_{NO}(T, f) p(T) p(f) \mathrm{d}T \mathrm{d}f \Big/ \int_{f_1}^{f_2} p(f) \mathrm{d}f] \mathrm{d}T \mathrm{d}f
\tag{7.32}
$$

The other approach is using the product of the PDF for the dimensionless temperature fluctuation with PDF for the oxygen mass fraction fluctuation. The time-averaged reaction rate is

$$
\overline{w}_{NO} = \int_0^1 \int_0^1 w_{NO}(\theta, Y_{O_2}) (\overline{\rho}/\rho) p(\theta) p(Y_{O_2}) \mathrm{d}\theta \mathrm{d}Y_{O_2}
\tag{7.33}
$$

where $\theta = (T - T_{min})/(T_{max} - T_{min})$. Alternatively, only the PDF for the mixture fraction fluctuation is used, i.e.,

$$
\overline{w}_{NO} = \int_0^1 w_{NO}(f) p(f) \mathrm{d}f
\tag{7.34}
$$
$$
f = (X - X_2)/(X_1 - X_2); \quad X = Y_f - Y_{O_2}/\beta
$$

For all of the above-stated versions, the β function PDF is used

$$
p(\phi) = [\phi^{a-1}(1-\phi)^{b-1}]/[\int_0^1 \phi^{a-1}(1-\phi)^{b-1} \mathrm{d}\phi]
\tag{7.35}
$$

where ϕ may be temperature, oxygen mass fraction, or mixture fraction. For the RMS values, three models [9,10] can be used. The PDF-1 model is an empirical algebraic expression

$$\overline{\phi'^2} = s\overline{\phi}(1 - \overline{\phi}) \tag{7.36}$$

where $s = 0.6$. The PDF-2 model is an algebraic second-order moment expression

$$\overline{\phi'^2} = c_1 \mu_T \left(\frac{\partial \overline{\phi}}{\partial x_j}\right)^2 /(c_2 \rho \varepsilon / k) \tag{7.37}$$

The PDF-3 model is a transport equation of the second-order moment

$$\frac{\partial}{\partial x_j}(\rho v_j g) = \frac{\partial}{\partial x_j}\left(\frac{\mu_T}{\sigma_\phi}\frac{\partial g}{\partial x_j}\right) + c_1 \mu_T \left(\frac{\partial \overline{\phi}}{\partial x_j}\right)^2 - c_2 \rho \varepsilon g / k + 2\overline{\phi' S'_\phi}, \quad g = \overline{\phi'^2} \tag{7.38}$$

where the last term on the right-hand side of Eq. (7.38) is the source term due to radiation and reaction. The presumed PDF-finite-rate model was used to simulate the NO formation in methane−air diffusion combustion [9,10]. The prediction results are given in Figs. 7.8 and 7.9 [6]. It can be seen that in most cases the NO and HCN concentration is underpredicted, and the PDF-3 model is somewhat better. The reason is that the product of separate PDFs, instead of a joint PDF, neglects the correlation between the temperature and concentration fluctuations, hence certainly underpredicting the reaction rate. Fetting et al. adopted a presumed joint PDF [11] for two species concentrations and temperature to simulate CO−air diffusion combustion. The instantaneous CO−O_2 reaction rate is

$$w_{co} = 3.98 \times 10^{17} \rho^{1.75} Y_{co} (Y_{H_2O}/M_{H_2O})^{0.5} (Y_{O_2}/M_{O_2})^{0.25} \exp(-E/RT) \tag{7.39}$$

The species mass fraction and temperature in terms of the mixture fraction and enthalpy are

$$Y_{H_2O} = (1 - f)Y_{H_2O,in} \quad Y_{O_2} = (1 - f)Y_{O_2,in} - \beta(fY_{CO,in} - Y_{CO})$$
$$T = (h - h_f Y_{CO})/c_p$$

and $w_{co} = w_{co}(Y_{CO}, f, h)$

The time-averaged reaction rate is obtained by using a joint PDF

$$\overline{w}_{co} = \int_0^1 \int_0^1 \int_{h_{1,in}}^{h_{2,in}} w_{co}(Y_{CO}, f, h)p(Y_{CO}, f, h)dY_{CO}dfdh \tag{7.40}$$

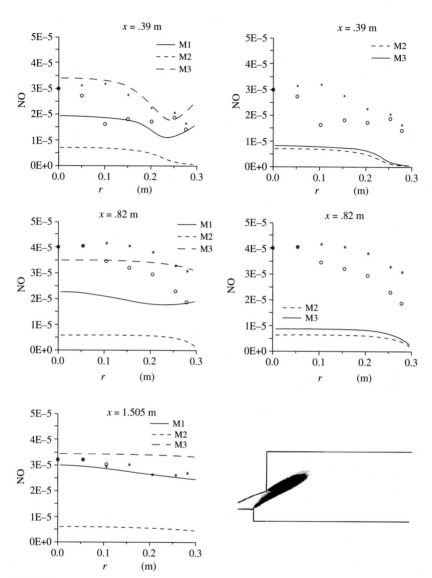

FIGURE 7.8   NO formation in methane−air diffusion combustion (•, ○ Exp; — Pred.).

The joint PDF is presumed to be a Gaussian distribution as

$$p(Y, f, h) = C\exp(-Q/2) \qquad (7.41)$$

where C and Q are the functions of variance, covariance, and the time-averaged Y, f, h. The variance and covariance are determined by their transport equations. This model was compared with other three models

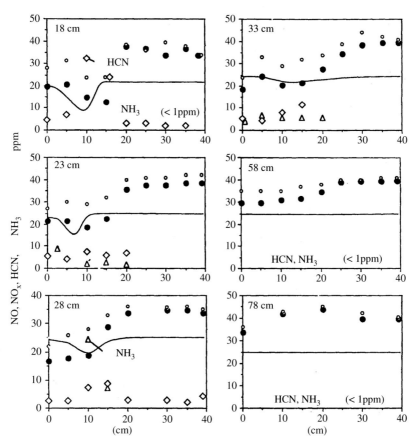

FIGURE 7.9 NOx formation in methane−air diffusion combustion (Dots, Exp.; Line, Pred.).

$$\overline{w}_{CO} = w_{CO}(\overline{Y}_{CO}, \overline{f}, \overline{h})$$

$$\overline{w}_{CO} = \int_0^1 \int_0^1 \int_{h_{1,in}}^{h_{2,in}} w_{co} p(Y_{CO}) p(f) p(h) \mathrm{d}h \mathrm{d}f \mathrm{d}Y_{CO}$$

$$\overline{w}_{CO} = w_{CO}(\overline{Y}_{CO}, \overline{f}, \overline{h})[1 + f(\overline{\phi'^2}, \overline{\phi'\psi'}, \ldots)]$$

Fig. 7.10 gives the predicted CO molar fraction in a CO−O_2 jet diffusion flame [6], where different models, including the reaction rate in terms of time-averaged values, the model using a product of three 1-D PDFs and the model using a joint PDF, are adopted. It can be seen that these four models do not give the results in agreement with the experimental results.

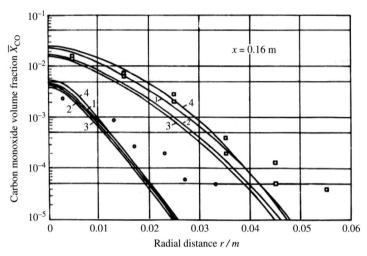

FIGURE 7.10 The CO concentration in a $CO-O_2$ jet diffusion flame.

7.5 THE PDF TRANSPORT EQUATION MODEL

The shortcomings of the presumed PDF models are: (1) the PDF in the whole flow field is changeable, e.g., it approaches to a two-δ PDF at the outer boundary of a jet, and approaches a Gaussian PDF at the axis of a jet; (2) the detailed chemical kinetics cannot be incorporated into the presumed PDF models. Hence the PDF transport equation model was proposed by Pope [12]. There are different versions of this model. One is called the "composition PDF equation model," in which the velocity field is obtained by using the ordinary turbulence models—the statistic moment models—and the temperature and species concentrations are obtained using the PDF equations. The transport equation of joint PDF in the phase space of geometric, species, and enthalpy coordinates is

$$\frac{\partial}{\partial t}(\rho P) + \frac{\partial}{\partial x_j}(\rho v_j P) + \frac{\partial}{\partial \psi_k}(\rho S_k P) = \frac{\partial}{\partial \psi_k}\left(\rho < \frac{\partial J_{j,k}}{\rho \partial x_j}\bigg|\psi > P\right) - \frac{\partial}{\partial x_j}\left(\rho < v''_j \bigg|\psi > P\right)$$

(7.42)

where the first two terms on the left-hand side are the time-changing rate and the convection of PDF in the geometrical space, the third term on the left-hand side is the source term due to chemical reactions, S_k is the reaction rate of species k, and the symbol ψ denotes the composition space coordinate. v''_j is the velocity fluctuation, the notation $< >$ denotes expectations, and $<A|B>$ is the conditionally averaged value of event A, given that event B occurs. All the terms on the left-hand side are exact, and do not need models. The advantage of the PDF equation model is that the highly nonlinear reaction term is the exact one, and does not need models. The two terms on the right-hand side represent the molecular mixing/diffusion and the scalar convection by turbulence (turbulent scalar flux) in the geometrical space, respectively, and need models. The turbulent scalar flux is modeled by gradient diffusion assumption

$$-\frac{\partial}{\partial x_j}\left[\rho<v''_j\Big|\psi>P\right] = -\frac{\partial}{\partial x_j}\left(\frac{\mu_T}{\rho Sc_T}\frac{\partial P}{\partial x_j}\right) \tag{7.43}$$

The weakest point of the PDF equation model is the modeling of the mixing term. Pope gave a simple description of the PDF method [13]. This method can be considered as a grid "particle" method, or a generalized Lagrangian method in the phase space. There is a large amount of computational "particles" (fluid elements) in each grid of the flow field. Each "particle" has a different location X^* (t), velocity U^* (t), species composition ϕ^* (t), and fluctuation frequency ω (t). Its mass is unchanged, and other variables change in time. The "particle" random position in the stochastic trajectories of the phase space is

$$\frac{dX^*(t)}{dt} = U^*(t) \tag{7.43}$$

The particle random species composition equation is

$$\frac{d\phi^*(t)}{dt} = -\frac{1}{2}C_\phi<\varpi>\left[\phi^*(t) - <\phi>\right] + S(\phi^*) \tag{7.44}$$

where the first term on the right-hand side is the molecular mixing term closed by the IEM (Interaction by Exchange with the Mean) model, reflecting the random value of the species composition relaxed to its local averaged value, and is proportional to the local averaged fluctuation frequency. The second term on the right-hand side is the reaction term, which is exact and does not need models. For the stochastic velocity, a simplified Langevin equation model (SLM) is used as

$$\frac{dU_i^*(t)}{dt} = -\frac{1}{<\rho>}\frac{\partial p}{\partial x_i} - \left(\frac{1}{2}+\frac{3}{4}C_0\right)<\varpi>(U_i^* - <U_i>)$$
$$+ (C_0<\varpi>k)^{1/2}\frac{dW_i}{dt} \tag{7.45}$$

The first term on the right-hand side of Eq. (7.45) is the averaged pressure gradient, the second term, like the IEM model, expresses the "particle" velocity relaxed to its local averaged value, and the last term includes the Wiener process, causing the diffusion in the velocity space. k is the turbulent kinetic energy, $<\omega>k$ is the time-averaged dissipation rate ε. $C_0\varepsilon$ is the diffusion coefficient. Eqs. (7.43)−(7.45) are solved by the Monte-Carlo method. Smoot et al. [14] simulated a bluff-body stabilized premixed methane−air flame using the composition PDF equation model together with a $k-\varepsilon$

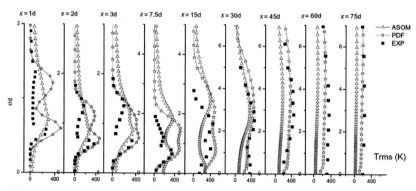

FIGURE 7.11   RMS value of temperature fluctuation.

turbulence model, IEM mixing model, and a five-step reduced chemistry mechanism. The predicted methane concentration and temperature are in general agreement with the measurement results, but there is some discrepancy in some locations due to either the overprediction of mixing by the $k - \varepsilon$ turbulence model, or the simplified five-step chemistry. Wang and Zhou et al. [15] used the composition PDF equation model together with the Reynolds stress turbulence model to simulate the Sandia methane–air piloted jet Flame C, the predicted RMS value of temperature fluctuation (Fig. 7.11) is in good agreement with the measured results. The ASOM model in this figure is explained in the following text.

7.6   THE BRAY–MOSS–LIBBY (BML) MODEL

The so-called "Bray–Moss–Libby (BML) model" was proposed [16] based on the concept of Damkoher–Shelkin's winkled flame surface for a pre-mixed turbulent flame [17]. In this model, a premixed combustion progress variable is defined as

$$c = (T - T_u)/(T_b - T_u) = Y_F/Y_{F,b} \qquad (7.46)$$

The Favre-averaged transport equation for c, when neglecting the molecular transport, is

$$\bar{\rho}\frac{\partial \bar{\bar{c}}}{\partial t} + \bar{\rho}\bar{\bar{u}}_j\frac{\partial \bar{\bar{c}}}{\partial x_j} = -\frac{\partial}{\partial x_j}\left(\overline{\rho u''_j c''}\right) + \bar{w}_c \qquad (7.47)$$

where the superscript double bars denote the Favre-averaged values, and the two primes symbols denote the fluctuation values in the Favre averaging, w is the reaction rate, which is closed by

$$\bar{w}_c = \rho_u S_L^0 l_0 \sum \qquad (7.48)$$

where l_0 is the local strain (stretch) factor of the laminar flame velocity, S_L^0 is the unstretched laminar flame velocity, and $\sum$ is the flame surface density (flame area per volume), determined either by an algebraic expression, or by a transport equation.

7.7   THE CONDITIONAL MOMENT CLOSURE (CMC) MODEL

The conditional moment closure (CMC) model was proposed by Bilger and Klimenko [18]. The conditional averaged value of a scalar is defined as

$$Q = \langle Y | \xi = \eta \rangle \tag{7.49}$$

The time-averaged value of mass fraction is obtained by integration over PDF

$$\langle Y \rangle = \int_{-\infty}^{\infty} \langle Y | \xi = \eta \rangle P(\eta) \mathrm{d}\eta \tag{7.50}$$

The transport equation of the conditional moment Q is

$$
\begin{aligned}
\frac{\partial Q P(\eta) \rho_\eta}{\partial t} &+ div(\rho_\eta \langle VY | \eta \rangle P(\eta)) = \langle w | \eta \rangle P(\eta) \rho_\eta \\
&+ \frac{\partial}{\partial \eta} \left(\langle N | \eta \rangle P(\eta) \rho_\eta \frac{\partial Q}{\partial \eta} - \frac{\partial \langle N | \eta \rangle P(\eta) \rho_\eta}{\partial \eta} Q \right)
\end{aligned}
\tag{7.51}
$$

where η is the mixture fraction, and $P(\eta)$ is the PDF. The scalar dissipation rate is modeled as $N = D(\nabla \xi)(\nabla \xi)$.

It was found by experiment that the second-order conditional moment is negligible; hence the conditionally averaged reaction rate in the conditionally averaged species equation can be solved without closure models. To find the time-averaged value from the conditionally averaged value can be done by using either a presumed PDF, or solving the $P(\eta)$ equation. Fig. 7.12 shows

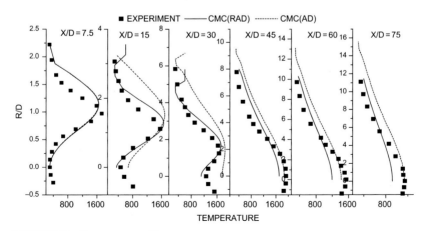

FIGURE 7.12   Temperature profiles.

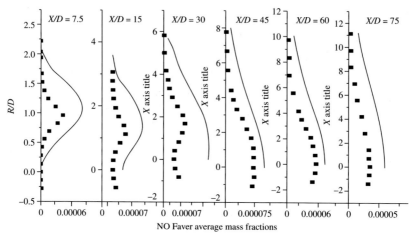

FIGURE 7.13   NO mass fraction.

the predicted temperature profiles of the Sandia methane−air jet diffusion flame using the CMC model with a transport equation of $P(\eta)$ [19]. The predictions are in good agreement with the experimental results. Fig. 7.13 gives the predicted NO mass fraction. It can be seen that the agreement with experiments is not good, and that the NO formation is overpredicted. The reason for this is that for NO formation the second-order conditional moment cannot be neglected, and besides it is necessary to close the scalar dissipation rate equation, and the model is too complex.

7.8   THE LAMINAR-FLAMELET MODEL

The laminar-flamelet model was proposed by Peters [20] to simulate nonpremixed flames. Its basic idea is to assume that there are laminar diffusion flamelets in a turbulent diffusion flame. By using the mixture fraction and flamelet equations the modeling of turbulence effect and finite-rate chemical kinetics can be decoupled. For a hydrocarbon−oxygen reaction,

$$C_m H_n + \nu'_{O_2} O_2 \to \nu''_{CO_2} CO_2 + \nu''_{H_2O} H_2O$$

a mixture fraction (a dimensionless combined mass fraction) can be defined as

$$Z = \frac{\xi_c/(mM_c) + \xi_H/(nM_H) + 2(Y_{O_2,2} - \xi_o)/(\nu'_{O_2} M_{O_2})}{\xi_{c,1}/(mM_c) + \xi_{H,1}/(nM_H) + 2Y_{O_2,2}/(\nu'_{O_2} M_{O_2})} \tag{7.52}$$

where the subscripts 1 and 2 denote the fuel and oxidizer inlet parameters, ξ_c, ξ_H, and ξ_o denote the mass fractions of the elements C, H, and O,

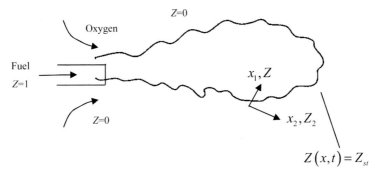

FIGURE 7.14   The isosurface of the mixture fraction in a turbulent jet diffusion flame.

respectively, M_c, M_H, and M_o express their atom weights, respectively, and ν_{o2} denotes the stoichiometric coefficient. The scheme of the concept of laminar flamelets is shown in Fig. 7.14.

The value of Z indicates the position of the flamelet. When taking the equal diffusivity of each species, the transport equation of Z can be obtained as

$$\rho\frac{\partial Z}{\partial t} + \rho u_j\frac{\partial Z}{\partial x_j} = \frac{\partial}{\partial x_j}\left(D\rho\frac{\partial Z}{\partial x_j}\right) \tag{7.53}$$

From Eq. (7.53), the averaged and RMS fluctuation value of Z in the whole flow field can be obtained by solving their transport equations, which reflect the effect of turbulence on the distribution of Z. For a presumed PDF of Z, the averaged species mass fraction and temperature can be obtained by

$$\tilde{Y} = \int Y(Z)p(Z)dZ; \tilde{T} = \int T(Z)p(Z)dZ \tag{7.54}$$

where $Y(Z)$ and $T(Z)$ are the instantaneous functional relationships between Y and Z or T and Z, obtained by solving the following flamelet equations, including the effect of finite-rate chemistry.

$$\rho\frac{\partial Y_\alpha}{\partial t} + \rho u_1\frac{\partial Y_\alpha}{\partial x_1} = \frac{\partial}{\partial x_1}\left(\rho D\frac{\partial Y_\alpha}{\partial x_1}\right) + \rho S_\alpha \tag{7.55}$$

$$\rho\frac{\partial T}{\partial t} + \rho u_1\frac{\partial T}{\partial x_1} = \frac{\partial}{\partial x_1}\left(\rho D\frac{\partial T}{\partial x_1}\right) + \rho S_T \tag{7.56}$$

$$\rho\frac{\partial Z}{\partial t} + \rho u_1\frac{\partial Z}{\partial x_1} = \frac{\partial}{\partial x_1}\left(\rho D\frac{\partial Z}{\partial x_1}\right) \tag{7.57}$$

where the source terms on the right-hand side of these equations are exact reaction terms, so the detailed chemistry can be incorporated into the flamelet model. The β-PDF is frequently taken as the presumed $p(Z)$. The laminar-flamelet model works well for simulating diffusion flames, such as the Sandia piloted methane−air jet diffusion flame.

7.9 THE SECOND-ORDER MOMENT COMBUSTION MODEL

As discussed above, in a simple model, like the EBU or EDM model, the effect of turbulence is usually dominant and the effect of chemical kinetics is significantly weakened. In the simplified PDF models, including the fast-chemistry, finite-rate, and local-equilibrium models, either the effect of chemical kinetics is still not taken into account, or the assumed PDF is frequently a product of several one-variable PDFs instead of the joint PDF, leading to underprediction of the time-averaged reaction rate. More advanced models, like the conditional moment closure (CMC) model and the flamelet model, may be a good choice, but they give good results only for certain flame types (e.g., nonpremixed flames) and flame structures (wrinkled flamelets, corrugated flamelets, broken reaction zones, etc.). The BML model and the G-equation model are limited to premixed flames and are good for certain flame structures. A general model for all flame types and flame structures is the PDF transport equation model, where the reaction term is an exact one and does not need closure models. However, its small-scale mixing closure still has some uncertainty, and due to using the Monte-Carlo method to solve the PDF equations, it is computationally very expensive, needs a computation time about two orders of magnitude greater than that for simple models, in particular when using in LES, and hence is not convenient for solving engineering problems. Therefore, engineers are still seeking an economical and reasonable turbulent combustion model. One choice is to return to the moment method.

7.9.1 The Early Developed Second-Order Moment Model

The idea of an early developed second-order moment combustion model is using an approximation of series expansion of the exponential term in the reaction rate expression, assuming $(E/RT)\cdot T'/T < <1$ and neglecting higher-order terms [21]. For a one-step second-order reaction, the instantaneous reaction rate is

$$w_s = B\rho^2 Y_1 Y_2 \exp(-E/RT) = B\rho^2 k Y_1 Y_2 \qquad (7.58)$$

After taking Reynolds decomposition $k = \bar{k} + k'$, $Y_1 = \bar{Y}_1 + Y'_1$, $Y_2 = \bar{Y}_2 + Y'_2$, $T = \bar{T} + T'$, the averaged reaction rate becomes

$$\overline{w}_s = \overline{B\rho^2 Y_1 Y_2 \exp(-E/RT)} = \overline{B\rho^2 k Y_1 Y_2} = B\rho^2 \overline{(\bar{k} + k')(\bar{Y}_1 + Y'_1)(\bar{Y}_2 + Y'_2)}$$

or

$$\overline{w}_s = B\rho^2 \overline{(\overline{Y}_1 + Y'_1)(\overline{Y}_2 + Y'_2)\exp\left[-\frac{E}{R(\overline{T} + T')}\right]} \tag{7.59}$$

where $k = \exp(-E/RT)$

The early developed second-order model uses a series expansion for the nonlinear exponential term as

$$\exp\left(-\frac{E}{RT}\right) = \exp\left[-\frac{E}{R(\overline{T} + T')}\right] = \exp\left[-\frac{E}{R\overline{T}}\left(1 + \frac{T'}{\overline{T}}\right)^{-1}\right]$$

Assuming $T'/\overline{T} << 1$, it becomes

$$\exp\left(-\frac{E}{RT}\right) \approx \exp\left[-\frac{E}{R\overline{T}}\left(1 - \frac{T'}{\overline{T}}\right)\right] = \exp\left(-\frac{E}{R\overline{T}}\right)\exp\left(\frac{E}{R\overline{T}^2}T'\right)$$

Expanding the exponential term by a series

$$\exp\left(-\frac{E}{R\overline{T}^2}T'\right) = 1 + \frac{E}{R\overline{T}}\frac{T'}{\overline{T}} + \frac{1}{2}\left(\frac{E}{R\overline{T}}\frac{T'}{\overline{T}}\right)^2 + \cdots$$

After a further assumption $\dfrac{E}{R\overline{T}}\dfrac{T'}{\overline{T}} << 1$ and neglecting the higher-order terms

$$\exp\left(-\frac{E}{R\overline{T}^2}T'\right) \approx 1 + \frac{E}{R\overline{T}}\frac{T'}{\overline{T}} + \frac{1}{2}\left(\frac{E}{R\overline{T}}\frac{T'}{\overline{T}}\right)^2$$

the time-averaged reaction rate becomes

$$\overline{w}_s = B\rho^2(\overline{Y}_1 + Y'_1)(\overline{Y}_2 + Y'_2)\exp\left(-\frac{E}{R\overline{T}}\right)\left[1 + \frac{E}{R\overline{T}}\frac{T'}{\overline{T}} + \frac{1}{2}\left(\frac{E}{R\overline{T}}\frac{T'}{\overline{T}}\right)^2\right]$$

or

$$\overline{w}_s = B\rho^2\overline{Y}_1\overline{Y}_2\exp\left(-\frac{E}{R\overline{T}}\right)\left[1 + \frac{\overline{Y'_1 Y'_2}}{\overline{Y}_1\overline{Y}_2} + \frac{E}{R\overline{T}}\left(\frac{\overline{T'Y'_1}}{\overline{T}\overline{Y}_1} + \frac{\overline{T'Y'_2}}{\overline{T}\overline{Y}_2}\right) + \frac{1}{2}\left(\frac{E}{R\overline{T}}\right)^2\frac{\overline{(T')^2}}{\overline{T}^2}\right] \tag{7.60}$$

However, in practical combustion processes E/RT is the order of 5−10 and the relative temperature fluctuation is not much less than unity, so this approximation leads to remarkably underpredicting the time-averaged reaction rate. This early developed model was adopted to predict methane−air diffusion combustion, taking 50 species and 300 elementary reactions into account [6]. Although the predicted temperature and main species concentration are close to the experimental results, the predicted NO concentration is much lower than the experimental results (Fig. 7.15), showing the serious shortcomings of this model.

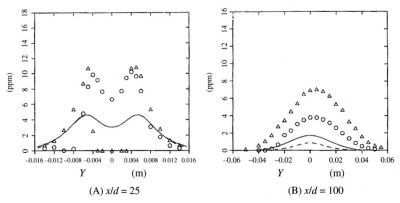

FIGURE 7.15   NO mass fraction in a methane−air diffusion flame ($\triangle$, $\bigcirc$ Exp; — Pred.).

An attempt was made by the present author to use only the concentration−concentration correlation $\overline{Y'_1 Y'_2}$, combined with the presumed PDF for the concentration-reaction rate coefficient correlation $\overline{k'Y'}$[22,23], i.e.,

$$\overline{k'Y'} = \overline{kY} - \overline{k}\,\overline{Y} = \iint kYp(T)p(Y)\mathrm{d}T\mathrm{d}Y - \overline{Y}\int kp(T)\mathrm{d}T$$

$$\overline{w}_s = B\rho^2[(\overline{Y}_1\overline{Y}_2 + \overline{Y'_1 Y'_2})\int kp(T)\mathrm{d}T + \overline{Y}_1(\iint kY_2 p(T)p(Y_2)\mathrm{d}T\mathrm{d}Y_2 - \overline{Y}_2\int kp(T)\mathrm{d}T)$$
$$+ \overline{Y}_2(\iint kY_1 p(T)p(Y_1)\mathrm{d}T\mathrm{d}Y_1 - \overline{Y}_1\int kp(T)\mathrm{d}T)]$$

When taking the top-hat PDFs for $p(T)$, $p(Y_1)$, and $p(Y_2)$, the time-averaged reaction rate becomes

$$\overline{w}_s = B\rho^2\overline{Y}_1\overline{Y}_2\exp(-E/R\overline{T})Z \tag{7.61}$$

$$Z = ch\left(\frac{E}{R\overline{T}}\frac{g_T^{1/2}}{\overline{T}}\right)\left[1 + \frac{\overline{Y'_1 Y'_2}}{\overline{Y}_1\overline{Y}_2} + \left(\frac{g_{Y1}^{1/2}}{\overline{Y}_1} + \frac{g_{Y2}^{1/2}}{\overline{Y}_2}\right)th\left(\frac{E}{R\overline{T}}\frac{g_T^{1/2}}{\overline{T}}\right)\right] \tag{7.62}$$

where

$$g_T = \overline{T'^2}, \ g_Y = \overline{Y'^2} \ ch(x) = (e^x + e^{-x})/2, \ th(x) = \frac{sh(x)}{ch(x)} \ sh(x) = (e^x - e^{-x})/2$$

Subsequently, it was found that this is not a good closure for $\overline{k'Y'}$. An updated second-order moment (SOM) combustion model, also called a unified second-order moment (USM) model, was proposed by the present author [24,25]. This is an unconditional second-order moment model, based fully on the idea of a second-order moment turbulence model (Reynolds stress equation model).

7.9.2 An Updated Second-Order Moment (SOM) Model

As mentioned above, when neglecting the density fluctuation and the third-order correlation, the time-averaged reaction rate is

$$\overline{w_s} = B\rho^2 \left[\left(\overline{Y_1 Y_2} + \overline{Y_1' Y_2'} \right) \overline{k} + \overline{Y_1 k'Y_2'} + \overline{Y_1 k'Y_1'} \right]$$

where $k = \exp(-E/RT)$ is the reaction rate coefficient, $\overline{k} = \int \exp(-E/RT)p(T)dT$ is the time-averaged value of k, and $p(T)$ is the probability density distribution function of temperature. The basic idea of constructing a second-order moment turbulent combustion model is to close the second-order moment terms using an approach like that in the turbulence modeling of Reynolds stresses [25]. The instantaneous conservation equations for two species mass fractions are

$$\frac{\partial}{\partial} (\rho Y_1) + \frac{\partial}{\partial x_j} (\rho v_j Y_1) = \frac{\partial}{\partial x_j} \left(\rho D \frac{\partial Y_1}{\partial x_j} \right) - w_1$$

$$\frac{\partial}{\partial} (\rho Y_2) + \frac{\partial}{\partial x_j} (\rho v_j Y_2) = \frac{\partial}{\partial x_j} \left(\rho D \frac{\partial Y_2}{\partial x_j} \right) - \beta w_1$$

where β is the stoichiometric ratio. The instantaneous energy equation for low Mach-number constant-specific heat, when neglecting the pressure change in time and the work done by body forces, is

$$\frac{\partial}{\partial} (\rho c_p T) + \frac{\partial}{\partial x_j} (\rho v_j c_p T) = \frac{\partial}{\partial x_j} \left(\lambda \frac{\partial T}{\partial x_j} \right) - w_1 Q_1$$

where λ is the heat conductivity, and Q is the heat of reaction. Starting from these equations, the exact transport equations of second-order moments $\overline{Y_1' Y_2'}$ and $\overline{k' Y_1'}$ can be obtained as:

$$\frac{\partial}{\partial t} (\rho \overline{Y_1' Y_2'}) + \frac{\partial}{\partial x_j} (\rho \overline{v_j}\, \overline{Y_1' Y_2'}) = \frac{\partial}{\partial x_j} \left(\rho D \frac{\partial \overline{Y_1' Y_2'}}{\partial x_j} \right) - \left(\rho \overline{v_j' Y_1'} \frac{\partial \overline{Y_2}}{\partial x_j} + \rho \overline{v_j' Y_2'} \frac{\partial \overline{Y_1}}{\partial x_j} \right)$$

$$- \rho \frac{\partial \overline{v_j' Y_1' Y_2'}}{\partial x_j} - 2\rho D \frac{\overline{\partial Y_1' \partial Y_2'}}{\partial x_j\, \partial x_j} - \left[\overline{w_1 (Y_2 + \beta Y_1)} - \overline{w_1} \left(\overline{Y_2} + \beta \overline{Y_1} \right) \right]$$

$$\tag{7.63}$$

$$\frac{\partial \rho \overline{k' Y_1'}}{\partial t} + \rho \overline{v_j} \frac{\partial \overline{k' Y_1'}}{\partial x_j} = \frac{\partial}{\partial x_j} \left(\rho D \frac{\partial \overline{k' Y_1'}}{\partial x_j} \right) - \rho \overline{v_j' k'} \frac{\partial \overline{Y_1}}{\partial x_j} - \rho \overline{v_j' Y_1'} \frac{\partial \overline{k}}{\partial x_j} - \frac{\partial \overline{v_j' k' Y_1'}}{\partial x_j}$$

$$- 2\rho D \frac{\overline{\partial Y_1' \partial k'}}{\partial x_j\, \partial x_j} - \frac{\lambda}{c_p} \frac{E}{R} \frac{\overline{Y_1'}}{T^2} \frac{\partial k}{\partial x_j} \frac{\partial T}{\partial x_j} - \frac{2\lambda}{c_p} \frac{\overline{Y_1'}}{T} \frac{\partial k}{\partial x_j} \frac{\partial T}{\partial x_j} - \overline{K' w_1} + \frac{EQ_1}{c_p R} \frac{\overline{Y_1'\, k w_1}}{T^2}$$

$$\tag{7.64}$$

The left-hand-side terms in Eqs. (7.63) and (7.64) are exact time-changing rates and convection terms, and do not need to be closed. The first terms on the right-hand side of Eqs. (7.63) and (7.64) are molecular diffusion terms, which also do not need to be closed. The physical meanings of the other terms on the right-hand side of these equations are production due to mean flows, turbulent diffusion, and viscous dissipation and reaction terms, that need to be closed. By taking the gradient modeling for the production and diffusion terms, assuming that the scalar dissipation rates and reaction terms are proportional to the dissipation rate of turbulent kinetic energy with a two-times scale, the generalized form of the closed transport equations of the second-order moments can be obtained as

$$
\frac{\partial}{\partial t}\left(\rho \overline{\phi' \psi'}\right) + \frac{\partial}{\partial x_j}\left(\rho v_j \overline{\phi' \psi'}\right) = \frac{\partial}{\partial x_j}\left(\frac{\mu_e}{\sigma_g}\frac{\partial \overline{\phi' \psi'}}{\partial x_j}\right) + c_{g1}\mu_T \left(\frac{\partial \phi}{\partial x_j}\right)\left(\frac{\partial \psi}{\partial x_j}\right)
$$
$$
- c_{g2}\rho\left(\frac{a}{\tau_c} + \frac{b}{\tau_T}\right)\overline{\phi' \psi'}
$$

$$(7.65)$$

where ϕ and ψ denote the mass fractions Y_1, Y_2 or the reaction-rate coefficient k, and c_{g1}, c_{g2} are empirical constants, τ_c, τ_T denote the reaction time scale and turbulence time scale, defined as $\tau_T = k/\varepsilon$;

$$
\tau_c = \left[B\rho\left(\overline{Y_2} + \beta\overline{Y_1}\right)\exp\left(-\frac{E}{R\overline{T}}\right)\right]^{-1}.
$$

Usually we have $c_{g1} = 0.01$, $c_{g2} = 1.4$, $a = 0.9$, $b = 0.05$. The second-order moment (SOM) combustion model is thus closed. Its physical meanings are: (1) the time-averaged reaction rate depends on reaction kinetics (preexponential factor B and activation energy E) and turbulence properties (second-order moments); (2) turbulence enhances the reaction rate; and (3) the transport of scalar correlation moments likes that of Reynolds stresses. For strongly shear flows, neglecting the convection and diffusion terms in Eq. (7.65), an algebraic second-order moment (ASOM) model can be obtained as

$$
\overline{\phi' \psi'} = c\mu_T \left(\frac{\partial \phi}{\partial x_j}\right)\left(\frac{\partial \psi}{\partial x_j}\right)\bigg/\left[\rho\left(\frac{b}{\tau_T} + \frac{a}{\tau_c}\right)\right]
$$

$$(7.66)$$

7.9.3 Application of the SOM Model in RANS Modeling

For RANS modeling, a basic version of the Reynolds stress model is used for modeling turbulence. For turbulent combustion models beside the above-stated SOM model, an EBU-Arrhenius (E-A) model and a presumed PDF model are adopted as:

$$
\overline{w}_s = \min[\overline{w}_{sA}, \overline{w}_{sT}]
$$

where $\overline{w}_{sA} = \overline{\rho^2 k Y_1 Y_2} = B\overline{\rho}^2 \overline{Y_1} \overline{Y_2}\exp(-E/R\overline{T})$

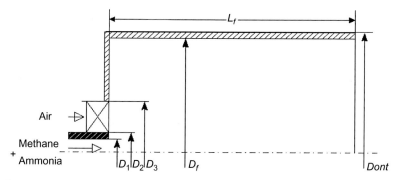

FIGURE 7.16 A swirl combustor for methane−air diffusion combustion.

$$\overline{w}_{sT} = -A_{EBU}\overline{\rho}\frac{\varepsilon}{k}\min\left\{Y_{CH4}, \frac{Y_{O_2}}{\beta}, B_{EBU}\frac{Y_P}{1+\beta}\right\}$$

The time-averaged reaction rate for the simplified PDF model of NO formation is:

$$\overline{w_{NO}} = \iint w_{NO}(T, Y_{O_2})p_1(T)p_2(Y_{O_2})dTdY_{O_2}$$

Assuming a PDF of β-function gives:

$$p(\phi) = \frac{\phi^{a-1}(1-\phi)^{b-1}}{\displaystyle\int_0^1 \phi^{a-1}(1-\phi)^{b-1}d\phi}$$

where $a = \overline{\phi}\left[\frac{\overline{\phi}(1-\overline{\phi})}{\overline{\phi'^2}}\right]$, $b = a\left[\frac{1-\overline{\phi}}{\overline{\phi}}\right]$, $\overline{\phi'^2} = \frac{C_1\mu_t\left(\frac{\partial\overline{\phi}}{\partial x_j}\right)\left(\frac{\partial\overline{\phi}}{\partial x_j}\right)}{C_2\overline{\rho}\frac{\varepsilon}{k}}$

The empirical constants C_1 and C_2 are taken as 2.86 and 2.0, respectively. The first case of RANS modeling was conducted for turbulent methane−air swirling diffusion combustion in a swirl combustor with a central fuel inlet [26], as shown in Fig. 7.16. The predicted and measured temperature profiles are shown in Fig. 7.17. Clearly, in most regions the SOM model predictions are in good agreement with the experimental results, while the predicted temperature profiles using the E-A model show a noticeable qualitative difference from the experimental results, particularly in the upstream region. The E-A model overpredicts the temperature. Since the E-A model exaggerates the turbulence effect and neglects or underestimates the chemical kinetic effect.

Fig. 7.18 gives the predicted thermal NO profiles using the SOM model and the E-A + simplified PDF model and their comparison with the measurement results. Generally speaking, the E-A + simplified PDF model remarkably underpredicts the thermal NO formation and the predicted NO

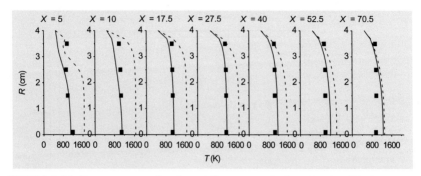

FIGURE 7.17 Temperature profiles for swirling nonpremixed combustion (■ Exp.; ——— SOM; ------- E-A) [13].

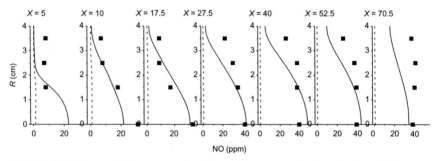

FIGURE 7.18 Thermal NO concentration for swirling nonpremixed combustion (■ Eep.; ——— SOM; ---- E-A and PDF).

concentration is nearly one order of magnitude lower than that measured. As indicated above, the simplified PDF model uses a product of two PDFs to approximate the joint PDF, hence underpredicting the reaction rate. Comparatively, the predictions using the SOM model are much better and are close to the experimental results. There is still some discrepancy between predictions using the SOM model and experiments, because the simplified chemical mechanism was used.

The second case of RANS modeling is the piloted jet methane−air nonpremixed flame measured in the Sandia Laboratory, as shown in Fig. 7.19.

The pilot jet methane−air nonpremixed flame was simulated simultaneously using the alegebraic SOM (ASOM) model and the composition PDF equation model [15,27]. The composition PDF equation model is one of different versions of the PDF transport equation models, where the flow is simulated using the Reynolds stress model, and all scalars are simulated by solving the scalar PDF equations using the Monte-Carlo method. For the composition PDF equation model, an IEM mixing model and a detailed chemistry mechanism of 23 species with 102 elementary reactions are

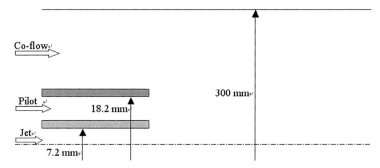

FIGURE 7.19 A piloted jet flame (Sandia Flame C).

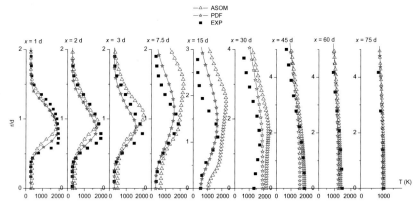

FIGURE 7.20 Time-averaged temperature for Sandia Flame C.

adopted. Fig. 7.20 gives the predicted time-averaged temperature. The PDF equation modeling results are in good agreement with the experimental data; while the ASOM modeling results are also in general agreement with the experimental data, except at cross-sections of $x = 7.5d$ and $x = 15d$, where the temperature is overpredicted. So, generally speaking, both models give good results and the PDF model is somewhat better. Fig. 7.21 shows the time-averaged methane concentration. At all cross-sections both models give results close to the experimental data. Again, the PDF equation modeling results are somewhat better at some cross-sections. Considering that the PDF equation model uses detailed reaction mechanism and solves Lagrangian equations using the Monte-Carlo algorithm, free of numerical diffusion; whereas the ASOM model uses a one-step global reaction mechanism and finite difference method with numerical diffusion, but the average discrepancies between experimental data and PDF equation model results are about 10%, and are about 17% for the ASOM model, and the ASOM model needs only 1/300 of the computation time that is needed by the PDF model, we can

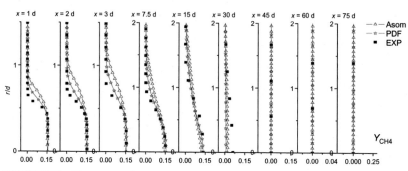

FIGURE 7.21 Time-averaged methane concentration for Sandia Flame C [16].

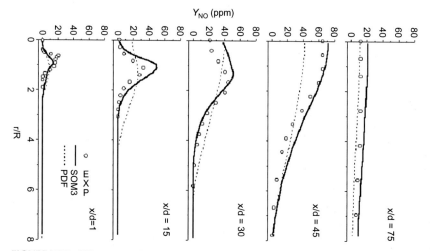

FIGURE 7.22 NO concentration (o Exp.; — SOM; ... Presumed PDF).

conclude, that the ASOM model is a reasonable and economical one and it is suggested to be used in simulating practical large-size engineering facilities. Fig. 7.22 gives the NO concentration predicted by both the SOM model and the presumed PDF model [24]. It can be seen that the SOM model gives much better results than the presumed PDF model, which is in qualitative disagreement with the experimental results.

7.9.4   Validation of the SOM Model by DNS

In order to validate the SOM model, the isothermal fully developed turbulent reacting channel flows were simulated by direct numerical simulation (DNS) [28]. The adopted numerical method is a spectral method. Fig. 7.23 shows the comparison between the exact values given by DNS and modeled values

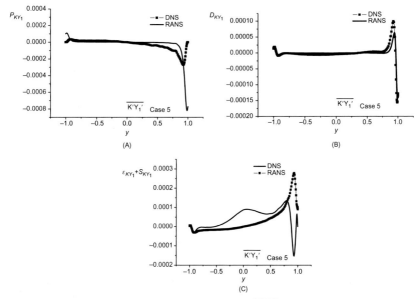

FIGURE 7.23 Exact and modeled values of terms in the $\overline{K'\,Y'_1}$ equation.

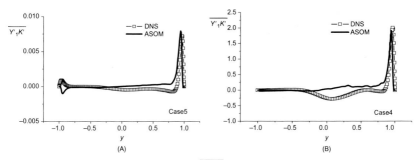

FIGURE 7.24 Exact and modeled values of $\overline{K'\,Y'_1}$ for the ASOM model.

of production, diffusion, dissipation, and reaction terms in the $\overline{K'Y'}$ equation. It is seen that in most regions the modeled values of the production and diffusion terms are in good agreement with the exact values, except in the near-wall region, where some modification should be made. The closure models for the dissipation and reaction terms are not good. This implies that the modeling of dissipation and reaction terms remains to be improved. However, in general, the SOM combustion model for RANS modeling is validated by the DNS of isothermal turbulent reacting flows. The DNS data can be used to validate the ASOM combustion model. Fig. 7.24 shows the exact value of $\overline{K'Y'}$ given by DNS and the modeled value given by the ASOM model. It is interesting to note that the agreement is sufficiently good.

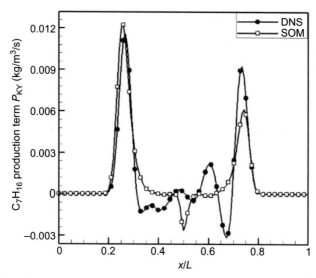

FIGURE 7.25 Production term in the SOM model.

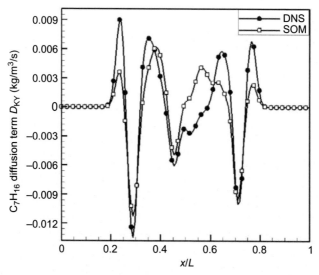

FIGURE 7.26 Diffusion term in the SOM model.

Recently, the SOM model was also validated by DNS of heptane−air turbulent combustion [29]. Figs. 7.25 and 7.26 show the DNS obtained and SOM modeled production terms and diffusion terms, respectively. The agreement is fairly good. Fig. 7.27 shows that after taking some dynamically determined coefficients, the agreement between the DNS obtained and SOM modeled reaction terms is in sufficiently good agreement.

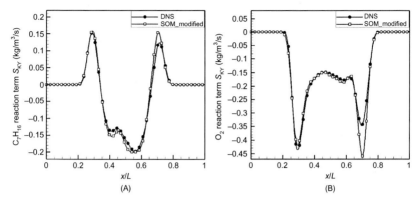

FIGURE 7.27 Reaction terms in a dynamic SOM model.

7.10 MODELING OF TURBULENT TWO-PHASE COMBUSTION

Turbulent two-phase combustion (spray combustion and solid-fuel combustion) is widely encountered in gas-turbine combustors, ramjet and rocket engine combustors, internal engine combustors, and utility and industrial furnaces. Two approaches—Eulerian–Lagrangian and Eulerian–Eulerian (two-fluid) modeling—have been developed for simulating turbulent two-phase combustion. Currently, two-fluid modeling has been developed mainly for dense gas-particle flows and combustion in fluidized beds, where the particle-flow behavior, like the particle stress, is determined only by interparticle collision based on the kinetic theory or its modified form [30,31]. For spray and pulverized solid-fuel combustion, where the droplet/particle concentration is low and the droplet/particle dispersion is determined only by its turbulent fluctuation, the most commonly used approach is the Lagrangian treatment of the dispersed phase—trajectory models [7,32]—which needs large computational effort to give detailed information of the droplet/particle flow field in the three-dimensional space, and has some difficulty in treating combusting droplets/particles entering or leaving recirculation zones. An alternative approach is two-fluid modeling of turbulent two-phase combustion. However, to take this approach it is necessary to solve the problems of modeling the particle turbulence (particle dispersion) and the particle history effect. As discussed in Chapter 6, Modeling of Turbulent Dispersed Multiphase Flows, a $k - \varepsilon - k_p$ and a USM two-phase turbulence model were proposed by the present author to simulate the particle turbulence. To simulate the particle history effect, a two-fluid-trajectory (TFT) model [33,34,35] and a full two-fluid model (FTF model) [36,37] were developed by the present author. Both TFT and FTF models use Eulerian partial differential equations of the particle phase to give the particle number density and particle velocity. The FTF model uses Eulerian partial differential equations to describe all particle history effects—the particle mass change due to moisture evaporation, devolatilization, and char combustion, and the particle temperature change due to convection, diffusion, and heat transfer between two phases; whereas the TFT model uses Lagrangian ordinary differential equations to describe these effects.

7.10.1 Two-Fluid Modeling of Turbulent Two-Phase Combustion

In the framework of two-fluid modeling of two-phase combustion, the volume-averaged and time-averaged gas-phase continuity, momentum, energy, and species equations should be

$$\frac{\partial \rho}{\partial t} + \frac{\partial}{\partial x_j}(\rho v_j) = -\sum_p n_p \dot{m}_p = S \tag{7.67}$$

$$\frac{\partial}{\partial t}(\rho v_i) + \frac{\partial}{\partial x_j}(\rho v_j v_i) = -\frac{\partial p}{\partial x_i} + \mu\left(\frac{\partial v_j}{\partial x_i} + \frac{\partial v_i}{\partial x_j}\right) + \rho g_i$$
$$+ \sum_p \frac{m_p}{\tau_r}\left[n_p(v_{pi} - v_i) + \overline{n_p v_{pi}}\right] + v_i S + F_{mi} - \frac{\partial}{\partial x_j}(\overline{\rho v_j v_i}) \tag{7.68}$$

$$\frac{\partial}{\partial t}(\rho h) + \frac{\partial}{\partial x_j}(\rho v_j h) = \frac{\partial}{\partial x_j}\left(\lambda \frac{\partial T}{\partial x_j}\right) - q_r$$
$$+ \sum_p n_p Q_p - \sum_k n_p \dot{m}_p C_{pp} T_p + W_s Q_s - \frac{\partial}{\partial x_j}(\overline{\rho v_j h}) \tag{7.69}$$

$$\frac{\partial}{\partial t}(\rho Y_s) + \frac{\partial}{\partial x_j}(\rho v_j Y_s) = \frac{\partial}{\partial x_j}\left(D\rho \frac{\partial Y_s}{\partial x_j}\right) - W_s - \alpha_s \sum_p n_p \dot{m}_p - \frac{\partial}{\partial x_j}(\overline{\rho Y_s v_j}) \tag{7.70}$$

In order to simulate the particle history effect due to moisture evaporation, devolatilization, and char combustion and heat transfer between two phases, four particle-phase continuity equations, one momentum, and one energy equations should be used in the FTF model [36,37]

$$\frac{\partial n_p}{\partial t} + \frac{\partial}{\partial x_j}(n_p v_{pj}) = -\frac{\partial}{\partial x_j}(\overline{n_p v_{pj}}) \tag{7.71}$$

$$\frac{\partial(n_p m_p)}{\partial t} + \frac{\partial}{\partial x_j}(n_p m_p v_{pj}) = -m_p \frac{\partial}{\partial x_j}(\overline{n_p v_{pj}}) + n_p \dot{m}_p \tag{7.72}$$

$$\frac{\partial(n_p m_c)}{\partial t} + \frac{\partial}{\partial x_j}(n_p m_c v_{pj}) = -m_c \frac{\partial}{\partial x_j}(\overline{n_p v_{pj}}) + n_p \dot{m}_c \tag{7.73}$$

$$\frac{\partial(n_p\,m_w)}{\partial t} + \frac{\partial}{\partial x_j}\left(n_p\,m_w\,v_{pj}\right) = -m_w\frac{\partial}{\partial x_j}\left(\overline{n_p\,v_{pj}}\right) + n_p\,\dot{m}_w \tag{7.74}$$

$$\frac{\partial}{\partial t}\left(n_p v_{pi}\right) + \frac{\partial}{\partial x_j}\left(n_p v_{pj} v_{pi}\right) = n_p\,g_i + \frac{1}{\tau_r}\left[n_p(v_i - v_{ki}) - \overline{n_p\,v_{pi}}\right]$$

$$+ \frac{n_p\dot{m}_p}{m_p}\left(v_i - v_{pi}\right) - \frac{\partial}{\partial x_j}\left(n_p\overline{v_{pj}\,v_{pi}} + v_{pj}\overline{n_p\,v_{pi}} + v_{pi}\overline{n_p\,v_{pj}}\right) \tag{7.75}$$

$$\frac{\partial}{\partial t}\left(n_p h_p\right) + \frac{\partial}{\partial x_j}\left(n_p v_{pj} h_p\right) = \frac{n_p}{m_p}\left(Q_h - Q_p - Q_{rp}\right)$$

$$+ \frac{n_p\dot{m}_p}{m_p}\left(h - h_p\right) - \frac{\partial}{\partial x_j}\left(n_p\overline{v_{pj}h_p} + v_{pj}\overline{n_p h_p} + h_p\overline{n_p v_{pj}}\right) \tag{7.76}$$

The particle mass changing rate $\dot{m}_p$ due to moisture evaporation, pyrolization, and char combustion is determined by the following equations

$$\dot{m}_p = \dot{m}_w + \dot{m}_v + \dot{m}_h \tag{7.77}$$

$$\dot{m}_w = \pi d_p\,NuD\,\rho\,\ln\left(1 + \frac{Y_{ws} - Y_{wg}}{1 - Y_{ws}}\right),$$

$$Y_{ws} = B_w\exp\left(-\frac{E_w}{RT_p}\right) \tag{7.78}$$

$$\dot{m}_v = m_c\,\alpha B_v\exp\left(-\frac{E_v}{RT_p}\right) \tag{7.79}$$

$$\dot{m}_c = -m_c\,B_v\exp\left(-\frac{E_v}{RT_p}\right) \tag{7.80}$$

$$\dot{m}_p = \pi d_p NuD\rho\,\ln\left(\frac{\dot{m}_s/\dot{m}_p - Y_s}{\dot{m}_s/\dot{m}_p - Y_{ss}}\right) \tag{7.81}$$

$$\dot{m}_s = \pi d_p{}^2\rho Y_{ss}B_s\exp\left(-\frac{E}{RT_p}\right) \tag{7.82}$$

$$\dot{m}_h = \sum \dot{m}_s \tag{7.83}$$

In the TFT model, the particle number density and velocity are determined by continuity and momentum equations in Eulerian coordinates (continuum description), but the particle mass and temperature changes due to evaporation, devolatilization, char combustion, and particle heat transfer are determined by ordinary differential equations together with a set of algebraic expressions, and a particle energy equation in Lagrangian coordinate is used instead of Eq. (7.76) [33– 35].

$$m_p C \frac{dT_p}{dt} = \pi d_p^2 \varepsilon \sigma \left(T^4 - T_p^4 \right) + \dot{m}_p C_p \left(T - T_p \right) \left[\exp \left(\dot{m}_p C_p / \left(\pi d_p Nu \lambda \right) \right) - 1 \right]^{-1}$$
$$- \dot{m}_w L_w - \dot{m}_v q_v + \dot{m}_h Q_c$$

$$(7.84)$$

For volatile and CO combustion, the conventional EBU-Arrhenius model is used. The time-averaged reaction rate is determined by:

$$W_s = \min(W_E, W_A)$$

where

$$W_E = c_E \, \rho \frac{\varepsilon}{k} \min \left(Y_s, \, Y_{ox}/\beta \right); W_A = B_s \rho^2 \, Y_s \, Y_{ox} \exp \left(-E/RT \right)$$

For NOx formation, the Zel'dovich mechanism of thermal NO and the DeSoete mechanism of fuel NO are considered, and an algebraic second-order moment model is used for the turbulence-chemistry model of NO formation [38].

$$\overline{W}_s = B \, \rho^2 \left[\left(\overline{Y}_1 \, \overline{Y}_2 + \overline{Y'_1 \, Y'_2} \right) \overline{K} + \overline{Y}_1 \, \overline{K' \, Y'_2} + \overline{Y}_2 \, \overline{K' \, Y'_1} \right]$$
$$K = \exp \left(-E/RT \right)$$

$$(7.85)$$

$$\overline{K' \, Y'_1} = C_1 \frac{k^3}{\varepsilon^2} \frac{\partial \overline{K}}{\partial x_j} \frac{\partial \overline{Y}_1}{\partial x_j} \quad \overline{Y'_1 \, Y'_2} = C_1 \frac{k^3}{\varepsilon^2} \frac{\partial \overline{Y}_1}{\partial x_j} \frac{\partial \overline{Y}_2}{\partial x_j}; \quad \overline{Y'_2 \, K'} = C_1 \frac{k^3}{\varepsilon^2} \frac{\partial \overline{Y}_2}{\partial x_j} \frac{\partial \overline{K}}{\partial x_j}$$

$$(7.86)$$

For the numerical procedure, both gas-phase equations and Eulerian particle-phase equations in the FTF model and the TFT model are integrated in the computational cell using the upwind scheme to obtain finite difference equations. The gas-phase equations are solved using the SIMPLE algorithm. The Eulerian particle-phase equations are solved in a similar way, but without $p - v$ corrections. For the TFT model the Lagrangian particle equations are solved to predict particle temperature and mass along the trajectories, obtained from Eulerian predictions. Therefore, multiple iterations have been made based on a two-way coupling (Eulerian gas–Eulerian particle) for the FTF model, and a four-way coupling (Eulerian gas–Eulerian particle–Lagrangian particle) for the TFT model. The algorithm is discussed in detail in Chapter 8.

7.10.2 Two-Fluid-Simulation of Coal Combustion in a Combustor with High-Velocity Jets

A sudden-expansion coal combustor with high-velocity jets discharged from two holes at the front surface, as shown in Fig. 7.28, is used to stabilize the coal flame in the boiler furnace.

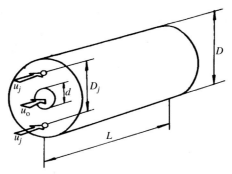

FIGURE 7.28   A coal combustor with high-velocity jets.

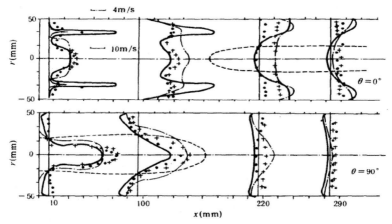

FIGURE 7.29   Two-phase velocities in the coal combustor (● gas, exp; + particle, exp; — gas, pred., -.-particle, pred.).

The coal combustion in this combustor was simulated using both TFT and FTF models [34,37]. The predicted and measured two-phase velocities are shown in Fig. 7.29. It can be seen that the predictions are in agreement with the measured results. A large-size recirculation zone, connecting the corner and near-axis zones is induced by the high-velocity jets. The predicted and measured particle mass flux is given in Fig. 7.30. Good agreement is obtained. High particle concentration exists in the recirculation zone, and is favorable to flame stabilization.

The predicted gas temperature by the FTF model and the TFT model is given in Figs. 7.31 and 7.32, respectively. Both prediction results show that the high temperature develops right at the stagnation point of the recirculation zone, indicating the important role of the high-velocity jets. It can be seen that there are some differences between these two modeling results. The FTF model gives higher gas temperature and more rapid temperature change along the radial direction than those given by the TFT model.

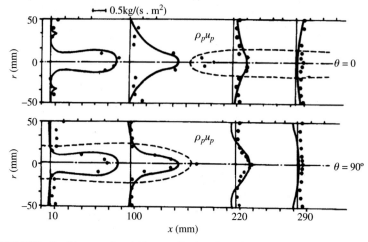

FIGURE 7.30 Particle mass flux (● Exp., — Pred. --- Boundary of recirculation zone).

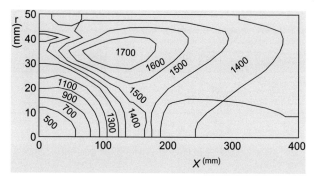

FIGURE 7.31 Gas temperature (FTF model).

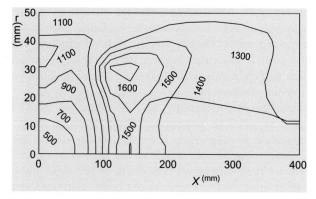

FIGURE 7.32 Gas temperature (TFT model).

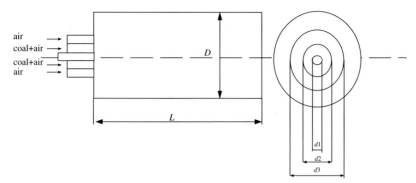

FIGURE 7.33 The swirl coal combustor.

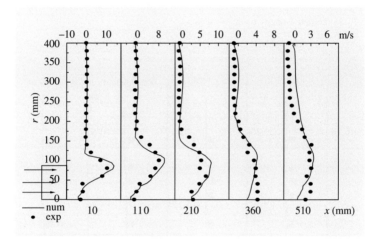

FIGURE 7.34 Gas velocity (— Pred., ● Exp.).

7.10.3 Two-Fluid Modeling of Coal Combustion and NO Formation in a Swirl Combustor

The swirl burners are widely adopted in the utility boilers of power stations. Different techniques have been used to improve flame stabilization, reduce NO formation, and simultaneously keep the high combustion efficiency in operating these burners. To simulate the coal combustion and NO formation in these burners, the FTF model for coal combustion and ASOM turbulence-chemistry model for NO formation were adopted [38]. The simulated swirl combustor was measured by Abbas et al. [39]. Its geometrical configuration is shown in Fig. 7.33. The predicted and measured gas and particle velocities are given in Figs. 7.34 and 7.35, respectively. Good agreement is obtained. The predicted and measured gas temperature and NO concentration are

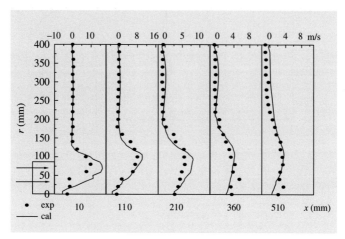

FIGURE 7.35   Particle velocity (— Pred., ● Exp.).

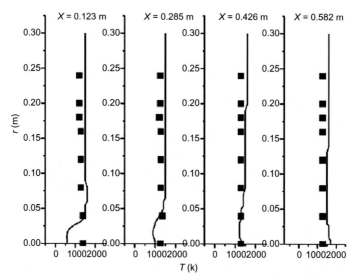

FIGURE 7.36   Gas temperature (— Pred., ■ Exp.).

shown in Figs. 7.36 and 7.37, respectively. The agreement here is also good. The effect of swirl number on the averaged NO emission at the exit is shown in Fig. 7.38. Both predictions and measurements indicate that as the swirl number increases, the NO emission at the exit at first decreases, and then increases. There is an optimal swirl number, at which the NO emission is lowest.

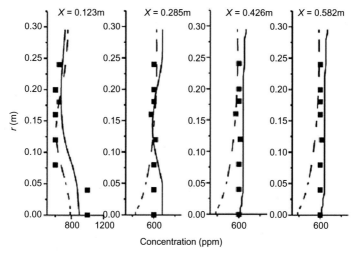

FIGURE 7.37 NO concentration (− − ASOM; — original SOM; ■ exp.).

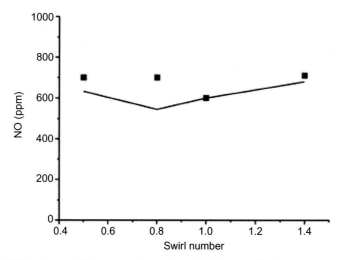

FIGURE 7.38 Averaged NO concentration at the exit (■ Exp., — Pred.).

7.10.4 Eulerian−Lagrangian Modeling of Two-Phase Combustion

The above-discussed two-fluid modeling of two-phase combustion was developed by the author. Most existing simulations of spray and pulverized-coal combustion take the Eulerian−Lagrangian (E-L) approach. The E−L two-phase flow equations have already been given in Chapter 6, and the submodels, as droplet evaporation, coal-particle devolatilization, and char combustion models are the same as those in two-fluid modeling. The application of these E−L simulations for two-phase combustion will be given in Chapter 9.

7.11 LARGE-EDDY SIMULATION OF TURBULENT COMBUSTION

Large-eddy simulation (LES) of single-phase and two-phase turbulent combustion has attracted more and more attention, as it can give instantaneous flow and flame structures leading to a better understanding of the turbulence−chemistry interaction and give statistical results, better than Reynolds-averaged N-S (RANS) modeling results. As stated in Chapter 5, Modeling of Single-Phase Turbulence and Chapter 6, Modeling of Turbulent Dispersed Multiphase Flows, LES is used to solve filtered 3-D instantaneous continuity, momentum, energy, and species equations, using subgrid scale (SGS) models for unresolved scales. The SGS stress models can be the Smagorinsky model, dynamic Smagorinsky model, and SGS k-equation model. Presently, the adopted SGS combustion models are: filtered PDF equation model, G-equation model; linear-eddy model, laminar-flamelet (LFM) model; EBU (EDM) model, the general tendency is extending RANS models to SGS models. The most popular are the laminar-flamelet model and the EDM (or EBU) model [40,41]. However, these combustion models either do not always give satisfactory results, or are insufficiently validated by experiments. In this section, large-eddy simulation (LES) of swirling diffusion combustion, jet diffusion combustion, bluff-body stabilized premixed combustion, and liquid-spray combustion are presented. In addition, more recently, LES of pulverized-coal combustion was studied by some investigators [42,43]. However, either the LES statistical results of temperature and species concentration are of less experimental validation [42], or both LES and RANS modeling methods remarkably overpredict the temperature and LES results are not better than RANS modeling results [43]. In this section, a LES of swirling coal combustion made by the author and his colleagues is also reported.

7.11.1 LES Equations and Closure Models for Simulating Gas Turbulent Combustion

The filtered continuity, momentum, energy, and species equations for LES of turbulent gas combustion are:

$$\frac{\partial \rho}{\partial t} + \frac{\partial}{\partial x_i}(\rho \bar{u}_i) = 0 \qquad (7.87)$$

$$\frac{\partial}{\partial t}(\rho \bar{u}_i) + \frac{\partial}{\partial x_j}(\rho \bar{u}_i \bar{u}_j) = \frac{\partial}{\partial x_j}\left(\mu \frac{\partial \bar{u}_i}{\partial x_j}\right) - \frac{\partial \bar{p}}{\partial x_i} - \frac{\partial \tau_{ij}}{\partial x_j} \tag{7.88}$$

$$\frac{\partial \rho \bar{h}}{\partial t} + \frac{\partial}{\partial x_j}(\rho \bar{h} \bar{u}_j) = \frac{\partial}{\partial x_j}\left(\frac{\mu}{\text{Pr}} \frac{\partial \bar{h}}{\partial x_j}\right) - \frac{\partial q_{u_j h}}{\partial x_j} \tag{7.89}$$

$$\frac{\partial \rho \bar{Y}_s}{\partial t} + \frac{\partial}{\partial x_j}(\rho \bar{u}_j \bar{Y}_s) = \frac{\partial}{\partial x_j}\left(\frac{\mu}{\text{Sc}} \frac{\partial \bar{Y}_s}{\partial x_j}\right) - \bar{w}_s - w_{sgs} - \frac{\partial g_{u_j Y_s}}{\partial x_j} \tag{7.90}$$

In these equations, the subgrid stress, heat flux, mass flux, and reaction rate need closure models. The subgrid scale stress τ_{ij} is defined by

$$\tau_{ij} \equiv \rho \overline{u_i u_j} - \rho \bar{u}_i \bar{u}_j \tag{7.91}$$

For the SGS stress, the Smagorinsky–Lilly (SL) eddy-viscosity model is adopted as

$$\tau_{ij} - \frac{1}{3}\tau_{kk}\delta_{ij} = -2\mu_t \bar{S}_{ij}, \quad \bar{S}_{ij} \equiv \frac{1}{2}\left(\frac{\partial \bar{u}_i}{\partial x_j} + \frac{\partial \bar{u}_j}{\partial x_i}\right), \quad \mu_t = \rho L_s^2 |\bar{S}|, \quad |\bar{S}| \equiv (2\bar{S}_{ij}\bar{S}_{ij})^{1/2} \tag{7.92}$$

where $L_s = \min(\kappa d, C_s V^{1/3})$

The subgrid scale mass flux and heat flux are closed by gradient modeling as

$$g_{jsgs} = \rho(\overline{u_j Y_s} - \bar{u}_j \bar{Y}_s) = \frac{\mu_t}{\sigma_Y} \frac{\partial \bar{Y}_s}{\partial x_j} \tag{7.93}$$

$$q_{j\,sgs} = \rho(\overline{u_j T} - \bar{u}_j \bar{T}) = \frac{\mu_t}{\sigma_T} \frac{\partial \bar{T}}{\partial x_j} \tag{7.94}$$

where σ_Y and σ_T are model constants, $\sigma_Y = \sigma_T = 1.0$. $\bar{w}_s, w_{sgs}$ are the filtered reaction rate and the SGS reaction rate of s species, respectively. Two combustion models are used in LES. The first is the second-order moment (SOM) combustion model, proposed by the author. The SOM-SGS combustion model, using a gradient modeling and expressing the effect of small-scale temperature and species fluctuations on the SGS reaction rate, is given by

$$w_{sgs} = \rho^2 \left[\bar{K}(\overline{Y_{ox}Y_{fu}} - \bar{Y}_{ox}\bar{Y}_{fu}) + \bar{Y}_{ox}(\overline{KY_{fu}} - \bar{Y}_{fu}\bar{K}) + \bar{Y}_{fu}(\overline{KY_{ox}} - \bar{Y}_{ox}\bar{K})\right] \tag{7.95}$$

and the subgrid scale correlation terms are given by the algebraic expressions

$$\overline{\Phi\Psi} - \bar{\Phi}\bar{\Psi} = c\mu_t \left(\frac{\partial \bar{\Phi}}{\partial x_j}\right)\left(\frac{\partial \bar{\Psi}}{\partial x_j}\right) / \left[\rho\left(\frac{a}{\tau_T} + \frac{(1-a)}{\tau_C}\right)\right] \tag{7.96}$$

where Φ and Ψ denote Y_1 or Y_2 or K, τ_C is the chemical reaction time, τ_T is the turbulent diffusion time, and a and c are model constants. The reaction time and fluctuation time are given by

$$\tau_C = \left[B\rho(\overline{Y}_{O_2} + \beta\overline{Y}_{CH_4}) \exp\left(-\frac{E}{R\overline{T}} \right) \right]^{-1}, \quad \tau_T = 1/|\overline{S}|$$

where β is the stoichiometric coefficient. The second combustion model is the LES version of EBU model, where the reaction rate is given by

$$\overline{w}_s + w_{sgs} = c_1\rho|\overline{S}|\min\left\{ \overline{Y_{CH_4}}, \frac{\overline{Y_{O_2}}}{\beta}, c_2\frac{\overline{Y_P}}{1+\beta} \right\} \tag{7.97}$$

where $\beta = 4$. For the methane−air and propane−air reaction mechanisms, the global one-step reaction rates are given as:

$$w_{fu} = 2.119 \times 10^{11} Y_{ox}^{1.3} Y_{fu}^{0.2} \exp(-2.027 \times 10^8/RT) \tag{7.98}$$

$$w_{fu} = 1.0 \times 10^{10} \rho^2 Y_{ox} Y_{fu} \exp(-1.84 \times 10^4/T) \tag{7.99}$$

7.11.2 LES of Swirling Diffusion Combustion, Jet Diffusion Combustion, and Bluff-Body Premixed Combustion

Fig. 7.39 gives the LES-obtained time-averaged temperature for the swirling methane−air diffusion combustion in Fig. 7.16, using both the LES-SOM model and LES-EBU model and also RANS-SOM modeling results [44]. It can be seen that both LES-SOM and RANS-SOM results are in good agreement with the experimental results, and the LES-SOM results are better than the RANS-SOM results at the cross-sections of $x = 5$ and $x = 10$. Obviously, in most regions the LES-SOM model is much better than the LES-EBU

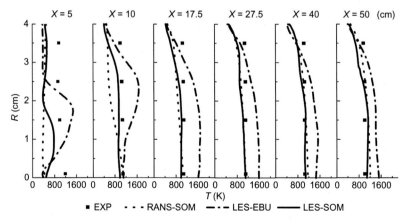

FIGURE 7.39 Time-averaged temperature for the swirling diffusion combustion.

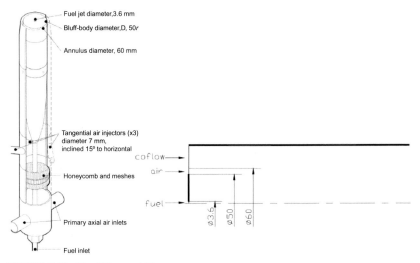

FIGURE 7.40   The Sidney swirl burner.

model, which remarkably overpredicts the temperature. There is still some discrepancy between the modeling results and the experiments near the inlet, which is caused by the oversimplified one-step global chemical kinetics.

A bluff-body stabilized swirling diffusion flame in a Sydney swirl burner, measured in Reference [45], shown in Fig. 7.40, was simulated using LES with different SGS stress and combustion models [46]. The predicted time-averaged temperature is shown in Fig. 7.41. It can be seen that in this case the difference between different models is not obvious.

The propane−air diffusion flame, as shown in Fig. 7.42, was studied by LES [47] using a SOM-SGS combustion model. Figs. 7.43 and 7.44 give the time-averaged axial velocity and temperature, respectively. In most regions the predictions are in good agreement with the experimental results. This is another example indicating the feasibility of the SOM-SGS model. The discrepancy between predictions and experiments in some regions may be caused by the oversimplified global chemical kinetics.

The methane−air jet flame shown in Fig. 7.19 was also studied using large-eddy simulation with an SOM combustion model (LES-SOM) and Reynolds-averaged modeling with SOM combustion model (RANS-SOM) [48]. The predicted statistically averaged temperature is given in Fig. 7.45. It can be seen that better agreement is obtained between LES-SOM results and experiments. At the first five cross-sections the RANS-SOM modeling results are close to the LES-SOM and experimental results, but in the downstream region, the LES-SOM results are better than the RANS-SOM results. RANS-SOM modeling gives more uniform distribution and faster temperature reduction than the measurement does, while the LES-SOM modeling gives obviously better results.

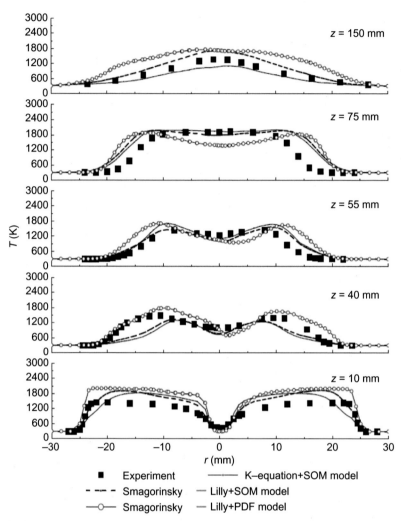

FIGURE 7.41 Time-averaged temperature for the Sidney swirling flame.

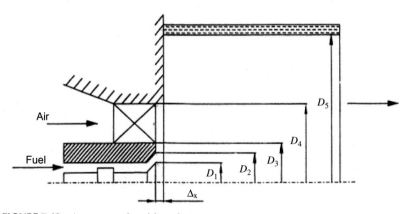

FIGURE 7.42 A propane−air swirl combustor.

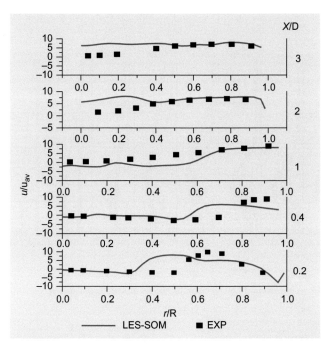

FIGURE 7.43   Time-averaged axial velocity.

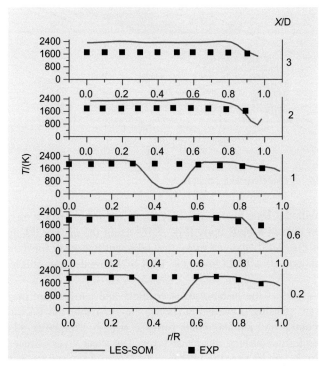

FIGURE 7.44   Time-averaged temperature.

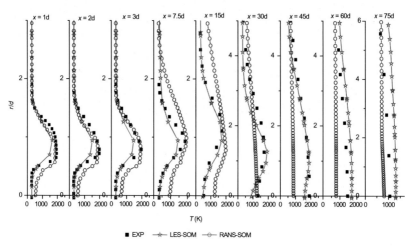

FIGURE 7.45   Time-averaged temperature for the jet diffusion combustion.

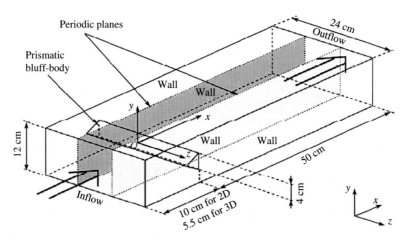

FIGURE 7.46   A combustor with a bluff body.

The predicted time-averaged temperature for the propane—air premixed combustion behind a bluff body (Fig. 7.46) using the LES-SOM model [49] is given in Fig. 7.47. One can make the judgment that even for the premixed combustion behind a bluff body the LES-SOM model can still work very well.

Fig. 7.48 shows the instantaneous vorticity and temperature maps for the methane—air swirling combustion. It can be seen that the oncoming flow from the central and annular inlets forms a strong shear layer. Many small vortices are formed around this shear layer. The shear layer diffuses quickly; the vorticity is largest near the inlet and the large vortex structures formed in

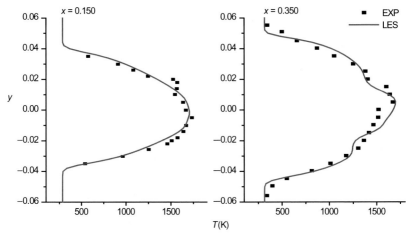

FIGURE 7.47   Time-averaged temperature for bluff-body premixed combustion.

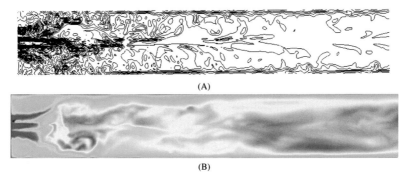

FIGURE 7.48   Instantaneous vorticity (A) and temperature maps (B) for swirling diffusion combustion.

the upstream region are quickly broken up. Comparison of the instantaneous temperature maps with the vortex structures shows that the flame is located in the region of high shear. Obviously, the chemical reaction is intensified by the large-eddy structures in swirling flows. However, no distinct thin flame surface is observed. This implies that swirl thickens the flame front. Fig. 7.49 gives the instantaneous vorticity surface and temperature map for the methane−air piloted jet combustion. The typical strong coherent structures of jet flows are observed. For jet combustion, unlike swirling combustion, a thin flame-front structure is obvious. In the upstream region, there is a thin flame front, whereas the vortices are rather small, and in the downstream region the large vortices are formed where the reaction is completed.

All of the above-stated examples of experimental validation of LES statistical results convince us that, generally speaking, the SOM

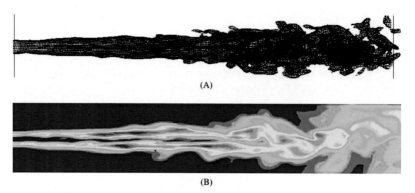

FIGURE 7.49 Instantaneous vorticity surface (A) and temperature maps (B) for jet diffusion combustion.

combustion model can always give better results in LES of both nonpremixed and premixed combustion, while the traditional EBU combustion model cannot do so.

7.11.3 LES of Ethanol−Air Spray Combustion

The ethanol−air spray combustion was studied by LES with an algebraic SOM-SGS combustion model [50]. The filtered gas-phase equations for LES of two-phase combustion are

$$\frac{\partial \rho}{\partial t} + \frac{\partial}{\partial x_i}(\rho \bar{u}_i) = \bar{S} \tag{7.100}$$

$$\frac{\partial}{\partial t}(\rho \bar{u}_i) + \frac{\partial}{\partial x_j}(\rho \bar{u}_i \bar{u}_j) = \frac{\partial}{\partial x_j}\left(\mu\left(\frac{\partial \bar{u}_i}{\partial x_j} + \frac{\partial \bar{u}_j}{\partial x_i}\right) - \frac{2}{3}(\mu \frac{\partial \bar{u}_j}{\partial x_j})\delta_{ij}\right) \\ - \frac{\partial \bar{p}}{\partial x_i} - \frac{\partial \tau_{ij}}{\partial x_j} + \sum_k \overline{\frac{\rho_k}{\tau_{rk}}(u_{ki} - u_i)} + \overline{u_i S} \tag{7.101}$$

$$\frac{\partial \rho \bar{Y}_s}{\partial t} + \frac{\partial}{\partial x_j}(\rho \bar{u}_j \bar{Y}_s) = \frac{\partial}{\partial x_j}\left(\frac{\mu}{Sc}\frac{\partial \bar{Y}_s}{\partial x_j}\right) - \bar{w}_s - w_{sgs} - \frac{\partial g_{u_j Y_s}}{\partial x_j} + \alpha_s S \tag{7.102}$$

$$\frac{\partial \rho \bar{h}}{\partial t} + \frac{\partial}{\partial x_j}(\rho \bar{h} \bar{u}_j) = \frac{\partial}{\partial x_j}\left(\frac{\mu}{Pr}\frac{\partial \bar{h}}{\partial x_j}\right) - \frac{\partial q_{u_j h}}{\partial x_j} - q_r + S_h \tag{7.103}$$

The SGS stress is closed using a SGS k-equation model. The SGS mass flux and heat flux are closed using the gradient modeling. For the SGS combustion model, two options were taken. One choice is the algebraic

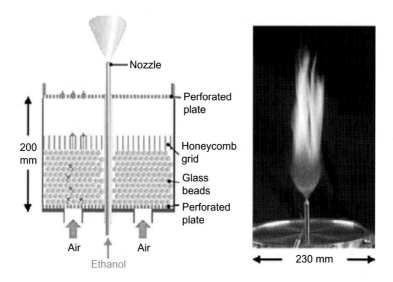

FIGURE 7.50   A spray combustor.

SOM-SGS model, the other is $w_{sgs} = 0$ (finite-rate model). For ethanol−oxygen combustion, a finite-rate global reaction rate is taken into account

$$2C_2H_5OH + 3O_2 \rightarrow 4CO_2 + 6H_2O \tag{7.104}$$

The Arrhenius expression of the one-step global reaction kinetics for w_s is

$$w_s = 8.345 \times 10^9 \rho^2 Y_{fu} Y_{ox} \exp[-1.26 \times 10^8/(RT)] \tag{7.105}$$

For droplet motion and evaporation, the Lagrangian approach was used. The LES was made for a spray combustor measured at UC Berkeley [51], as shown in Fig. 7.50.

Fig. 7.51 gives the LES statistically averaged gas temperature distribution for ethanol spray combustion, given by LES-SOM, LES-FA (finite-rate combustion model), and RANS modeling using the PDF equation model reported in the literature, and its comparison with the measurement results. It can be seen that the LES-SOM results are better than the LES-FA results, indicating that the small-scale temperature and concentration fluctuations do have an obvious effect on combustion. In particular, the LES-SOM results give the temperature peak at $r = 0.016$ m observed in experiments, which cannot be given by the LES-FA results. Furthermore, the LES results are in much better agreement with the experiments than those obtained by RANS modeling using the most complex PDF equation model. Therefore, in general, LES can give more exact statistical results than RANS modeling, even when using very simple one-step global kinetics.

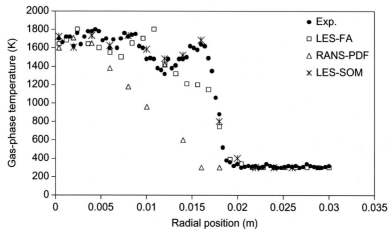

FIGURE 7.51 Gas temperature for spray combustion.

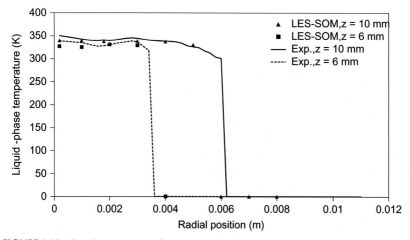

FIGURE 7.52 Droplet temperature for spray combustion.

Fig. 7.52 shows the statistically averaged droplet temperature and its comparison with the measurement results. The agreement is good, with only some minor differences, possibly owing to inaccurately specified spray inlet conditions due to the uncertainty given in experimental conditions. Fig. 7.53 gives the predicted instantaneous vorticity map. The coherent structures initially are produced in the near-inlet shear region of fuel jet flows, then are intensified to become large vortices, and finally are weakened in the downstream region. The instantaneous gas temperature map is shown in Fig. 7.54. It can be seen that the high temperature is developed in the high shear region and some flame islands are observed, indicating droplet-group combustion.

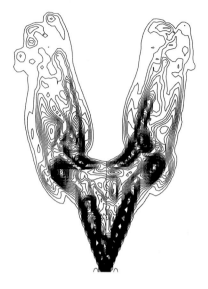

FIGURE 7.53   Instantaneous vorticity map.

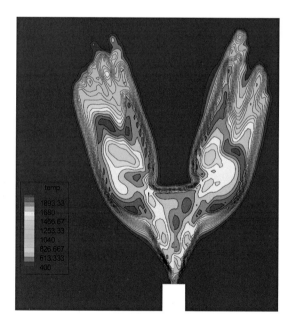

FIGURE 7.54   Instantaneous gas temperature (K).

7.11.4   LES of Swirling Coal Combustion

Large-eddy simulation of coal combustion has been reported recently. Some investigators reported a LES of coal-air jet combustion, but LES statistical results are less of detailed experimental validation. Up to now, no studies on

LES of swirling coal combustion have been reported. To use LES for coal combustion, a Eulerian–Lagrangian large-eddy simulation with Smagorinsky–Lilly's subgrid scale stress model, presumed-PDF fast-chemistry and EBU gas combustion models, particle devolatilization and combustion models are used to study swirling coal combustion [52]. The LES statistical results are validated by the measurement results for swirling coal combustion made by us and those from the literature. The instantaneous LES results are used to study the turbulence and flame structures and their interaction.

The filtered gas continuity, momentum, energy, and species equations for Eulerian–Lagrangian LES of reacting gas-particle flows and combustion are the same as those listed in the previous section. For the subgrid scale stress model, the Smagorinsky–Lilly model is adopted as

$$\tau_{ij} - \frac{1}{3}\tau_{kk}\delta_{ij} = -2\mu_t \overline{S}_{ij}, \ \overline{S}_{ij} \equiv \frac{1}{2}\left(\frac{\partial \overline{u}_i}{\partial x_j} + \frac{\partial \overline{u}_j}{\partial x_i}\right), \ \mu_t = \rho L_s^2 |\overline{S}|, \ |\overline{S}| \equiv \sqrt{2\overline{S}_{ij}\overline{S}_{ij}}$$

$$L_s = \min(\kappa d, C_s V^{1/3}) \tag{7.106}$$

The gradient modeling is taken for the subgrid scale heat flux and mass flux. Two gas combustion models are used. One is the eddy-break-up (EBU) model

$$\overline{w}_s + w_{sgs} = c_1 \rho |\overline{S}| \min\left\{\overline{Y}_f, \frac{\overline{Y}_{ox}}{\beta}, c_2 \frac{\overline{Y}_P}{1+\beta}\right\} \tag{7.107}$$

where $c_1 = 4.0$, $c_2 = 0.5$

The gas-phase combustion includes two reactions:

$$CH_4 + 2O_2 \rightarrow CO_2 + 2H_2O \ (\text{reaction 1}) \tag{7.108}$$

$$2CO + O_2 \rightarrow 2CO_2 \ (\text{reaction 2}) \tag{7.109}$$

Alternatively, a presumed-PDF fast-chemistry gas combustion model is used. In the latter case a mixture fraction equation should be solved

$$\frac{\partial}{\partial t}(\rho \overline{f}) + \frac{\partial}{\partial x_j}(\rho \overline{u}_j \overline{f}) = \frac{\partial}{\partial x_j}\left(\frac{\mu}{Sc}\frac{\partial \overline{f}}{\partial x_j}\right) - \frac{\partial m_{jsgs}}{\partial x_j} \tag{7.110}$$

where $m_{jsgs} = \rho\left(\overline{u_j f} - \overline{u}_j \overline{f}\right) = \frac{\mu_t}{\sigma_f}\frac{\partial \overline{f}}{\partial x_j}$; $\overline{ff} - \overline{f}\,\overline{f} = \frac{1}{2}L_s^2 |\nabla \overline{f}|^2$

For radiative heat transfer for the P1 model is used as:

$$-\overline{q}_r = aG - 4a\sigma \overline{T}^4 \tag{7.111}$$

For comparison, simultaneously, an RANS modeling using a Reynolds stress equation model (RSM) and the same combustion model as that used in LES are used to simulate swirling gas combustion. For particle combustion, neglecting moisture evaporation and taking the submodels of two-equation

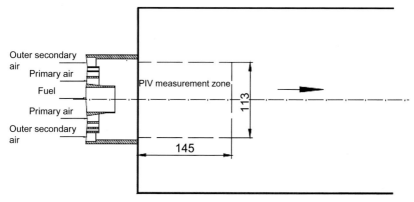

FIGURE 7.55   A swirl coal combustor.

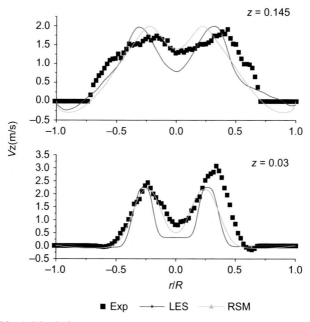

FIGURE 7.56   Axial velocity.

devolatilization and diffusion-kinetic particle oxidation model with total formation of CO are used. The Lagrangian particle continuity, momentum, and energy equations are similar to those used in spray combustion. The simulation was made in a swirl coal combustor, as shown in Fig. 7.55. Figs. 7.56—7.59 are LES-predicted time-averaged velocity, temperature, oxygen, and CO concentrations, respectively, for coal combustion using the simplified PDF and fast-chemistry model. The agreement between prediction

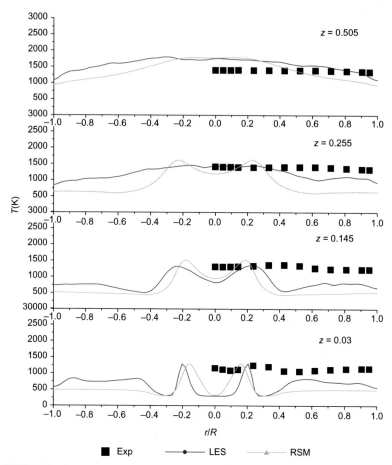

FIGURE 7.57 Temperature.

and measurement is not good. The temperature is underpredicted and CO concentration is overpredicted. This discrepancy may be caused by the shortcomings of the simplified PDF fast-chemistry model.

Alternatively, the LES of coal combustion was also made in a 1/10 model of the swirl coal combustor measured by Abbas et al. [39] as shown in Fig. 7.60.

Figs. 7.61 to 7.63 are LES-predicted temperature, CO_2, and oxygen concentrations, respectively, for coal combustion using the EBU combustion model. It can be seen that except in the near-axis region of upstream cross-sections where the temperature and species concentrations are overpredicted, in most regions the agreement is good. The reason for this is that in the case of coal combustion, char oxidation plays a dominant role, and the gas combustion model has a minor effect. The discrepancy between predictions and

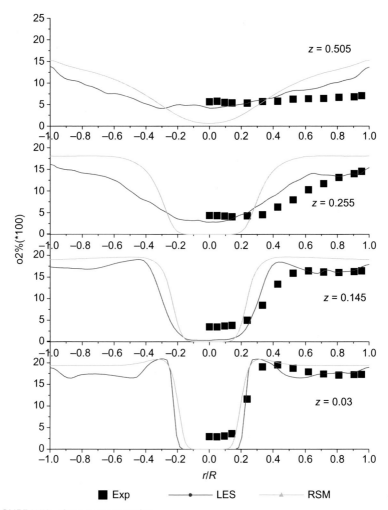

FIGURE 7.58 Oxygen concentration.

experiments may be caused by the subgrid scale stress model, the gas combustion model, as well as the particle devolatilization and char oxidation models.

Fig. 7.64 gives the LES-obtained instantaneous vorticity map for swirling coal combustion and swirling gas combustion in the same combustor. The vorticity is higher at the near-inlet region and shows its strip structures. In the downstream region the vorticity lowers. In the near-wall zone there are wavy coherent structures. It is unexpected that the vorticity in the case of coal combustion develops faster than that in the case of gas combustion, showing the effect of combusting particles on the gas turbulence. Fig. 7.65

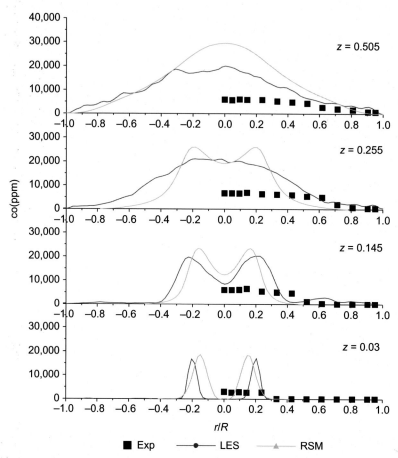

FIGURE 7.59 CO concentration.

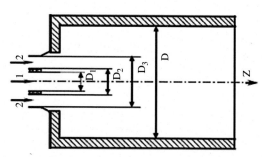

FIGURE 7.60 A model of a swirl coal combustor (1, primary air + coal; 2, secondary air with swirl).

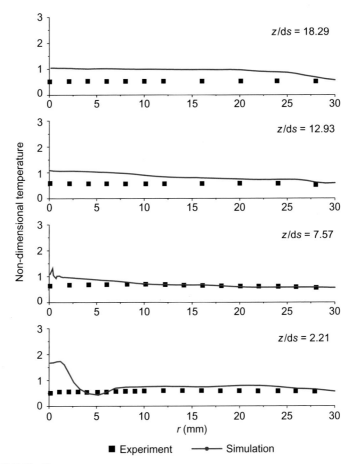

FIGURE 7.61   Temperature.

shows the LES-predicted instantaneous particle concentration map. Obviously, in the near-axis inlet region the vorticity is low and the particle concentration is also low. Near the inlet the particles show strip-shaped structures. A large amount of particles is concentrated in the periphery of gas coherent structures. In the downstream region the particle concentration tends to become more uniform and subsequently somewhat higher in the near-wall region. Fig. 7.66 gives the LES-predicted instantaneous temperature map for swirling coal and gas combustion. There are close interactions between the coherent structures and combustion. The high-temperature flame zone is located at the near-inlet high particle concentration and high vorticity region with strip-shaped coherent structures. In the downstream region, where the vorticity and particle concentration are low, the temperature becomes lower. The high-temperature zone for coal combustion develops

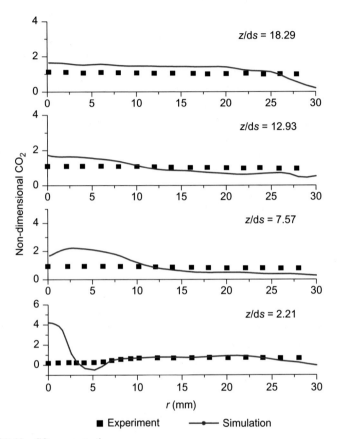

FIGURE 7.62 CO_2 concentration.

earlier than that for the cases of gas combustion. Here a question may arise, why is the coal combustion faster than the gas combustion? This is because, in the present case, the vorticity is stronger in the near-inlet region than that in the case of gas combustion and the particle-released volatile and CO intensively burn in this region.

7.12   DIRECT NUMERICAL SIMULATION OF TURBULENT COMBUSTION

As discussed in Chapter 5, Modeling of Single-Phase Turbulence and Chapter 6, Modeling of Turbulent Dispersed Multiphase Flows the direct numerical simulation (DNS) is used to solve 3-D instantaneous equations in Kolmogorov dissipation scales, and hence can give the information for all scales without using any models. Its accuracy is determined by numerical methods and periodic boundary conditions. The spectral methods

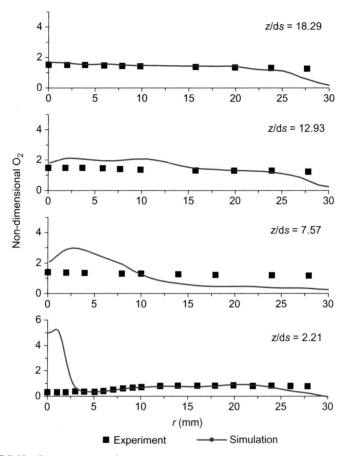

FIGURE 7.63 Oxygen concentration.

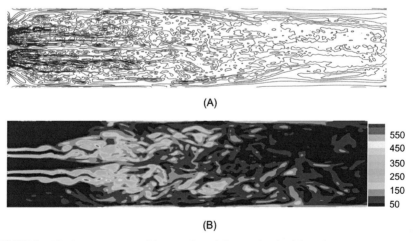

FIGURE 7.64 Instantaneous vorticity map for swirling combustion (A, coal; B, gas).

FIGURE 7.65 Instantaneous particle concentration map for swirling coal combustion.

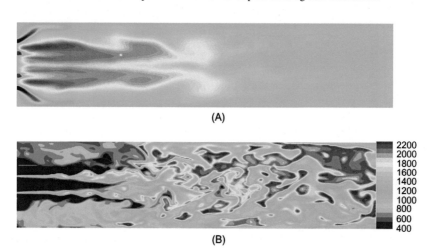

FIGURE 7.66 Instantaneous temperature map for swirling combustion (A, coal; B, gas).

(for time-developing flows) and higher-order finite difference methods (for space-developing flows) can be used. It is recognized that the DNS of turbulent combustion is a breakthrough in the history of combustion science. The DNS of turbulent combustion aims to solve three kinds of problems: (1) the detailed structure of turbulent flames, such as nonpremixed or premixed flames, wrinkled or broken flames; (2) the effect of combustion on turbulence, e.g., flame induces or reduced turbulence; (3) to validate the RANS and SGS combustion models using the DNS database. Many DNS studies give the flame structures, and told us that the flame structure is very complex; there are no pure nonpremixed or premixed flames. DNS gives new findings in the flame structures of gas and two-phase combustion. The structure of a lifted H_2−air nonpremixed flame was obtained by Takeno et al. [53] using DNS. It is interesting to note that in fact it is not a pure diffusion flame, but consists of a rich premixed flame, many diffusion flame islands and an edge flame. Domingo et al. give the instantaneous vorticity and temperature isosurfaces of a partially premixed swirling flame using DNS [54], showing the complex structures of turbulence and flame. It was found that in the upstream regions the flame is mainly premixed, in the downstream region the flame is partially premixed, and there is a local diffusion flame in the

FIGURE 7.67 Temperature and heat release maps.

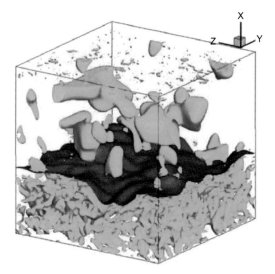

FIGURE 7.68 Vorticity (green) and temperature (red) isosurfaces.

core region. Luo et al. studied the swirling hydrogen−air premixed flame using DNS [55]. The temperature and heat release maps are shown in Fig. 7.67, and the vorticity and temperature isosurfaces are shown in Fig. 7.68. The flame front is affected by the local turbulent eddies and becomes wrinkled. The contour of high heat release rate exhibits a bowl shape, which suggests that the flame front is stabilized around the recirculation zone in the swirling flame. As the flow traverses downstream, the effects of forcing diminish, while instabilities such as Kalvin−Helmholtz instability, centrifugal instability, and diffusive-thermal instability develop.

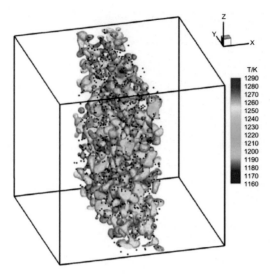

FIGURE 7.69 The mixture fraction and temperature isosurfaces.

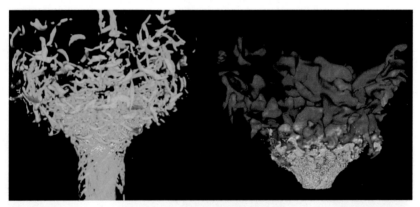

FIGURE 7.70 The vorticity (left) and temperature (right) isosurfaces of a swirling spray flame.

The mixture fraction and temperature isosurfaces (Fig. 7.69) of a heptane−air flame in homogeneous and isotropic turbulence obtained by Luo et al. using DNS [56], show many flame islands, instead of a continuous flame surface.

Luo et al. used DNS to study a swirling spray flame [57]. The results give the instantaneous vorticity and temperature isosurfaces, shown in Fig. 7.70. The effect of different inlet conditions, including lean coswirling, rich coswirling, lean counterswirling, and rich counterswirling, was studied. It was found that the spray flame consists of both premixed and diffusion flames.

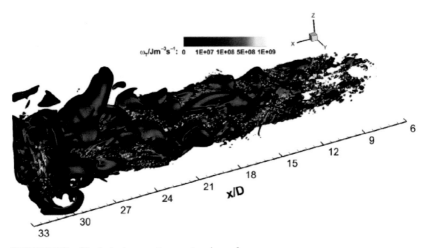

FIGURE 7.71 The instantaneous temperature isosurface.

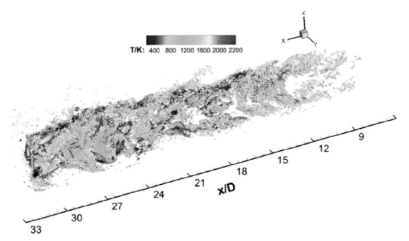

FIGURE 7.72 The heat release rate isosurface.

For DNS of coal combustion, Luo et al. studied the coal-jet flame by DNS [58]. Figs. 7.71 and 7.72 give the instantaneous temperature isosurface with the heat release rate superimposed and the heat release rate isosurface, respectively. It is seen that the jet flame has fully developed and has a complicated flame structures. Some high heat release rate regions are scattered on the flame surface. The reaction zones have different structures from upstream to downstream regions.

The flame index map is given in Fig. 7.73. It indicates that in different axial positions the premixed and nonpremixed flames coexist but occupy different parts.

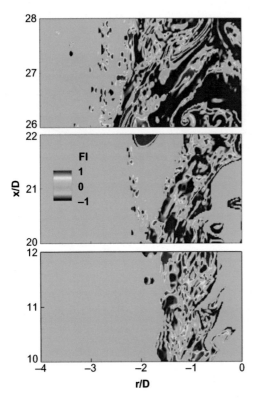

FIGURE 7.73 The Flame Index.

In the validation of RANS and SGS combustion models using DNS, in the 1990s, Pope et al. [59] used DNS of isothermal reacting flows to validate the mixing model in the PDF equation, the CMC model, and the flamelet model. Subsequently, Domingo et al. [60] proposed a flame-surface density-simplified PDF (FSD-PDF) SGS combustion model using DNS of a methane-air V-shaped flame. Hawkes and Chen [61] validated the flamelet model using DNS of a methane−air flame. Poinsot et al. [62] used DNS of a propane−air flame to validate the flamelet model and the CMC model. Recently, Minamoto et al. [63] used the DNS of a swirling hydrogen−air premixed flame to validate the flamelet and EDC models, and show that the combination of these two models gives better results. Chan et al. [64] used DNS to access the model assumptions and budget terms of the unsteady flamelet equations. A typical example is using DNS to validate the mixing models in the PDF transport equation by Krisman et al. [65]. The previous section discussed the second-order moment combustion model being validated by DNS of both isothermal reacting flows. A recent review of combustion model development using DNS was given by Trisjono and Pitsch [66], showing different methods of validation and error analysis and giving some representative examples.

As for the effect of combustion on turbulence, Zhang and Rutland [67] used DNS of a premixed flame to study the effect of combustion on several terms in the Reynolds stress equation. In general, the validation of RANS models using DNS is limited mainly to the flamelet, CMC, and PDF transport equation models. The future development in DNS of turbulent combustion should be studies on fully resolved DNS (FDNS) of spray/coal two-phase combustion.

REFERENCES

[1] L. Landau, E. Lifshitz, Mekhanica Sploshneih Sred (Continuum Mechanics) (in Russian). Moscow, Gostsekhizdat (State Technology Press), 1944.

[2] D.B. Spalding. Development of eddy-break-up model of turbulent combustion. In: Combustion Inst. Pittsburgh PA, Eds. 16th Symposium (International) on Combustion, Pittsburgh: The Combustion Institute, pp. 1657−1669, 1976.

[3] B.F. Magnussen, B.H. Hjertager. On mathematical modeling of turbulent combustion with special emphasis on soot formation and combustion. in: Combustion Inst. Pittsburgh PA, eds. 16th Symposium (International) on Combustion. Pittsburgh: The Combustion Institute, 1976, pp. 719−729.

[4] FLUENT User's Guide, 2003.

[5] D.B. Spalding, Concentration fluctuations in a round turbulent free jet. Chem. Eng. Sci. 26 (1971) 95−107.

[6] L.X. Zhou, Dynamics of Multiphase Turbulent Reacting Fluid Flows (in Chinese). Defense Industry Press, 2002.

[7] L.D. Smoot, P.J. Smith, Coal Combustion and Gasification., Plenum Press, New York, 1985.

[8] L.X. Zhou, Advances in modeling NO_x formation in turbulent flows. Adv. Mech. (in Chin.) 30 (2000) 77−82.

[9] A. Okasanon, E. Maki-Mantila, Use of PDF in modeling of NO formation in methane combustion. Proceedings of 3rd International Conference on Combustion Technology for a Clean Environment, Ed. Carvalho, Paper 18. 3, Lisbon, 1995.

[10] A. Beretta, N. Mancini, F. Podenzani, L. Vigevano, The influence of the temperature fluctuation variance on NO predictions for a gas flame. in: Proceedings of 3rd International Conference on Combustion Technology for a Clean Environment, Ed. Carvalho, Paper 18.4, Lisbon, 1995.

[11] E. Gutheil, H. Bockhorn, F. Fetting, Elements of modeling of turbulent diffusion flames., Proceedings of 1st International Symposium on Coal Combustion, Hemisphere, 1988, pp. 181−188.

[12] S.B. Pope, PDF methods for turbulent reactive flows. Progress Energy Combustion Sci. 11 (1985) 119−192.

[13] S.B. Pope, New developments in PDF modeling of non-reactive and reactive turbulent flows. Proc. 2nd Inter. Symp. on Turbulence, Heat and Mass Transfer, Delft University Press, 1997, pp. 35−45.

[14] S.M. Cannon, B.S. Brewster, L.D. Smoot, PDF modeling of lean premixed combustion using in situ tabulated chemistry. Combustion Flame 119 (1999) 233−252.

[15] F. Wang, L.X. Zhou, C.X. Xu, G.M. Goldin, Comparison between a composition PDF transport equation model and an ASOM model for simulating a turbulent jet flame. Int. J. Heat Mass Transfer 51 (2008) 136−144.

[16] P. Libby, F.A. Williams, Turbulent Reacting Flows. Academic Press, London, 1994.

[17] Zhou Lixing, Combustion Theory and Reacting Fluid Dynamics (in Chinese). Science Press, Beijing, 1986.

[18] A.Y. Klimenko, R.W. Bilger, Conditional moment closure for turbulent combustion. Progress Energy Combustion Sci. 25 (1999) 595−687.

[19] C. Zou, C.G. Zheng, L.X. Zhou, Conditional moment closure model for simulation of turbulent diffusion combustion and nitric oxides formation. Acta Mech. Sin. (in Chin.) 34 (2002) 969−977.

[20] N. Peters, Turbulent Combustion. Cambridge University Press, 2000.

[21] E.E. Khalil, On the prediction of reaction rates in turbulent premixed confined flames. AIAA Paper, 1980.

[22] L.X. Zhou, A new second-order moment model for turbulent combustion. J. Eng. Thermophys. (in Chin.) 17 (1996) 353−356.

[23] X.L. Chen, L.X. Zhou, J. Zhang, Numerical simulation of methane-air turbulent jet flame using a new second-order moment model. Acta Mech. Sin. 16 (2000) 41−47.

[24] L.X. Zhou, L. Qiao, J. Zhang, A unified second-order moment turbulence-chemistry model for simulating turbulent combustion and NO_x formation. Fuel 81 (2002) 1703−1709.

[25] F. Wang, Studies on the second-order moment turbulent reaction model by RANS, LES and DNS, Ph.D. Dissertation (in Chinese). Department of Engineering Mechanics, Tsinghua University, 2006.

[26] L.X. Zhou, F. Wang, J. Zhang, Simulation of swirling combustion and NO formation using a USM turbulence-chemistry model. Fuel 82 (2003) 1579−1586.

[27] L.X. Zhou, Development of SOM combustion model for Reynolds-averaged and large-eddy simulation of turbulent combustion and its validation by DNS. Sci. China E-51 (2008) 1073−1086.

[28] F. Wang, C.X. Xu, L.X. Zhou, C.K. Chan, DNS-LES validation of an algebraic second-order-moment combustion model. Numer. Heat Transfer B55 (2009) 523−532.

[29] Luo Kun, Bai Yun, Yang Jianshan, Wang Haiou, Zhou Lixing, Fan Jianren, A-priori validation of a second-order moment combustion model via DNS database. Int. J. Heat Mass Transfer 86 (2015) 415−425.

[30] Y.W. Zhang, F.L. Lei, Y.H. Xiao, CFD simulation and parametric study of coal gasification in a circulating fluidized bed. Asia-Pacific J. Chem. Eng. 10 (2015) 307−317.

[31] G.B. Yu, J.H. Chen, J.R. Li, T. Hu, S. Wang, H.L. Lu, Analysis of SO_2 and NO_x emissions using two-fluid method coupled with eddy dissipation concept reaction sub-model in CFB combustors. Energy Fuels 28 (2014) 2227−2236.

[32] L.D. Smoot, International research center's activities in coal combustion. Progress Energy Combustion Sci. 24 (1998) 409−501.

[33] L.X. Zhou, J. Zhang, A Lagrangian-Eulerian particle model for turbulent two-phase flows with reacting particles. in: F.G. Zhuang (Ed.), Proceedings of 10th International Conference on Numerical Methods in Fluid Dynamics. Springer-Verlag, 1986, pp. 705−709.

[34] L.X. Zhou, W.Y. Lin, J. Zhang, W.W. Luo, X.Q. Huang, Gas-particle flows and coal combustion in a burner/combustor with high-velocity jets. Combustion Flame 99 (1994) 669−678.

[35] L.X. Zhou, L. Li, R.X. Li, J. Zhang, Simulation of 3-D gas-particle flows and coal combustion in a tangentially fired furnace using a two-fluid-trajectory model. Powder Technol. 125 (2002) 226−233.

[36] L.X. Zhou, A multi-fluid model of two-phase flows with pulverized-coal combustion. in: J.K. Feng (Ed.), Proceedings of 1st International Symposium on Coal Combustion, Hemisphere, 1988, pp. 207−213.

[37] L.X. Zhou, Y.C. Guo, W.Y. Lin, Two-fluid models for simulating reacting gas-particle flows, coal combustion and NO_x formation. Combustion Sci. Technol. 150 (2000) 161−180.

[38] L.X. Zhou, Y. Zhang, J. Zhang, Simulation of swirling coal combustion using a full two-fluid model and an AUSM turbulence-chemistry model. Fuel 82 (2003) 1001−1007.

[39] T. Abbas, P. Costen, F.C. Lockwood, The influence of near burner region aerodynamics on the formation and emission of nitrogen oxides in a pulverized coal-fired furnace. Combustion Flame 91 (1991) 346−363.

[40] W. Kim, S. Menon, H.C. Mongia, Large-eddy simulation of a gas turbine combustor flow. Combustion Sci. Technol. 143 (1999) 25−62.

[41] C. Fureby, C. Lokstrom, Large-eddy simulation of bluff-body stabilized flames. Proceedings of 25th International Symposium on Combustion, The Combustion Institute, Bouder, Colorado, USA, 1994, pp. 1257−1264.

[42] R. Kurose, H. Makino, Large eddy simulation of a solid-fuel jet flame. Combustion Flame 135 (2003) 1−16.

[43] M. Gharebaghi, R.M.A. Irons, L. Ma, M. Pourkashanian, A. Pranzitelli, Large eddy simulation of oxy-coal combustion in an industrial combustion test facility. Int. J. Greenhouse Gas Control (2011). Available from: https://doi.org/10.1016/j.ijggc.2011.05.030.

[44] L.Y. Hu, L.X. Zhou, J. Zhang, Large-eddy simulation of a swirling diffusion flame using a SOM SGS combustion model. Numer. Heat Transfer B50 (2006) 41−58.

[45] University of Sydney, Swirl Flame Web Database. <http://www.aeromech.usyd.edu.au/thermofluids/swirl.htm>, 2002.

[46] L.Y. Hu, L.X. Zhou, Y.H. Luo, Large-eddy simulation of the Sydney swirling non-premixed flame and validation of several sub-grid scale models. Numer. Heat Transfer B53 (2008) 39−58.

[47] W.L. Wang, L.X. Zhou, R.X. Li, Large-eddy simulation of propane-air swirling diffusion combustion using SOM sub-grid scale combustion model. J. Combustion Sci. Technol. (in Chin.) 13 (2007) 97−100.

[48] F. Wang, L.X. Zhou, C.X. Xu, Large-eddy simulation of correlation moments in turbulent combustion and validation of the RANS-SOM combustion model. Fuel 85 (2006) 1242−1247.

[49] F. Wang, L.X. Zhou, C.X. Xu, Y. Huang, Large-eddy simulation of premixed combustion and validation of the combustion model., J. Propulsion Technol. (in Chin.) 29 (2008) 33−36.

[50] K. Li, L.X. Zhou, C.K. Chan, Studies of the effect of spray inlet conditions on flow and flame structures of ethanol-spray combustion by large-eddy simulation. Numer. Heat Transfer A62 (2012) 44−59.

[51] H.W. Ge, I. Düwel, H. Kronemayer, R.W. Dibble, E. Gutheil, C. Schulz, J. Wolfrum, Laser-based experimental and Monte Carlo PDF numerical investigation of an ethanol/air spray flame. Combustion Sci. Technol. 180 (2008) 1529−1547.

[52] L.Y. Hu, L.X. Zhou, Y.,H. Luo, C.S. Xu, Measurement and simulation of swirling coal combustion. Particuology 11 (2013) 189−197.

[53] T. Takeno, Y. Mizobuchi, Significance of DNS in combustion science. Comptes Rendus Mecanique 334 (2006) 517−522.

[54] L. Vervisch, P. Domingo, Two recent developments in numerical simulation of premixed and partially premixed turbulent flames. Comptes Rendus Mecanique 334 (2006) 523–530.

[55] H.O. Wang, K. Luo, et al., A DNS study of hydrogen/air swirling premixed flames with different equivalence ratios. Int. J. Hydrog. Energy 37 (2012) 5246–5256.

[56] H.O. Wang, K. Luo, J.R. Fan, Direct numerical simulation and CMC (conditional moment closure) sub-model validation of spray combustion. Energy 46 (2012) 606–617.

[57] K. Luo, H. Pitsch, et al., Direct numerical simulations and analysis of three-dimensional n-heptane spray flames in a model swirl combustor. Proc. Combustion Inst. 33 (2011) 2143–2152.

[58] Y. Bai, K. Luo, K.Z. Qiu, J.R. Fan, Numerical investigation of two-phase flame structures in a simplified coal jet flame. Fuel 182 (2016) 944–957.

[59] M.R. Overholt, S.B. Pope, Direct numerical simulation of a statistically stationary, turbulent reacting flow. Combustion Theory Model. 3 (1999) 371–408.

[60] P. Domingo, L. Vervisch, S. Payet, et al., DNS of a premixed turbulent V flame and LES of a ducted flame using a FSD-PDF sub-grid scale closure with FPI-tabulated chemistry. Combustion Flame 143 (2005) 566–586.

[61] E.R. Hawkes, J.H. Chen, Comparison of direct numerical simulation of lean premixed methane-air flames with strained laminar flame calculations. Combustion Flame 144 (2006) 112–125.

[62] C. Jimenez, B. Cuenot, T. Poinsot, et al., Numerical simulation and modeling for lean stratified propane-air flames. Combustion Flame 128 (2002) 1–21.

[63] Y. Minamoto, K. Aoki, et al., DNS of swirling hydrogen-air premixed flames., Int. J. Hydrog. Energy 40 (2015) 13604–13620.

[64] W.L. Chan, H. Kolla, J.H. Chen, M. Ihme, Assessment of model assumptions and budget terms of the unsteady flamelet equations for a turbulent reacting jet-in-cross-flow. Combustion Flame 161 (2014) 2601–2613.

[65] A. Krisman, J.C.K. Tang, E.R. Hawkes, D.O. Lignell, J.H. Chen, A DNS evaluation of mixing models for transported PDF modeling of turbulent nonpremixed flames. Combustion Flame 161 (2014) 2085–2106.

[66] P. Trisjono, H. Pitsch, Systematic analysis strategies for the development of combustion models from DNS: a review. Flow, Turbulence Combustion 95 (2015) 231–259.

[67] S. Zhang, C.J. Rutland, Premixed flame effects on turbulence and pressure-related terms. Combustion Flame 102 (1995) 447–461.

Chapter 8

The Solution Procedure for Modeling Multiphase Turbulent Reacting Flows

In the preceding chapters, the fundamentals, basic equations, closure models, and their validation are discussed. In general, the equations are simultaneous nonlinear partial differential equations, and these equations must be solved numerically, and discretized using finite element or finite difference methods. For computational fluid dynamics, the finite difference method is predominant. For single-phase flows the numerical method is described in many books, for example in Reference [1]. For simulating fluid flows, heat transfer, and combustion, the most widely used is the so-called SIMPLE (semi-implicit pressure-linked) algorithm. In this chapter, only a brief introduction is given to the specific problems in some numerical procedures for simulating multiphase turbulent reacting flows.

8.1 THE PSIC ALGORITHM FOR EULERIAN–LAGRANGIAN MODELS

The numerical method for Eulerian–Lagrangian simulation of dispersed multiphase flows is the PSIC (Particle Source in Cell) method, first proposed by Crowe et al. [2]. It is now widely used and included in commercial software, such as ANSYS-FLUENT, STAR-CD, etc. In the PSIC method, for the gas flow field, the usual SIMPLE algorithm (SIMPLE, SIMPLER, SIMPLEC) may be adopted, i.e., the $p - v$ correction, line-by-line or plane-by plane TDMA, and under-relaxation iterations. The particle source term in the computational cell for the gas continuity can be written as

$$-\Delta V \cdot S = -\Delta V \left(\sum_k n_k \dot{m}_k \right) = -\Delta V \left(\sum_k n_k \frac{\Delta m_k}{\Delta t} \right) = -\sum_k \frac{\Delta V \cdot n_k}{\Delta t} \Delta m_k$$
$$= \sum_{k,c} \sum_{j,c} N_{k,j}(m_{k,i} - m_{k,e})$$

where the first term on the far right-hand side is the summation of the mass of all particle-size trajectories crossing this cell, and the second term is the

Theory and Modeling of Dispersed Multiphase Turbulent Reacting Flows.
DOI: https://doi.org/10.1016/B978-0-12-813465-8.00008-9
253

summation of the mass of the same-size particles with different initial positions and directions of trajectories. $N_{k,j}$ is the total number flux of k-group particles, which keeps constant along the j-th trajectory. Similarly, the particle source terms in the momentum and energy equations are

$$\Delta V \cdot S_{pvi} = \sum_{k,c} \sum_{j,c} N_{k,j}(m_{k,e}v_{ik,e} - m_{k,i}v_{ik,i})$$

$$\Delta V \cdot S_{ph} = \sum_{k,c} \sum_{j,c} N_{k,j}[c_p T(m_{k,e} - m_{k,i}) - Q_k]$$

where the subscripts i and e express the values at the entrances and exits of trajectories in the gas computational cell, respectively, and Q_k is the convective heat transfer between the gas and particles. For the particle boundary conditions, the inlet particle concentration distribution, particle size, and velocity distribution, particle initial position and particle-wall collision conditions must be given. The gas pressure correction equation must be based on the gas continuity equation with particle source terms. To predict the particle trajectories and particle history along the trajectories, it is necessary to select an appropriate integral method for determining the crossing positions of trajectories with the walls of gas cells, including determining which wall to cross with, when the particle trajectory leaves a gas cell, and promptly stopping the particle trajectory computation in that cell. The particle trajectories can be obtained by solving the particle momentum equation using the Rung−Kutta method:

$$x_k = \int_0^t u_k dt, \quad r_k = \int_0^t v_k dt, \quad \theta_k = \int_0^t (w_k/r_k)dt$$

The problem is how to choose Δt. At first we may take

$$\Delta t^{(1)} = \min\left[\frac{\Delta x_i}{u_{ki}}, \frac{r_e - r_i}{v_{ki}}\right]$$

and use the Rung−Kutta method to calculate the particle position x_k and r_k in this time interval. If the point (x_k, r_k) cannot fall into the anticipated region in Fig. 8.1, then we should take

$$\Delta t^{(2)} = \min\left[\frac{\Delta x_i}{(u_{ki} - u_{ke})/2}, \frac{r_e - r_i}{(v_{ki} - v_{ke})/2}\right]$$

and repeat the trajectory computation, until the point (x_k, r_k) falls into the anticipated region in Fig. 8.1. Crowe et al. assume a constant gas velocity and a constant particle relaxation time in the gas cell, namely,

$$v_i = \text{const}, \quad \tau_{rk} = \text{const}$$

hence they give an analytical solution as

$$u_{ke} = u - (u - u_{ki})\exp(-\Delta t/\tau_{rk})$$
$$v_{ke} = v - (v - v_{ki})\exp(-\Delta t/\tau_{rk})$$

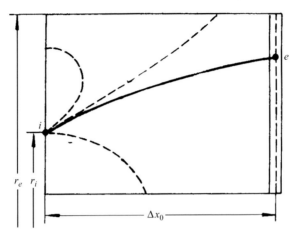

FIGURE 8.1   Method 1 for computing particle trajectories.

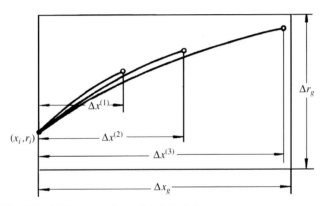

FIGURE 8.2   Method 2 for computing particle trajectories.

Finally, the particle trajectory is obtained by linear interpolation as

$$x_{ke} = x_{ki} + 0.5(u_{ki} + u_{ke})\Delta t$$
$$r_{ke} = r_{ki} + 0.5(v_{ki} + v_{ke})\Delta t$$

In fact, the particle mass, size, drag, and relative velocity change along the trajectories, therefore the present author suggests using an analytical solution of trajectories of particles with mass change [3]

$$\Delta x = x - x_i = t_1[uf_1(\bar{d}) - (u - u_i)f_2(\bar{d})]$$
$$\Delta r = r - r_i = t_1[vf_1(\bar{d}) - (v - v_i)f_2(\bar{d})]$$

where (x_i, r_i) is the initial position of trajectories entering the gas call (Fig. 8.2)

$$f_1(\bar{d}) = \int_{\bar{d}}^{1} \frac{\bar{d}}{1 + \Pi^{0.5(m+1)}} \mathrm{d}\bar{d}$$

$$f_2(\bar{d}) = \int_{\bar{d}}^{1} \frac{\bar{d}^{(m+1)}}{1 + \Pi^{0.5(m+1)}} \mathrm{d}\bar{d}$$

$$t_1 = \rho_s d_{ki}^2 / [4D\rho \ln(1 + B)]$$
$$= \rho_s d_{ki}^2 c_p / [4\lambda \ln(1 + B)]$$
$$\bar{d} = d_k / d_{ki}, \quad m = 3.24/B$$
$$\Pi = 0.25(\mathrm{Re}_{ki})^{1/2}$$

For the given particle inlet position, velocity and size in a certain gas cell $x_i, u_{ki}.v_{ki}, d_{ki}$, at first take the value $\bar{d}^{(1)}$ near unity, then calculate $f_1(\bar{d}), f_2(\bar{d})$, and compare the predicted $\Delta x^{(1)}$, $\Delta r^{(1)}$ with the sizes of the gas cell Δx_g, Δr_g. If $\Delta X^{(1)}$ is smaller than Δx_g and $\Delta r^{(1)}$ is smaller than Δr_g, may take $\bar{d}^{(2)}$ to be smaller than $\bar{d}^{(1)}$, and calculate $\Delta X^{(2)}$, $\Delta x^{(2)}$, until $\Delta X^{(n)}$ at first reaches ΔX_g or $\Delta x^{(n)}$ at first reaches Δx_g. Using such a method, the position of trajectories in the gas cell is determined. The flow chart of the PSIC method is shown in Fig. 8.3.

8.2   THE LEAGAP ALGORITHM FOR E−E−L MODELING

In simulating multiphase turbulent reacting flows, for example, pulverized-coal combustion, a two-fluid-trajectory (gas Eulerian and particle

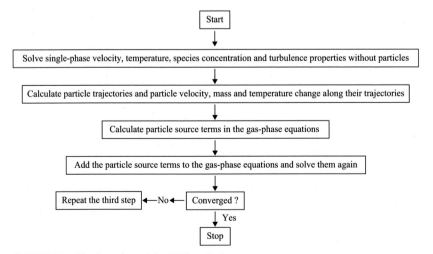

FIGURE 8.3   The flow chart of the PSIC method.

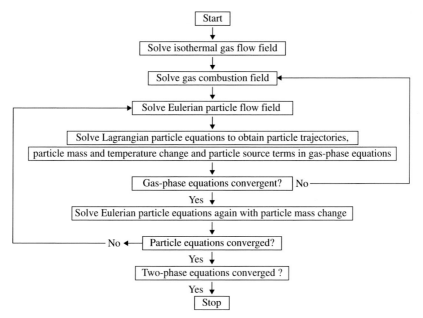

FIGURE 8.4 The LEAGAP flow chart.

Eulerian−Lagrangian, E−E−L) model was proposed by the author [4]; its algorithm is called LEAGAP (Lagrangian−Eulerian−Eulerian Algorithm of Gas-Particle Flows). In this algorithm, in addition to the iteration inside the gas phase, there are coupling and iterations among the gas Eulerian, particle Eulerian, and particle Lagrangian computations. The LEAGAP flow chart is shown in Fig. 8.4.

8.3 THE PERT ALGORITHM FOR EULERIAN−EULERIAN MODELING

A PERT algorithm (Pure Eulerian Algorithm of Reacting Two-phase Flows) was proposed by the author [4] for pure two-fluid or Eulerian−Eulerian modeling of multiphase turbulent reacting flows. In this algorithm, in addition to the iteration inside the gas phase, there are coupling and iterations between the Eulerian gas and Eulerian particle computations. The flow chart of the PERT algorithm is shown in Fig. 8.5.

8.4 THE GENMIX-2P AND IPSA ALGORITHMS FOR EULERIAN−EULERIAN MODELING

Spalding extended the SIMPLE algorithm to two-fluid modeling of two-phase flows [5]. At first the diffusion terms in each equation of each phase

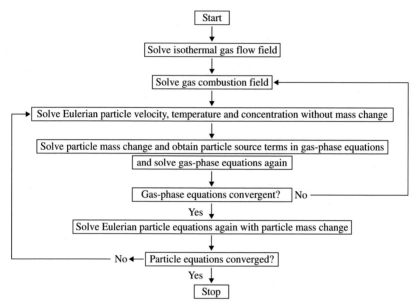

FIGURE 8.5   The flow chart of the PERT algorithm.

are modeled in the form of volume fraction gradient. So, the conservation equations of all the dependent variables of each phase are expressed in the following form:

$$\frac{\partial}{\partial t}(\alpha_k \rho_k \varphi_k) + \frac{\partial}{\partial x_j}(\alpha_k \rho_k v_{kj} \varphi_k) = \frac{\partial}{\partial x_j}\left(\Gamma_{\varphi k}\frac{\partial \varphi_k}{\partial x_j}\right) + \alpha_k S_{\varphi k}$$

where α_k is the volume fraction of phase k, ϕ is a property of phase k, $\Gamma_{\phi k}$ is the transport coefficient inside phase k, and $S_{\phi k}$ is the source term, including the source term inside phase k and the source term of interaction between phase k and other phases. Hence the original numerical methods, like GENMIX and SIMPLE algorithms, can be extended to two-phase flows, called GENMIX-2P and IPSA algorithms. The specific features of solving two-phase flows are: (1) each phase has a volume fraction (or mass fraction, apparent density); (2) need to give the turbulence properties of each phase and the interactions between two phases, like drag force and heat transfer between two phases; (3) the partial pressures of two phases and their pressure correction. The basic steps of the GENMIX-2P algorithm for two-dimensional parabolic flows (jets, boundary layers, pipe flows) are:

1. Introduce a stream function ψ, and transform the $x-y$ coordinate system into an $x-\omega$ coordinate system, where ω is a relative stream function;
2. Presume or guess the two-phase volume fractions;
3. Solve the gas momentum equation in the x direction and energy equation using the TDMA method;

4. Find the position in the y-coordinate of the boundary grid nodes;
5. Solve gas continuity equation to find the velocity v;
6. Solve the gas momentum equation in the y direction to find the pressure distribution in the y direction;
7. Solve particle-phase equations to find the particle velocity, temperature, and volume fraction;
8. Compare the predicted particle volume fraction with the initially pre-sumed value and make corrections, and repeat the computation until the correction is smaller than a certain value.

For the IPSA algorithm in modeling of two-dimensional elliptic two-phase flows, at first the two-phase differential equations are discretized into finite difference equations

$$\varphi_k = \left(\sum A_i^\varphi \varphi_i + b \right) / \sum A_i^\varphi$$

Let $g = \rho v a = $ density $\times$ velocity $\times$ cross-sectional surface be the mass flux across the wall of a cell, and the subscripts and e denote the inlet and outlet values, respectively, then the continuity equation of the first phase is

$$\varphi_1 = \left[\left(\sum \varphi_1 g_1 \right)_i + \Delta VS \right] / \sum g_{1,e}$$

To avoid divergence, the denominator of this equation can be timed by

$$\left\{ 1 + \left[\frac{\sum (\varphi_1 g_1)_i + \Delta VS}{\sum g_{1,e}} - \varphi_1 \right] + \left[\frac{\sum (\varphi_2 g_2)_i + \Delta VS}{\sum g_{2,e}} - \varphi_2 \right] \right\}$$

In general this term is unity, hence we have

$$\varphi_1 = \frac{\sum g_{2,e}[\sum (\varphi_1 g_1)_i + \Delta VS]}{\sum g_{2,e} \cdot \sum (\varphi_1 g_1)_i + \sum g_{1,e} \cdot \sum (\varphi_2 g_2)_i + (\sum g_{2,e} - \sum g_{1,e}) \Delta VS}$$

where $\sum$ expresses the summation over the walls of the cell. This expression is called the equation of volume fraction. The pressure correction accounts for two phases. If

$$\frac{\partial u_1}{\partial p}, \frac{\partial u_2}{\partial p}, \frac{\partial v_1}{\partial p}, \frac{\partial v_2}{\partial p}$$

is known, the error of u_1 and u_2 is e_1, and the error of v_1 and v_2 is e_2, then we can find $\frac{\partial e_1}{\partial p}, \frac{\partial e_2}{\partial p}$, and the pressure correction is

$$p' \left(\frac{\partial e_1}{\partial p} / \rho_1 + \frac{\partial e_2}{\partial p} / \rho_2 \right) = - \left(\frac{e_1}{\rho_1} + \frac{e_2}{\rho_2} \right)$$

where ρ_1 and ρ_2 are the material density of two phases, respectively.
The flow chart of the IPSA algorithm is shown in Fig. 8.6.

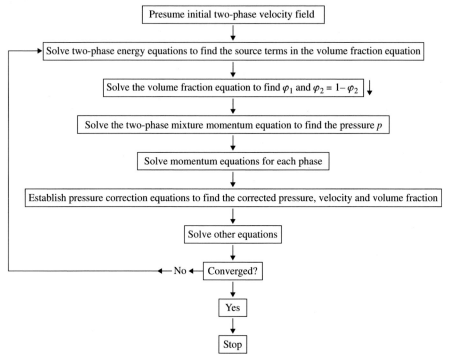

FIGURE 8.6 The flow chart of the IPSA algorithm.

The IPSA algorithm has been applied to simulate two-phase flows with large variation of two-phase volume fractions. However, due to lack of the dispersed-phase turbulence models it is not widely used in practical engineering applications.

REFERENCES

[1] S.V. Patankar. Numerical Heat Transfer and Fluid Flow. Hemisphere, New York, 1980.
[2] C.T. Crowe, M.P. Sharma, D.E. Stock, The particle-source-in-cell (PSIC) method for gas--droplet flows., J. Fluid Eng. 99 (1977) 325–332.
[3] L.X. Zhou, Combustion Theory and Reacting Fluid Dynamics (in Chinese)., Science Press, Beijing, 1986.
[4] L.X. Zhou, Dynamics of Multiphase Turbulent Reacting Fluid Flows (in Chinese)., Defense Industry Press, Beijing, 2002.
[5] D.B. Spalding, Numerical computation of multiphase fluid flow and heat transfer., in: C. Taylor (Ed.), Recent Advances in Numerical Mechanics, Pinerage Press, New Jersey, 1980.

Chapter 9

Simulation of Flows and Combustion in Practical Fluid Machines, Combustors, and Furnaces

The numerical simulation of multiphase turbulent reacting flows has made more and more progress and now has been widely applied in the design of flow, heat, and mass transfer and combustion/reaction equipment in power engineering, aeronautical and astronautical engineering, chemical and metallurgical engineering, petroleum engineering, hydraulic engineering, and nuclear engineering. Some examples have already been given in Chapter 7: e.g., the two-fluid modeling of pulverized coal−air two-phase combustion. In this chapter, more examples will be given to further illustrate the state of the art and the development of numerical simulation.

9.1   AN OIL-WATER HYDROCYCLONE

In petroleum pipelines the transport of oil, water, and sands are frequently encountered. Oil-water separation, including separation of water from the oil and separation of oil from the wasted water, is one of the important problems in petroleum engineering. Hydrocyclones, similar to gas-solid cyclones, are adopted for this purpose. However, due to the smaller difference between the oil and water densities and in order to give higher collection efficiency, stronger centrifugal forces and longer residence time are required for the hydrocyclones than for gas-solid cyclones. The predicted gas velocities in original and improved hydrocyclones [1] using the Reynolds-stress equation model are given in Figs. 9.1 and 9.2, respectively. It is seen that the two peaks of tangential velocity indicate a higher centrifugal force in the improved hydrocyclone than do those in the original hydrocyclone, and axial velocities are lower in the improved hydrocyclone than in the original hydrocyclone, indicating that the oil-droplet residence time in the former is longer than in the latter.

Further, the Eulerian-Lagrangian simulation of oil-water two-phase flows in these hydrocyclones was made using the Reynolds-stress turbulence model and the particle stochastic trajectory (ST) model. Figs. 9.3 and 9.4 give the

Theory and Modeling of Dispersed Multiphase Turbulent Reacting Flows.
DOI: https://doi.org/10.1016/B978-0-12-813465-8.00009-0
261

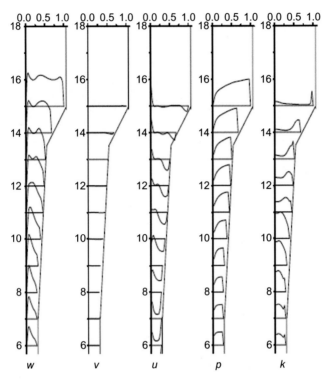

FIGURE 9.1   The velocity field in an original hydrocyclone.

predicted droplet trajectories in these hydrocyclones, respectively. Obviously, more droplets impinged on the walls of the improved hydrocyclone than in the original hydrocyclone, implying that the collection efficiency of the former is higher than that of the latter.

9.2   A GAS-SOLID CYCLONE SEPARATOR

Gas-solid cyclones are widely used for particle separation and collection in utility boiler furnaces, industrial furnaces, cement kilns, etc. Cyclonic flows are very strongly swirling flows, with swirl numbers much greater than unity, where the tangential velocity is an order of magnitude higher than the axial velocity, and the axial velocity is also higher than the radial velocity by an order of magnitude. The main problems are how to reduce the pressure drop and increase the collection efficiency. The geometrical configuration of a cyclone separator in our early studies [2] is shown in Fig. 9.5. The gas flow field was simulated using a k-ε turbulence model and measured using LDV (laser Doppler velocimetry). Fig. 9.6 gives the simulation results in comparison with the simplified analytical solution by S.L. Soo and the measurement results. The solid lines denote numerical results, the dashed lines express the

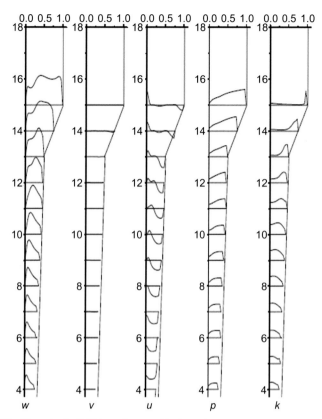

FIGURE 9.2 The velocity field in an improved hydrocyclone.

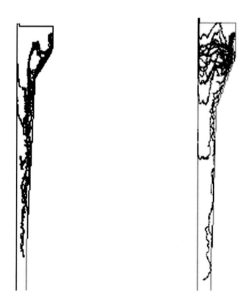

FIGURE 9.3 Oil-droplet trajectories in the original hydrocyclone (Left: $10-30\,\mu m$; Right: $30-100\,\mu m$).

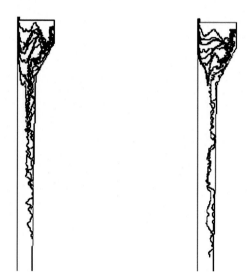

FIGURE 9.4   Oil-droplet trajectories in the improved hydrocyclone (Left: $10-30\,\mu m$; Right: $30-100\,\mu m$).

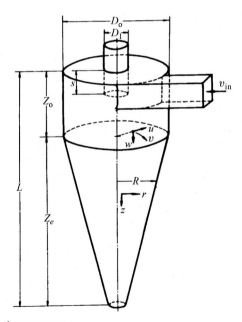

FIGURE 9.5   A cyclone separator.

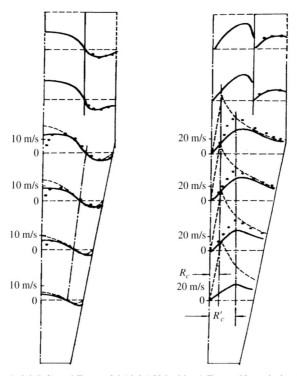

FIGURE 9.6   Axial (left) and Tangential (right) Velocities (●Exp, —Numerical, --Analytical).

analytical solution, and the dots are measurement data. It is seen that in most regions the numerically predicted velocity distribution agrees well with that given by the analytical solution, in particular for the zero-velocity positions. However, due to the shortcomings of the k-ε model for simulating strongly swirling flows in overestimating the turbulent mixing, the low axial velocity in the near-axis region and the peak of tangential velocity cannot be well predicted. According to the axial velocity distribution, there is a near-axis up-flow region and a near-wall down-flow region. The maximal up-flow velocity is much greater than the maximal down-flow velocity. The tangential velocity distribution shows the near-axis solid-body rotation zone and the near-wall potential vortex zone. At three cross sections of the conic part, the tangential velocity is almost unchanged. The large tangential velocity gradient in the near-axis region is the main source of turbulence production and mean energy dissipation, and hence is responsible for the pressure drop. The particles are concentrated mainly in the near-wall region, so the peak of tangential velocity does not play an important role in particle separation. Hence a central fin-cross-sectional body was mounted to make a more uniform tangential velocity distribution. The result is that when the inlet velocity is 21 m/s, the pressure drop is reduced by 37%, and simultaneously the collection efficiency is increased by 5%.

9.3 A NONSLAGGING VORTEX COAL COMBUSTOR

A nonslagging cyclone coal combustor, called a "vortex combustor," was studied cooperatively by the Catholic University of America and Tsinghua University, Beijing, China [3]. The geometrical configuration is shown in Fig. 9.7. Its special features are multiple tangential inlets and a central exit tube of reverse flows. The purposes of this design are to form more uniform tangential velocity distribution and multiple pulverized-coal suspension layers in the annular space and to prolong the particle residence time and increase the radiative heat transfer. The tested fuel is fine ash-free pulverized coal with averaged size of 40 μm. Gas-particle flows and combustion in this combustor were simulated using the algebraic stress turbulence model, particle ST model, EBU-Arrhenius CO and volatile combustion model, coal-particle moisture evaporation, devolatilization, and diffusion-kinetic char combustion model and four-flux radiation model.

Fig. 9.8 gives the predicted gas temperature fields of two cases for coal combustion. Fig. 9.9 shows the predicted gas temperature, and the volatile (simulated by ethylene) concentration and oxygen concentration fields. Due to strong mixing in the annular space, a high temperature of 1600K is formed near the tangential inlets; the coal particles are immediately ignited and burn, but in most regions the gas temperature is in the range of 1200−1400K, or

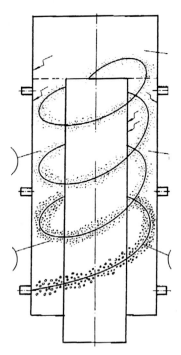

FIGURE 9.7 A vortex coal combustor.

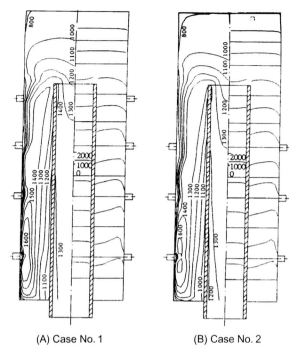

(A) Case No. 1   (B) Case No. 2

FIGURE 9.8   The gas temperature field.

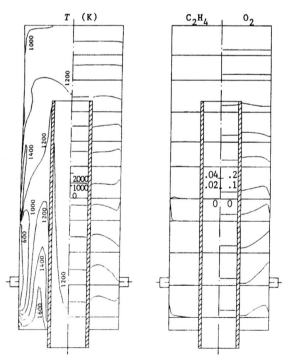

FIGURE 9.9   The gas temperature and species concentration.

900−1100°C, lower than the ash melting point, so it is nonslagging. Owing to lower temperature and high coal concentration, in experiments the nitric oxide formation is low, the oxygen consumption is exhausted at the top of the combustor, and the combustion efficiency reaches more than 95%. Therefore highly efficient and low pollutant combustion was obtained.

9.4 A SPOUTING-CYCLONE COAL COMBUSTOR

The spouting-cyclone combustor, proposed by the present author and his colleagues [4], is a two-staged nonslagging cyclone combustor, combining two perpendicularly rotating (vortex) flows, as shown in Fig. 9.10. Strongly recirculating gas-particle flows and swirling gas-particle flows with air staging in the combustor are expected to achieve highly efficient and clean combustion of ungrounded coal in domestic furnaces and small-size industrial combustors. The lower part of the combustor is a "spouting zone" as the main combustion zone, where an opposed-jet system of primary-air and bottom-air is used to induce large-size strongly recirculating flows (vortex flows with a horizontal axis) of gas and particles, allowing larger coal particles to circulate many times for increasing their residence time and burn-out rate. The combustion in this zone is fuel-rich to decrease the temperature and to provide nonslagging conditions for reducing NO_x emission. The upper part is a cyclone combustor (vortex flows with a vertical axis), which allows completion of the combustion of small-size particles coming from the lower zone and for the collection of particles. A tangential main stream and multiple tangential secondary-air jets are used to create strongly swirling flows, increasing the residence time of smaller particles and of particles that do not yet have complete combustion in the spouting zone. The air

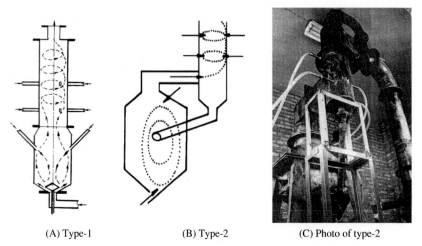

| (A) Type-1 | (B) Type-2 | (C) Photo of type-2 |

FIGURE 9.10 Two types of spouting-cyclone combustors (nonslagging two-staged cyclone combustors).

distribution and the heat exchangers can be arranged to control the temperature in this zone within a specified range. Collected particles can be sent back to the spouting zone to continue their burning. In order to develop these kinds of combustors, measurements and numerical simulation of gas-particle flows were carried out to make an optimal design, and coal combustion tests were conducted to measure the combustion temperature, efficiency and NO emission.

The isothermal gas-particle flows in the spouting zone and the cyclone zone were numerically simulated using a two-fluid modeling with both k-ε-A_p and k-ε-k_p two-phase turbulence models. Fig. 9.11 gives the longitudinal gas velocity in the spouting zone of the type-2 combustor, measured by the hot-ball probe. Fig. 9.12 show the velocity vectors in this zone obtained by numerical simulation. The predictions are in qualitative agreement with the measurement results. The large-size recirculating flow can be seen.

Fig. 9.13 shows the predicted gas and particle velocity vectors in the spouting zone of a type-1 combustor. It is seen that both gas (arrow) and particles (dash) have large-size recirculating flows, but the two-phase velocities are different in their magnitude and directions due to different inertia. By adjusting the angles of bottom jets and the ratio of primary-air/bottom-air velocity, the optimal recirculating flows can be obtained. Fig. 9.14 gives the tangential gas and particle velocities in the cyclone zone, measured by LDV, showing the features of strongly swirling flows—the Rankine-vortex structure, i.e., solid-body rotation plus potential vortex. The particle velocity lags behind the gas velocity due to larger inertia of particles. Fig. 9.15 shows the

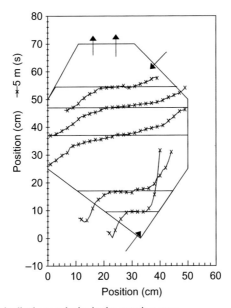

FIGURE 9.11   Longitudinal gas velocity in the spouting zone.

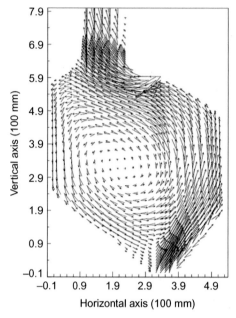

FIGURE 9.12 Two-phase velocity vectors in the spouting zone of the combustor of type 2 (arrows−gas; dash−particle).

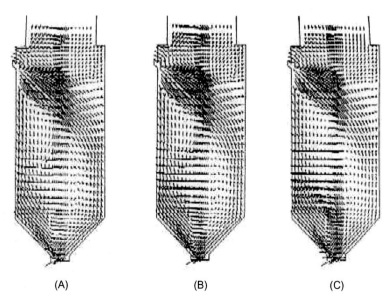

(A)   (B)   (C)

FIGURE 9.13 Two-phase velocity vectors in the spouting zone of the combustor of type 1 (arrows−gas; dash−particle) (A, B, C-for swirl numbers of s = 2.06, s = 2.50 and s = 3.15).

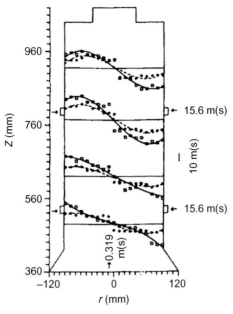

FIGURE 9.14 Two-phase tangential velocities ($\square$-Gas; ...Particle).

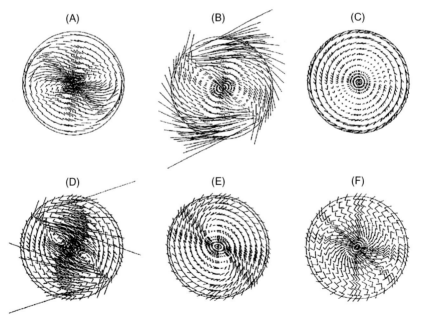

FIGURE 9.15 Two-phase tangential velocity vectors ($\rightarrow$Gas; —Particle) ($x = 600$, 665, 785, 935, 1015, 1200 mm from (A) to (F)).

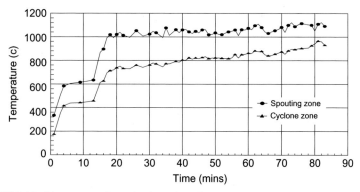

FIGURE 9.16   Temperature change in the coal combustor.

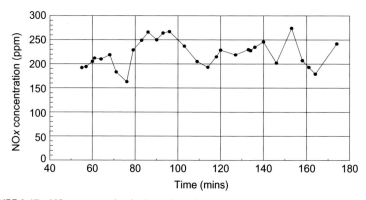

FIGURE 9.17 NO_x concentration in the coal combustor.

predicted tangential gas and particle velocity vectors at different cross sections of the cyclone zone. The strongly swirling flows look like 2-D axisymmetrical. Therefore, by adjusting the position and numbers of the tangential inlets, the optimal design of the cyclone part can be performed.

The coal combustion test was conducted in the combustor based on the optimal design from the two-phase flow measurements and modeling, using ungrounded Datong bituminous coal particles of 3- to 5-mm sizes. The inlet parameters are: bottom-air flow rate $30-45$ m^3/hour, primary-air flow rate $30-40$ m^3/hour, coal feeding rate $10-30$ kg/hour, secondary-air flow rate (for each inlet) $0-20$ m^3/hour. Fig. 9.16 shows the temperature change in the combustor, showing stable coal combustion in the spouting zone and the cyclone zone. The highest temperature is below the ash melting point, indicating the combustion is in a nonslagging mode. The combustion efficiency based on gas analysis reached 92%. Fig. 9.17 gives the measured NO_x

concentration, which is below 300 ppm. Therefore, it can be concluded that there is highly efficient and low-pollutant coal-particle combustion in the proposed nonslagging cyclone combustor.

9.5 PULVERIZED-COAL FURNACES

Some examples, such as the simulation of a coal combustor with high-velocity jets using the two-fluid trajectory (TFT) model, have already been discussed in Chapter 7. Here, more examples for the simulation of pulverized-coal furnaces are discussed. Possibly the best-known research center for coal combustion is the Advanced Combustion Engineering Research Center (ACERC) at Provo, Utah, USA, led by L.D. Smoot. A computer code named "PCGC3" was developed [5] using the k-ε turbulence model, the particle trajectory model, and the gas local instantaneous equilibrium combustion model, and it was used to simulate utility boiler furnaces. Fig. 9.18 is the grid system of a boiler furnace. Fig. 9.19 shows the predicted gas velocity vectors in comparison with some measured results. Fig. 9.20 gives the particle trajectories. Figs. 9.21 and 9.22 show the predicted and measured NO concentration. The agreement between predictions and measurements is good at some locations

A tangentially fired utility furnace was simulated by the present author and his colleagues using the TFT model [6]. The prediction results for

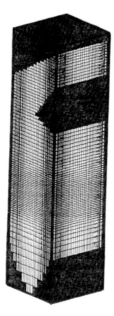

FIGURE 9.18 A grid system of simulated furnace.

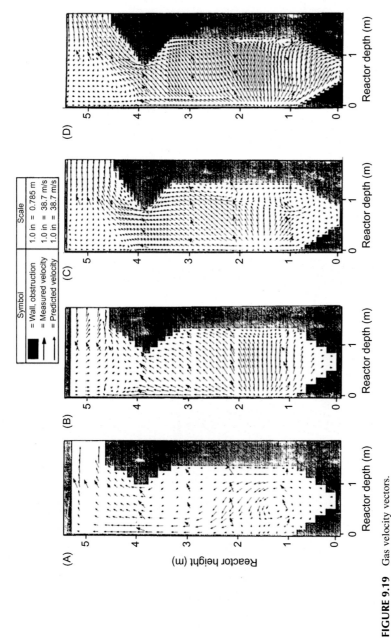

FIGURE 9.19 Gas velocity vectors.

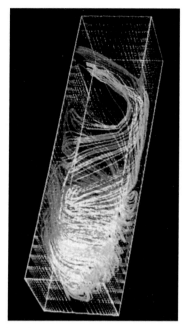

FIGURE 9.20 Particle trajectories.

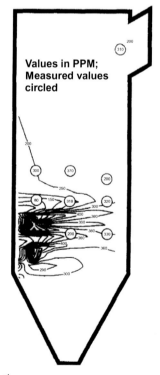

FIGURE 9.21 NO concentration map.

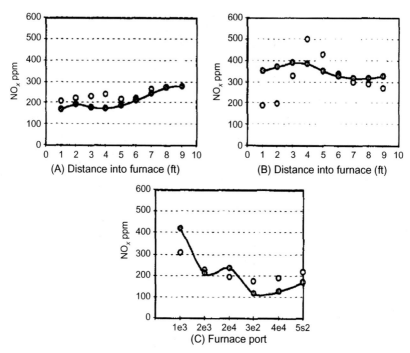

FIGURE 9.22   NO concentration profiles.

isothermal gas-particle flows in a cold model are validated by the PDPA (phase Doppler particle anemometer) measurement results and also compared with those obtained using the ST model in the conventional Eulerian-Lagrangian approach. The predictions for the isothermal gas-particle flows were made for a boiler cold model as shown in Figs. 9.23 and 9.24.

The predicted and measured particle concentration distribution in the isothermal cold model is shown in Fig. 9.25. It is seen that the TFT model gives much better agreement with the experiments than the ST model, while the ST model gives too high concentration in the near-wall region, which was not observed in the experiments. This is because the ST model underpredicts the particle turbulent dispersion. The predictions for the coal combustion case using the TFT approach with a k-ε-k_p two-phase turbulence model and an eddy break-up (EBU) gas combustion model were made for the furnace prototype shown in Fig. 9.26.

The grid nodes are $80 \times 25 \times 25$ for a domain of $40 \times 12.6 \times 12.6$ m. The pulverized-coal furnace is tangentially fired. It has four sets of burners near the corners, where each burner set consists of eleven pulverized-coal and air ports, including primary-air and secondary-air ports. The burner jets are directed horizontally at 39 degrees and 51 degrees to the normal of the wall to achieve a swirling flow. To reduce the numerical diffusion, the grid

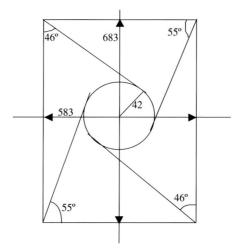

FIGURE 9.23 The lateral section of the furnace model.

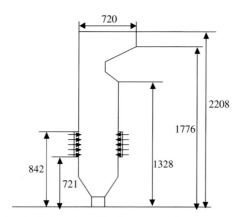

FIGURE 9.24 The longitudinal section of the furnace model.

system was rotated with an angle to be parallel with the direction of the inlet jets (Fig. 9.27). Fig. 9.28 shows the $v-w$ velocity vectors using three different grid systems for a rectangular computation domain of a furnace model with the sizes of $10 \times 2 \times 2$ m. The grid nodes for the grid system (a) are $21 \times 21 \times 21$, and for the grid system (b) $21 \times 42 \times 42$. The grid system (c) is a rotated one, with grid nodes of $21 \times 31 \times 31$. The inlet-flow direction is at an angle of $50°$ to the grid systems (a) and (b), and $0°$ to the grid system (c). It can be seen that the numerical diffusion for case (a) is very large; the inlet jet is soon immersed. Refining of the grid nodes, even twice, cannot give better results (case (b)). In case (c), the rotated grid system gives very good results, and the numerical diffusion is substantially reduced. Fig. 9.29

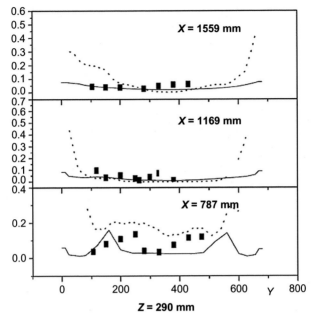

FIGURE 9.25 Particle concentration distribution (■ Measurements; —, FTF Model; ---, ST Model).

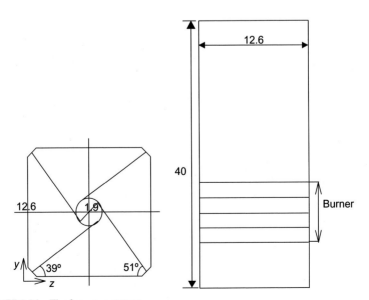

FIGURE 9.26 The furnace prototype.

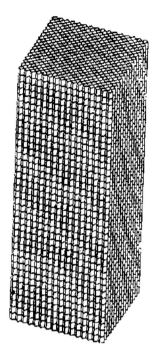

FIGURE 9.27 A rotated grid system.

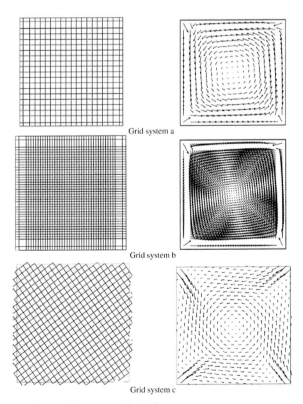

Grid system a

Grid system b

Grid system c

FIGURE 9.28 Different grid systems and velocity vectors.

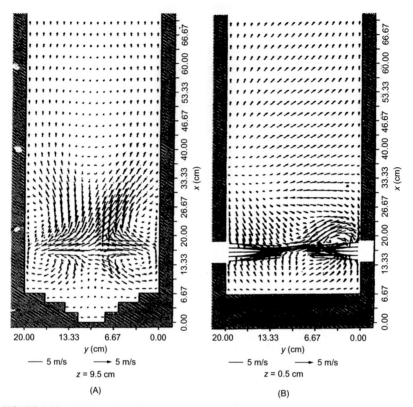

FIGURE 9.29   Isothermal two-phase velocity vectors (arrow−gas; dash−particle).

shows the predicted isothermal two-phase velocity vectors at each point of the furnace space, which can be obtained only by the two-fluid modeling, but is difficult to be obtained by Eulerian-Lagrangian modeling. It is seen that the gas velocity and particle velocity are different both in their magnitudes and directions due to different inertia. So, the no-slip approach is not appropriate. Fig. 9.30 gives the isothermal particle concentration isolines in the furnace. The particle concentration is high in the regions of inlet-jet impingement, and then gradually becomes more uniform and slightly higher near the wall due to the effect of centrifugal force.

Fig. 9.31 gives the predicted v-w gas velocity vectors at two cross sections of the furnace during coal combustion, using the TFT approach. The jet penetration is reasonable. Fig. 9.32 gives similar results using the Eulerian-Lagrangian approach by FLUENT code. Obviously, the predictions show that the burner jets have a very short penetration due to significant numerical diffusion, not observed in experiments. Fig. 9.33 shows the predicted gas temperature maps at a cross section of the furnace using both TFT and E-L modeling. It is seen that the unbelievably low temperatures, near 1273K, in

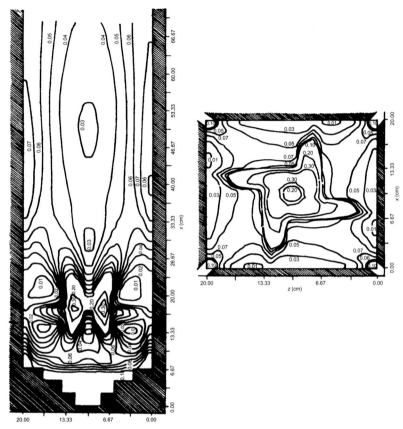

FIGURE 9.30 Particle concentration isolines (isothermal flows).

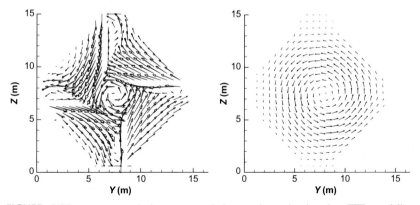

FIGURE 9.31 v-w gas velocity vectors during coal combustion by TFT modeling (Left: $x = 14.7$ m; Right: $x = 30$ m).

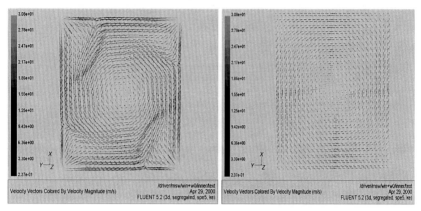

FIGURE 9.32 *v-w* gas velocity vectors during coal combustion by E-L modeling (Left: $x = 14.7$ m; Right: $x = 30$ m).

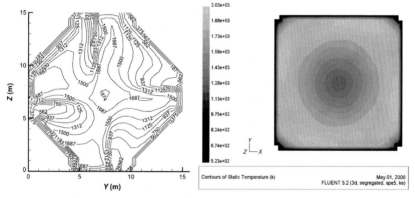

FIGURE 9.33 Temperature maps at a horizontal cross section ($x = 14.7$ m) (Left: TFT Model; Right: E-L Model).

the center region were obtained using the E-L model, since the E-L model results in too many particles moving in the near-wall region with very few particles in the near-center region. The TFT model predictions show the temperature near 1874K in the near-center region, which is reasonable in practical boiler furnaces.

Recently, Belosevic et al. [7] simulated a furnace tangentially fired by eight jet burners, with four tiers each: two lower-stage burners (the main burners for combustion of larger particle size classes) and two upper-stage burners, using the Eulerian-Lagrangian approach, k-ε turbulence model, coal-particle pyrolization model and char combustion model. The gas combustion model was not reported. The predicted gas temperature and oxygen concentration maps are shown in Figs. 9.34 and 9.35. No experimental validation was given.

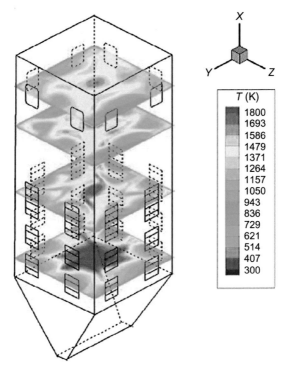

FIGURE 9.34 Gas temperature map.

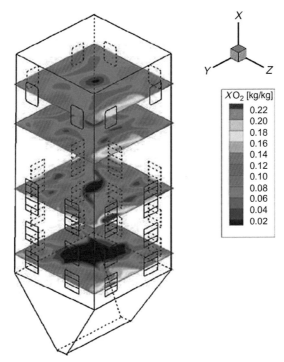

FIGURE 9.35 Oxygen concentration map.

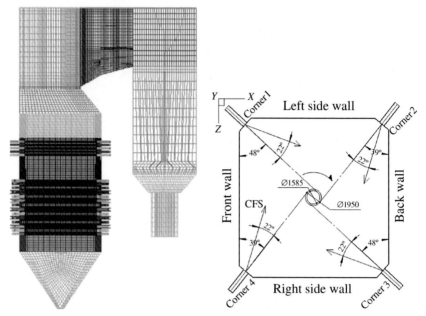

FIGURE 9.36 A 600 MW tangentially fired furnace.

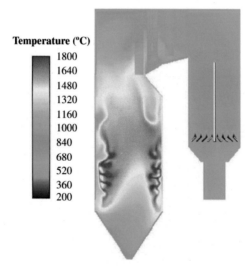

FIGURE 9.37 The gas temperature map.

Fan et al. [8] simulated a 600 MW tangentially fired furnace (Fig. 9.36) using a commercial code FLUENT with all models existing in it. The predicted gas temperature map is given in Fig. 9.37. The experimental validation was made only for a smaller down-fired combustor; no validation was made for the simulation of this large-size furnace.

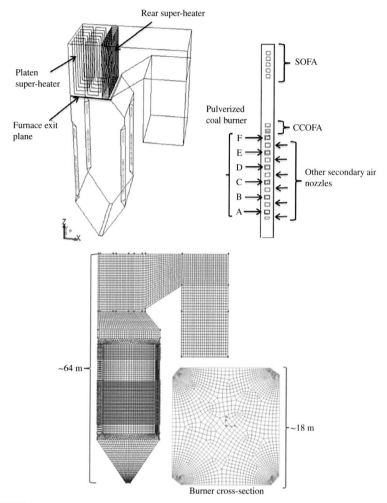

FIGURE 9.38 A tangentially fired furnace and its grid system.

Zhang et al. [9] simulated a 600 MW tangentially fired furnace (Fig. 9.38), using the commercial software ANSYS-FLUENT 15.0 with an Eulerian-Lagrangian approach. The predicted oxygen and NO$_x$ concentrations at a cross section were validated by experiments, as shown in Fig. 9.39. The agreement is fairly good. The particle trajectories, gas temperature and species concentration are shown in Figs. 9.40 and 9.41, respectively.

Liu et al. [10] simulated a 200 MW oxygen-fuel tangentially fired furnace (Fig. 9.42) using the commercial CFD code FLUENT 6.3. Figs. 9.43 and 9.44 give the predicted velocity vectors and gas temperature maps for different oxygen/air ratio, respectively. No detailed experimental validation was reported.

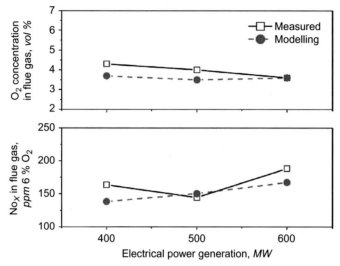

FIGURE 9.39   Oxygen and NO$_x$ concentrations.

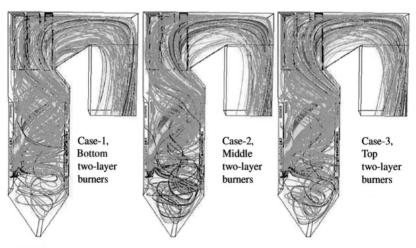

FIGURE 9.40   Particle trajectories in the furnace.

Khaldi et al. [11] simulated a 300 MW tangentially fired furnace. Its geometrical configuration and sizes are shown in Fig. 9.45. The commercial software FLUENT was used, adopting the Eulerian-Lagrangian approach with the standard k–ε model, the RNG k-ε model, the Reynolds-stress model and the EDC combustion model. The predicted isothermal tangential velocity at a cross section is shown in Fig. 9.46. It was reported that the RNG k–ε model gives the best results. Figs. 9.47 and 9.48 show the predicted velocity vectors and gas temperature maps at different cross sections. No experimental validation of the combustion cases was reported.

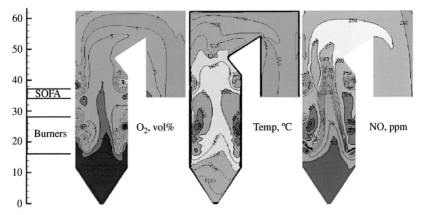

FIGURE 9.41 Gas temperature and species concentrations.

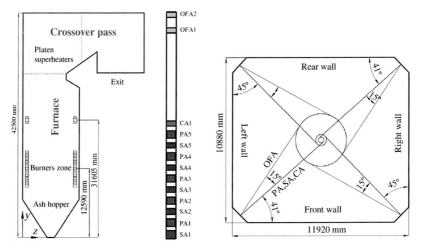

FIGURE 9.42 An oxygen-fuel tangentially fired furnace.

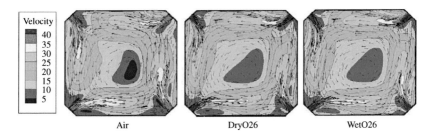

FIGURE 9.43 Velocity vectors at a cross section.

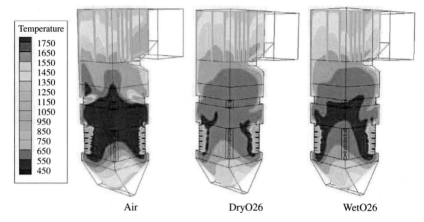

Air   DryO26   WetO26

FIGURE 9.44 Gas temperature along the height of the furnace.

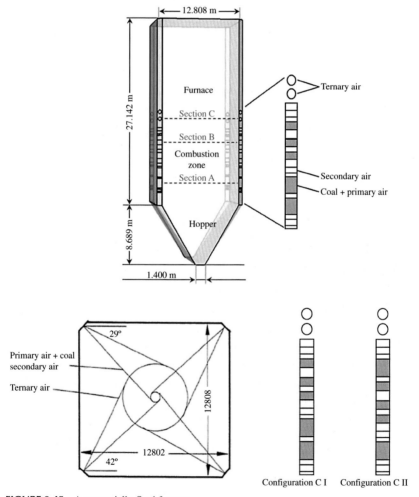

FIGURE 9.45 A tangentially fired furnace.

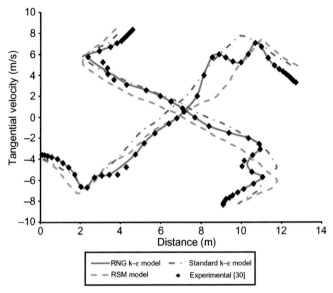

FIGURE 9.46 Tangential velocity predicted by different turbulence models.

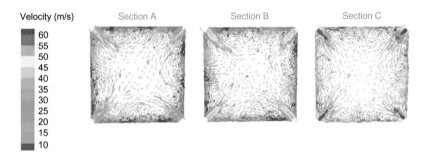

FIGURE 9.47 Velocity vectors at different cross sections.

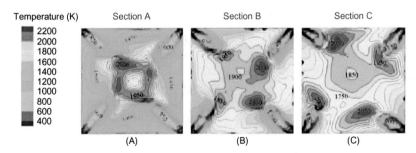

FIGURE 9.48 Gas temperature maps at different cross sections.

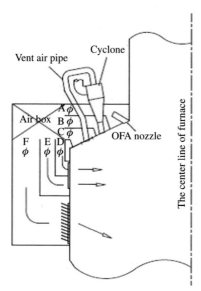

FIGURE 9.49 A down-fired furnace.

Li et al. [12] simulated a down-fired furnace, shown in Fig. 9.49. The predicted gas temperature along the distance to the primary-air nozzle is in fairly good agreement with the measurement results (Fig. 9.50). Fig. 9.51 gives the predicted temperature isolines for different OFA (over-fire air) ratios. It is seen that as the OFA ratio increases, the areas of the high-temperature region in the bottom of the upper furnace gradually increase.

9.6 SPRAY COMBUSTORS

Earlier simulations of spray combustion were conducted by Banhawy and Whitelaw [13] and Swithenbank et al. [14]. A sudden-expansion kerosene-spray combustor was simulated by Banhawy and Whitelaw using the k-ε turbulence model, droplet deterministic trajectory model and presumed PDF-fast chemistry combustion model. The predicted gas temperature and velocity are shown in Fig. 9.52. The predicted temperature is much lower than that measured, indicating the shortcomings of the deterministic trajectory model neglecting the droplet turbulent diffusion.

A gas-turbine combustor was simulated by Swithenbank et al. [14] using the k-ε turbulence model, droplet deterministic trajectory model and EBU-Arrhenius combustion model. The predicted droplet trajectories, gas velocity vectors and gas temperature are shown in Figs. 9.53–9.55. The droplet trajectories are concentrated in the head zone (primary zone) where a strong gas recirculation exists. The high temperature is developed in the head zone and near the wall. No detailed experimental validation was made for the

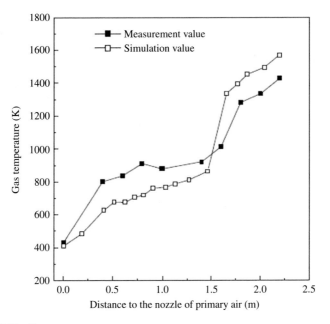

FIGURE 9.50 Gas temperature.

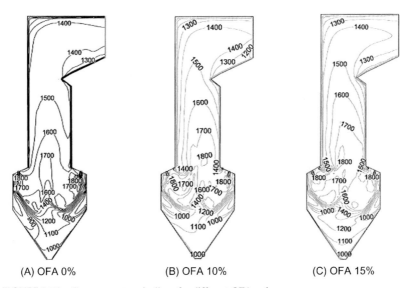

FIGURE 9.51 Gas temperature isolines for different OFA ratios.

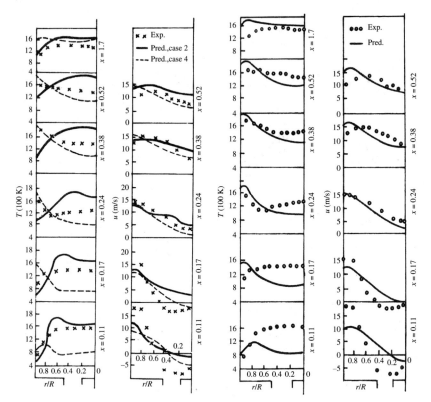

FIGURE 9.52   Temperature and velocity in spray combustion.

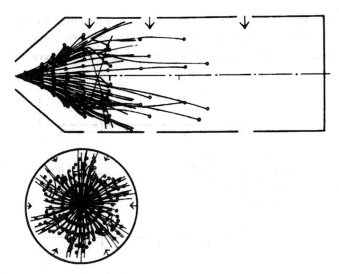

FIGURE 9.53   Droplet trajectories in a gas-turbine combustor.

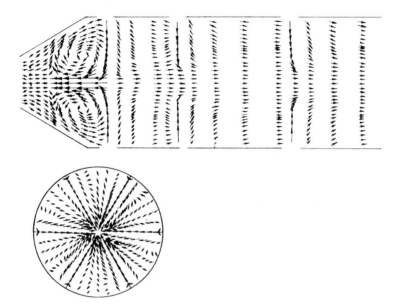

FIGURE 9.54   Gas velocity vectors in a gas-turbine combustor.

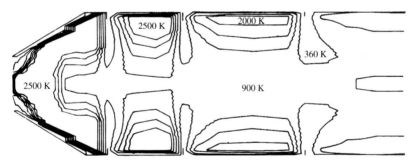

FIGURE 9.55   Gas temperature in a gas-turbine combustor.

predicted velocity, temperature and species concentration inside the combustor, although some comparison was made for the exit velocity and temperature profiles.

The research group led by the present author is beginning work on the simulation of spray-air flows caused by opposite injection of a swirl atomizer into heated air flow [15] (Fig. 9.56) using the k-ε turbulence model and droplet deterministic trajectory model. The predicted droplet trajectories, gas velocity and droplet mass flux and their comparison with experimental results are given in Figs.9.57–9.59. Predicted trajectories of 41.2–103 μm droplets are in fairly good agreement with experimental results (Fig. 9.57).

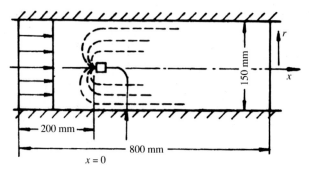

FIGURE 9.56   Opposite injection of a swirl atomizer.

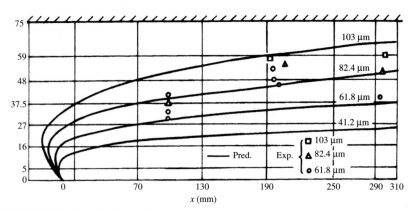

FIGURE 9.57   Droplet trajectories.

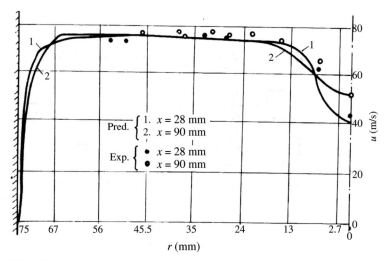

FIGURE 9.58   Gas velocity profile.

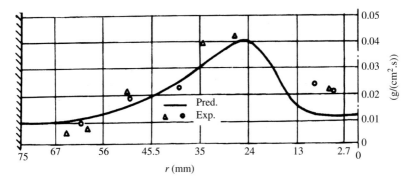

FIGURE 9.59 Spray mass flux.

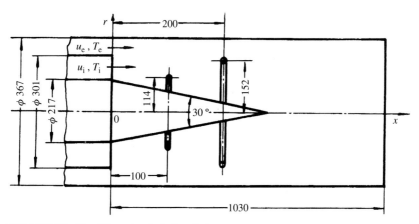

FIGURE 9.60 The diffuser in a turbofan afterburner.

The predicted gas velocity profiles are in good agreement with the experimental results (Fig. 9.58), indicating that the dense liquid spray near the nozzle of the atomizer, like a bluff body, causes a wake in the gas velocity profiles. The predicted liquid-spray mass flux at $x = 355$ mm behind the atomizer (Fig. 9.59) is in general agreement with the experimental results. In the near-axis region, the mass flux is underpredicted, since in fact there is turbulent diffusion of droplets, which is not taken into account in the deterministic trajectory model.

The evaporating spray-air-to-phase flows in the diffuser of a turbofan afterburner (Fig. 9.60) were simulated using a k-ε turbulence model and droplet deterministic trajectory model [16]. Predicted gas velocity (Fig. 9.61) and temperature (Fig. 9.62) in the absence of liquid spray are in good agreement with the experimental results. Predicted droplet trajectories (Fig. 9.63) and fuel vapor concentration (Fig. 9.64) indicate higher fuel concentration in the near-wall low-temperature zone, which is unfavorable to flame stabilization and complete combustion.

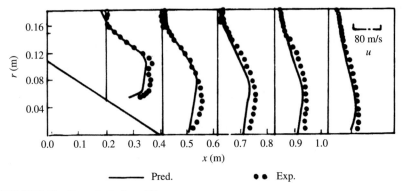

FIGURE 9.61   Gas velocity in a diffuser.

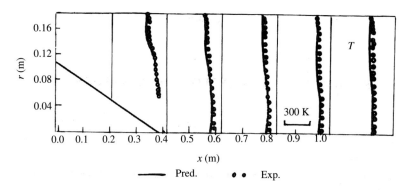

FIGURE 9.62   Gas temperature in a diffuser.

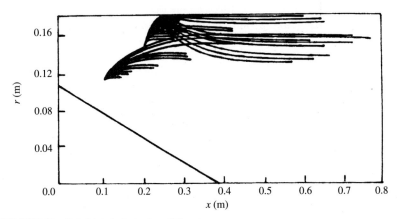

FIGURE 9.63   Droplet trajectories in a diffuser.

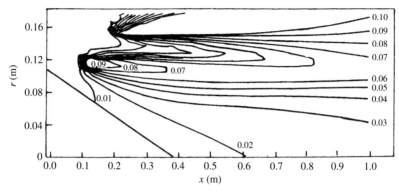

FIGURE 9.64 Fuel-vapor concentration in a diffuser.

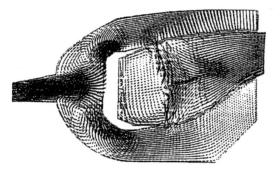

FIGURE 9.65 Gas velocity vectors.

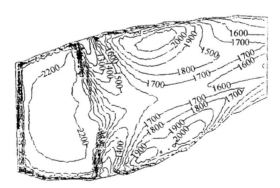

FIGURE 9.66 Temperature isolines.

An annular gas-turbine spray combustor was simulated by Zhao et al. [17] using a k-ε turbulence model, a deterministic droplet trajectory model and an EBU combustion model. The predicted gas velocity vectors and temperature are given in Figs. 9.65 and 9.66, respectively. The recirculating

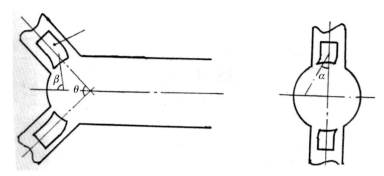

FIGURE 9.67   A ram-jet combustor with a central tube in the dual inlet.

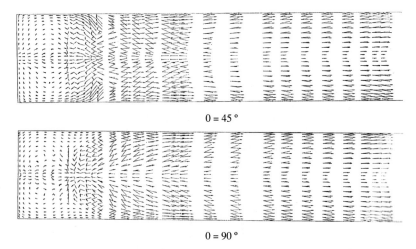

$\theta = 45\,^{\circ}$

$\theta = 90\,^{\circ}$

FIGURE 9.68   The two-phase velocity vectors in the ram-jet combustor (arrow—gas; dash—particle).

flows and high temperature in the near-wall region are seen clearly. No detailed validation of the prediction results inside the combustor was reported. There is only a comparison of the predicted temperature profile at the exit with the measured results.

A dual-inlet sudden-expansion spray combustor for ram-jet engines was simulated in the research group led by the present author [18, 19]. To improve the flame stabilization in the ram-jet combustor, a central tube with a tangential exit angle is mounted in the inlet tubes of the combustor for inducing the large-size recirculation zone, as shown in Fig. 9.67.

The isothermal two-phase flow field was simulated using a k-ε-A_p two-phase turbulence model. The predicted two-phase velocity vectors are given in Fig. 9.68. It is seen that large-size recirculating flows are formed in the

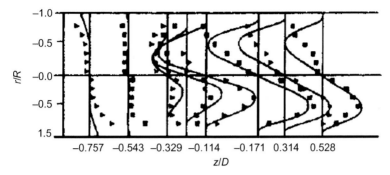

FIGURE 9.69 Gas tangential velocity in the ram-jet combustor (line—pred.; dots—exp.).

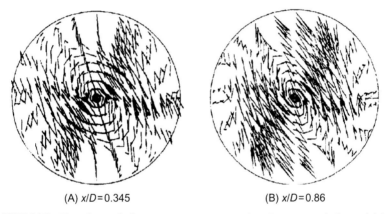

(A) x/D=0.345   (B) x/D=0.86

FIGURE 9.70 Two-phase velocity vectors at two cross sections (arrow—gas; dash—particle).

head zone of the combustor. The predicted gas tangential velocity is in agreement with the measurement results (Fig. 9.69). The predicted two-phase velocity vectors in the cross sections (Fig. 9.70) show gas and particle velocities, different in both their magnitudes and directions. The predicted gas velocity and droplet velocity during combustion, gas temperature, droplet temperature, fuel vapor concentration, and oxygen concentration, using the full two-fluid model with a k-ε-k$_p$ two-phase turbulence model, a droplet evaporation model, and an EBU combustion model, are shown in Figs. 9.71—9.76, respectively. It is seen that the two-phase velocities are different in their directions and magnitudes. The gas temperature is much higher than the droplet temperature and reaches 1200—1400K in the recirculation zone; the flame front is located at the boundary of the recirculation zone, indicating that the recirculation zone in the head part is favorable to flame stabilization. The fuel and oxygen concentrations are very low in the recirculation zone, where the main species are the combustion products. In the

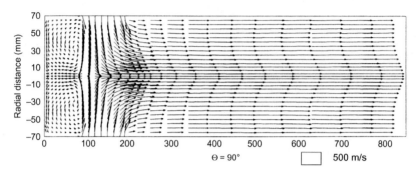

FIGURE 9.71 Gas velocity vectors during combustion.

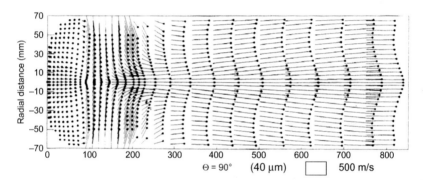

FIGURE 9.72 Droplet velocity vectors during combustion.

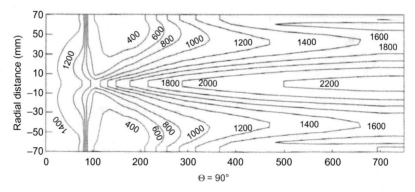

FIGURE 9.73 Gas temperature.

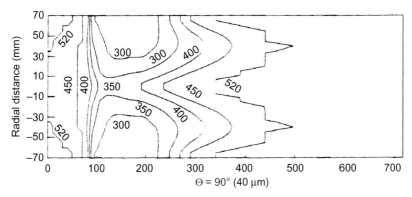

FIGURE 9.74 Droplet temperature.

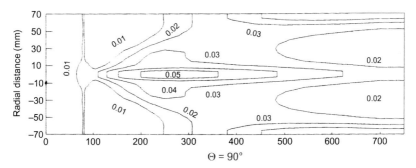

FIGURE 9.75 Fuel vapor concentration.

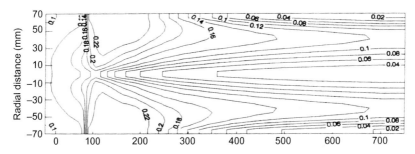

FIGURE 9.76 Oxygen concentration.

near-axis region of the combustor the fuel concentration gradually increases due to droplet evaporation. In the downstream region the fuel and oxygen concentrations decrease due to gas-phase combustion.

A gas-turbine combustor burning biofuel (Fig. 9.77) was simulated by Sallevelt et al. [20] using the commercial software ANSYS-FLUENT 14.5

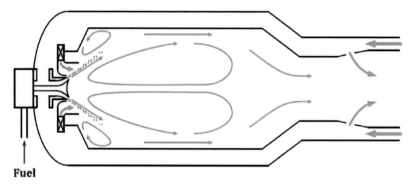

FIGURE 9.77 A biofuel gas-turbine combustor.

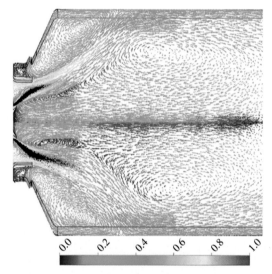

FIGURE 9.78 Velocity vectors (top–ethanol, bottom–pyrolysis oil).

with the SST-k-ω turbulence model, Lagrangian droplet motion and evaporation models, and both a "relaxation-to-equilibrium (RTE)" combustion model and a steady laminar flame-let combustion model.

The predicted velocity vectors and temperature maps for burning ethanol and pyrolysis oil are shown in Figs. 9.78 and 9.79, respectively. The combustion of pyrolysis fuel gives a smaller recirculation zone and lower temperature. No detailed quantitative validation of modeling results was reported.

The spray combustion in a diesel internal combustion engine was simulated by Vujanovic et al. [21] using a coupled Eulerian-Eulerian and Eulerian-Lagrangian method. The Eulerian-Eulerian (EE) method is used for

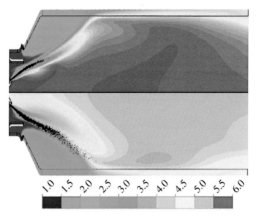

FIGURE 9.79 Temperature maps (top−ethanol, bottom−pyrolysis oil).

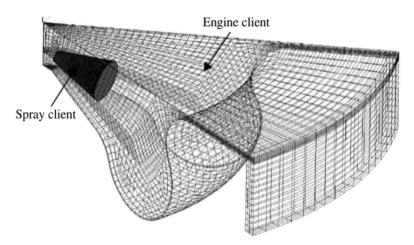

FIGURE 9.80 Simulated IC engine and computation domain.

spray modeling, including the fuel jet disintegration and droplet evaporation processes. The combustion process is modeled by employing the Eulerian-Lagrangian (EL) discrete droplet model. The simulated engine and computation domain are given in Fig. 9.80.

The predicted gas velocity map is shown in Fig. 9.81. The predicted temperature and NO concentration maps are given in Fig. 9.82.

The comparison between predicted and measured pressure and temperature changes with the crank angle is shown in Fig. 9.83. The agreement is good.

A NASA aircraft combustor was studied by Pitsch et al. [22] using LES with a multiregime flame-let combustion model, detailed chemical mechanism,

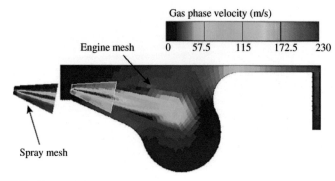

FIGURE 9.81 Temperature maps.

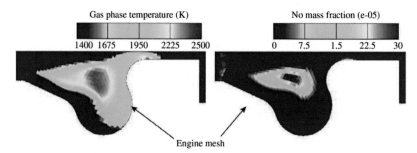

FIGURE 9.82 Temperature and NO concentration maps.

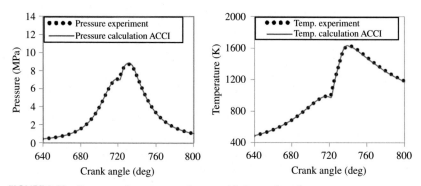

FIGURE 9.83 Pressure and temperature changes with the crank angle.

and droplet Lagrangian approach. Fig. 9.84 gives the predicted instantaneous droplet position and temperature. Fig. 9.85 shows the instantaneous axial velocity and temperature. The high temperature is developed near the corner recirculation zone and at the downstream region. Fig. 9.86 gives the time-

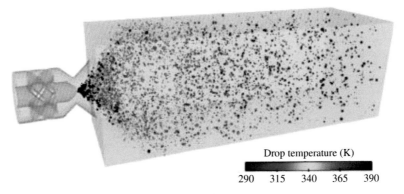

FIGURE 9.84 Instantaneous droplet position and temperature.

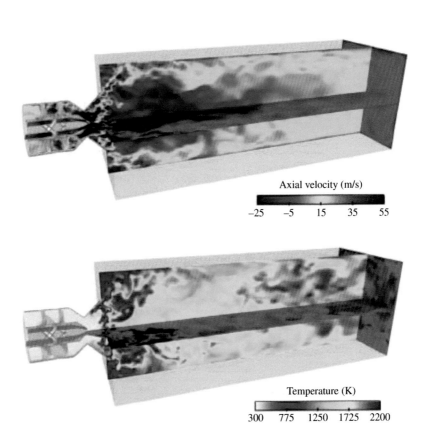

FIGURE 9.85 Instantaneous axial velocity (L) and temperature (R).

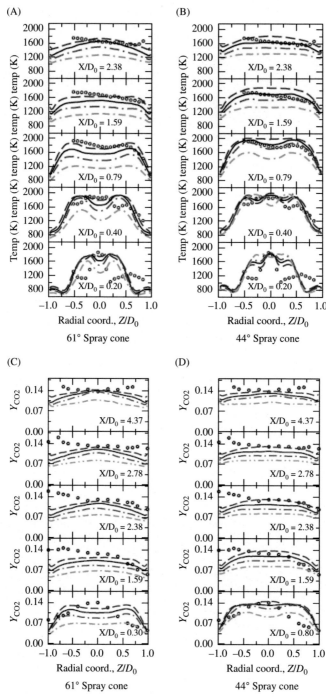

FIGURE 9.86 Temperature (L) and CO_2 concentration (R) (dots—exp.; lines—pred. using different gas properties).

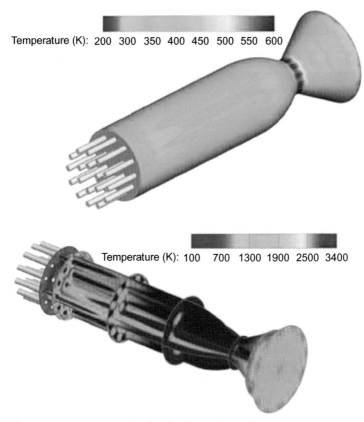

FIGURE 9.87 Temperature map in a liquid rocket engine combustor.

averaged temperature and CO_2 concentration. It is seen that by choosing appropriate gas thermodynamic properties the agreement between predictions and experimental results is better.

A rocket engine combustor burning liquid oxygen/methane was simulated by Song et al. [23] using the commercial software ANSYS-FLUENT, selecting the RANS modeling with the standard k-ε turbulence model, steady flamelet combustion model, a reduced chemical mechanism. Fig. 9.87 shows the predicted temperature map. However, no validation of the prediction results was made in this study.

9.7 CONCLUDING REMARKS

From the whole text of this book, it can be concluded that tremendous strides have been made in the theory and numerical modeling of multiphase turbulent reacting flows and their validation and application. However, it must be pointed out that many problems in different submodels still remain to be solved, such as the two-phase subgrid scale (SGS) stress models in two-fluid large-eddy simulation (LES) of two-phase flows and combustion, a reasonable and economic turbulence-chemistry model for simulating detailed reaction mechanism, turbulence modulation in dense gas-particle flows, etc. These models need validation by both DNS and experimental validation using PDPA, particle image velocimetry (PIV), coherent anti-Stokes Raman spectroscopy (CARS), and planar laser induced fluorescence (PLIF), etc. Although there are many difficulties ahead, it is without doubt that the theory and modeling of multiphase turbulent reacting flows will continue to progress and more engineering applications will be found in the 21st century. The phenomena involved are very complex; we sincerely hope that more investigators can be involved and contribute their efforts to this research field.

REFERENCES

[1] Y.J. Lu, Experimental and numerical studies on strongly swirling turbulent two-phase flows and separation in the liquid-liquid hydrocyclones, PhD thesis, Department of Engineering Mechanics, Tsinghua University, Beijing, China, 1997.

[2] L.X. Zhou, S.L. Soo, Gas-solid flow and collection of solids in a cyclone separator, Powder Technol. 63 (1990) 45−53.

[3] L.X. Zhou, S. Nieh, G. Yang, Modeling of vortexing gas-particle flows, combustion and flow field predictions, J. Eng. Thermop. English Edi. 2 (1990) 231−241.

[4] L.X. Zhou, Development of an innovative cyclone coal combustor, J. Combust. Sci. Technol. (China) 21 (2015) 287−292.

[5] L.D. Smoot (Ed.), Fundamentals of Coal Combustion, Elsevier, Amsterdam, 1993.

[6] L.X. Zhou, L. Li, R.X. Li, J. Zhang, Simulation of 3-D gas-particle flows and coal combustion in a tangentially fired furnace using a two-fluid-trajectory model, Powder Technol. 125 (2002) 226−233.

[7] S. Belosevic, I. Tomanovic, et al., Numerical study of pulverized coal-fired utility boiler over a wide range of operating conditions for in-furnace SO_2/NO_x reduction, Appl. Thermal Eng. 94 (2016) 657−669.

[8] Y. Liu, W. Fan, Y. Li, Numerical investigation of air-staged combustion emphasizing char gasification and gas temperature deviation in a large-scale, tangentially fired pulverized-coal boiler, Appl. Energy 177 (2016) 323−334.

[9] J. Zhang, Q.Y. Wang, Y.J. Wei, L. Zhang, Numerical modeling and experimental investigation on the use of brown coal and its beneficiated semicoke for coal blending combustion in a 600 MW utility furnace, Energy Fuels 29 (2015) 1196−1209.

[10] J.J. Guo, Z.H. Liu, P. Wang, X.H. Huang, J. Li, P. Xu, et al., Numerical investigation on oxy-combustion characteristics of a 200 MWe tangentially fired boiler, Fuel 140 (2015) 660−668.

[11] N. Khaldi, Y. Chouari, H. Mhiri, P. Bournot, CFD investigation on the flow and combustion in a 300 Mwe tangentially fired pulverized-coal furnace, Heat Mass Transfer 52 (2016) 1881−1890.

[12] G.K. Liu, Z.C. Chen, Z.Q. Li, G.P. Li, Q.D. Zong, Numerical simulations of flow, combustion characteristics, and NOx emission for down-fired boiler with different arch-supplied over-fire air ratios, Appl. Thermal Eng. 75 (2015) 1034–1045.

[13] Y.E.I. Banhawy, J.H. Whitelaw, Calculation of the flow properties of a confined kerosene-spray flame, AIAAJ 18 (1980) 1503–1510.

[14] F. Boysan, W.H. Ayers, J. Swithenbank, Z. Pan, 3-D model of spray combustion in gas-turbine combustors, J. Energy 6 (1982) 368–375.

[15] Zhou, L.X., Lin, W.Y., Jiang, Z., Numerical simulation of spray two-phase flows produced by opposite injection of a centrifugal atomizer into a pipe. Proc. First University Symposium on Engineering Thermophysics (in Chinese), Tsinghua University Press, 1985, pp. 401-410.

[16] L.X. Zhou, J. Zhang, Numerical modeling of turbulent evaporating gas-droplet two-phase flows in an afterburner diffuser of turbo-fan jet engines, Chinese J. Aeronautics 3 (1990) 258–265.

[17] J.X. Zhao, Combustion modeling, Science Press, Beijing, 2002.

[18] L.X. Zhou, W.Y. Lin, C.M. Liao, Modeling of 3-D turbulent swirling and recirculating gas-particle flows in a dual-inlet dump combustor, J. Eng. Thermop. (in Chinese) 15 (1994) 446–448.

[19] Y.C. Guo, L.X. Zhou, W.Y. Lin, Numerical modeling of spray combustion in a dual-inlet dump combustor, J. Eng. Thermop. (in Chinese) 18 (1997) 502–506.

[20] J.L.H.P. Sallevelt, A.K. Pozarlik, G. Brem, Numerical study of pyrolysis oil combustion in an industrial gas turbine, Energy Conv. Manag. 127 (2016) 504–514.

[21] M. Vujanovic, Z. Petranovic, W. Edelbauer, N. Duic, Modeling spray and combustion processes in diesel engine by using the coupled Eulerian–Eulerian and Eulerian–Lagrangian method, Energy Conv. Manag. 125 (2016) 15–25.

[22] E. Knudsen, Shashank, H. Pitsch, Modeling partially premixed combustion behavior in multiphase LES, Combustion Flame 162 (2015) 159–180.

[23] J. Song, B. Sun, Coupled numerical simulation of combustion and regenerative cooling in LOX/Methane rocket engines, Appl. Thermal Eng. 106 (2016) 762–773.

Index

Note: Page numbers followed by "*f*" refer to figures.

Printed in the United States
By Bookmasters